DIFFERENTIAL GEOMETRY AND RELATIVITY

MATHEMATICAL PHYSICS AND APPLIED MATHEMATICS

Editors:

VOLUME 3

ANDRÉ LICHNEROWICZ

DIFFERENTIAL GEOMETRY AND RELATIVITY

A Volume in Honour of André Lichnerowicz on His 60th Birthday

Edited by

M. CAHEN AND M. FLATO

D. REIDEL PUBLISHING COMPANY

DORDRECHT-HOLLAND / BOSTON-U.S.A.

Library of Congress Cataloging in Publication Data
Main entry under title:

Differential geometry and relativity.

(Mathematical physics and applied mathematics; v. 3)
'Collected works of A. Lichnerowicz': p.
Includes bibliographies and index.
1. Geometry, Differential—Addresses, essays, lectures. 2. Mathematical physics—Addresses, essays, lectures. 3. Relativity (Physics)—Addresses, essays, lectures. 4. Lichnerowicz, André, 1915– I. Cahen, Michel. II. Flato, Moshé. III. Lichnerowicz, André, 1915– IV. Series.
QC20.7.D52D54 516'.34 76–48299
ISBN 90–277–0745–6

Published by D. Reidel Publishing Company,
P.O. Box 17, Dordrecht, Holland

Sold and distributed in the U.S.A., Canada, and Mexico
by D. Reidel Publishing Company, Inc.
Lincoln Building, 160 Old Derby Street, Hingham
Mass. 02043, U.S.A.

CONTENTS

PREFACE

On the occasion of the sixtieth birthday of André Lichnerowicz a number of his friends, many of whom have been his students or coworkers, decided to celebrate this event by preparing a jubilee volume of contributed articles in the two main fields of research marked by Lichnerowicz's work, namely differential geometry and mathematical physics. Limitations of space and time did not enable us to include papers from all Lichnerowicz's friends nor from all his former students. It was equally impossible to reflect in a single book the great variety of subjects tackled by Lichnerowicz. In spite of these limitations, we hope that this book reflects some of the present trends of fields in which he worked, and some of the subjects to which he contributed in his long – and not yet finished – career.

This career was very much marked by the influence of his masters, Elie Cartan who introduced him to research in mathematics, mainly in geometry and its relations with mathematical physics, and Georges Darmois who developed his interest for mechanics and physics, especially the theory of relativity and electromagnetism. This particular combination, and his personal talent, made of him a natural scientific heir and continuator of the French mathematical physics school in the tradition of Henri Poincaré. Some of his works would even be best qualified by a new field name, that of physical mathematics: branches of pure mathematics entirely motivated by physics.

In differential geometry he continued the fundamental works of Elie Cartan, and greatly contributed to their formulation in a more precise and intrinsic manner. His books, "Théorie globale des connections et des groupes d'holonomie" and "Géométrie des groupes de transformations", are not only very clearly written in good French (a rarity nowadays): they have been basic textbooks for a whole generation of differential geometers, and they truly reflect the impact of Lichnerowicz on the field.

In contradistinction with the present tendency to over-algebrize everything, he succeeded in being precise, concise but clear, abstract and yet still concrete; he was always motivated by physics. In addition, Lichnerowicz is the author of numerous important contributions on Riemannian manifolds, Kaehlerian manifolds, and symplectic manifolds.

In general relativity the influence of Lichnerowicz was perhaps even greater, an influence which started from his Thesis in 1939. Here one must also mention his famous book, "Théories relativistes de la gravitation et de l'électromagnétisme". Through his many works and students he developed in France a school of rigorous relativistic theories which is still very much alive and meets yearly, together with other specialists, in the Journées Relativistes which he created. In general relativity one has to deal both with geometrical questions (space-time manifolds) and with difficult problems of mathematical analysis, problems of partial differential equations. In this field also he distinguished himself through works now classical such as solutions of the Einstein-Maxwell equations, study of the Cauchy problem for Einstein equations, and characterization of gravitational radiation through algebraic properties of the curvature tensor. He also dealt with problems of mechanics, in particular in relativistic magnetohydrodynamics or more recently shock waves.

All his works show a permanent interconnection between mathematics and mathematical physics, and are often a perfect illustration of what should be called physical mathematics. This is only one example of interdisciplinary research, in which André Lichnerowicz got interested long before it became fashionable. Among other fields he showed a strong interest for mathematical economics, a field in which his deeply mourned elder son Marc was about to make himself a leading name when he prematurely disappeared.

The content of this book will therefore be divided into two parts, differential geometry and mathematical physics. In each part the contributions will be ordered according to the name of the first mentioned contributor. The first part deals mainly with various problems of Riemannian manifolds. Other works in this part deal with symmetric spaces, almost-complex manifolds, Banach manifolds, and vector bundles. The second one deals with various problems of general relativistic systems, problems around invariance or covariance in classical or quantum special relativistic theories, or applications of symplectic geometry to mechanics. More details on the great variety of subjects treated can be found from the titles in the Contents.

Finally one cannot avoid mentioning the very important activity of Lichnerowicz in the reform of mathematics teaching. This will be explained in more detail in the paper following this preface, by R. Couty and A. Revuz.

This book is dedicated to André Lichnerowicz on the occasion of his sixtieth birthday with admiration and affection.

THE EDITORS

R. COUTY and A. REVUZ

THIRTY YEARS OF ACTIVITY IN THE RENOVATION OF MATHEMATICAL EDUCATION

Lichnerowicz's activity in the field of mathematical education both at the international level and in his own country was continued throughout his career. To mention only a few dates: 1956, first colloquium (Caen), 1962–1966, chairman of the I.C.E.M., 1966, second colloquium (Caen), 1966–1973, chairman of the French ministerial committee for mathematical education, chairman of the national committee of the I.R.E.M.s (Instituts de Recherche sur l'Enseignement Mathématique) since their creation (1968).

It is a remarkable feature of Lichnerowicz's activity in the field of mathematical education that he did not find it enough to utter sensible or bright ideas about the matter – and we may think that he has more confidence in sensible than in bright ideas – but he has given them practical support in a long and steadfast struggle, both at the international level and in his own country. There he has been the chairman of the ministerial committee which is everywhere designated by his name and he has given a powerful impulse to the transformation of French mathematical education.

The main incentive to his efforts has been the idea that mathematics should not be the private domain of some specialists, and of those who can't avoid using it, but is a common resource of the whole of mankind. He has considered that an educational system is a failure if it does not contrive to give to all its students the basic habits of a mathematical attitude of mind, in order that they may be able to develop them afterwards in their life and adapt them to the special problems they are bound to meet. In a well balanced education, mathematics must play an important part: the idea being not that everybody is going to become a mathematician, in the restricted sense, but that the possibility of a harmonious development of mankind depends not only on the goodwill of the peoples and their leaders, but also in the quick solution of very complex problems, where mathematics is by a long way not the only tool, but nevertheless a powerful, and in many cases, indispensable one. The sensible use of such a tool cannot be the prerogative of only a very few people without great social and political dangers.

Cahen & Flato (eds.), Differential Geometry and Relativity, 1–2.

Lichnerowicz has understood that we are confronted with a huge and urgent task, and that it cannot be accomplished without a tremendous collective effort: bright ideas and deep theories are useless if they don't quickly become the common property of every man. And that is why he so strongly supported in France the creation of the I.R.E.M.s (Instituts de Recherche sur l'Enseignement Mathématique), whose originality lies chiefly in their gathering, from university professors to kindergarten teachers, people who work at every level of mathematical education and who wish to cooperate in order to improve it both in the short and the long term. The tasks of the I.R.E.M.s are: fundamental and applied research in mathematical education; in-service training of teachers; cooperation in the initial training of teachers; and the gathering and diffusion among teachers of documents. These various tasks must not be considered as separate from each other: they must be closely linked in order to be effective. In order to maintain close contact with the real conditions of teaching, every one who works in an I.R.E.M. does half-time service in the I.R.E.M. and half-time service in his school. What is aimed at in the work of the I.R.E.M.s is a dialectical and simultaneous progress of both theory and practice: they try to avoid both theories that have no impact on practice, and practice which by lack of reflection degenerates into dead routine. According to this view, it is very important that as many people as possible work with the I.R.E.M.s in order to collect wide ranging observations, to experiment on a large scale, and to maintain permanent contact between didactical research and everyday life in the schools. The I.R.E.M.s, the number of which is now twenty-five, have made a decisive contribution to the carrying out in the schools of the last French reform of the mathematical curricula. They have tried – and in some cases they have succeeded – to break through the artificial barriers that traditionally separated the various levels of education from each other, or the teaching of mathematics from the teaching of science or the mother tongue. There remains a lot of work to do in these directions. Indeed the endeavour of the I.R.E.M.s to break through administrative barriers or inherited prejudices has brought them real difficulties. Nevertheless, their ideal of the sound teaching of mathematics in close connection with the teaching of all other human activities, is strong enough to give them the heart to face these difficulties; and they are sure that in this matter they can rely on the indefatigable help of Lichnerowicz who, since the creation of the I.R.E.M.s, has been the chairman of their national committee.

PART I

DIFFERENTIAL GEOMETRY

T. AUBIN

THE SCALAR CURVATURE

Introduction

We shall deal with some problems concerning the scalar curvature of compact riemannian manifolds. In particular we shall deal with the problem of Yamabe: Does there exist a conformal metric for which the scalar curvature is constant? And also problems posed by Chern, Nirenberg and others. All these problems are almost entirely solved, however there remain some open questions (see the conjectures).

The main theorems are Theorems 4 and 8.

1. The dimension two

The problems concerning scalar curvature turn out to be very particular when the dimension is two. And they can be completely solved. For compact riemannian manifolds M_2, the Euler-Poincaré characteristic is $\chi = \frac{1}{4\pi}\int_M R \, dv$. Hence the sign of the scalar curvature R has topological implications.

A conformal deformation of the metric $g' = e^{\varphi}g$, leads to a scalar curvature R' (related to g') according to

$$R' = e^{-\varphi}(R + \Delta\varphi), \tag{1}$$

where R is the scalar curvature of g. Let f be a C^∞ function on M_2; does there exist a conformal metric g' so that $R' = \chi e^f$? (and even, so that R' is a C^∞ function which changes sign, when $\chi = 0$).

This problem was studied by M. Berger [6] and solved if $\chi < 2$. Let h be the function verifying $\Delta h = -R + \int_M R \, dv$ and $\int_M h \, dv = 0$, when we assume $\int_M dv = 1$. Set $\psi = \varphi - h$, this implies

$$\Delta\psi + 4\pi\chi = \chi e^{f+h} e^{\psi}.$$

When $\chi \neq 0$, M. Berger considers the functionnal

$$I(\psi) = \int_M \nabla^{\nu}\psi \nabla_{\nu}\psi \, dv + 8\pi\chi \int_M \psi \, dv$$

Cahen & Flato (eds.), Differential Geometry and Relativity, 5–18.

with $\int_M e^{f+h+\psi}\,dv = 4\pi$ for all $\psi \in H_1$. One shows that the lower bound is finite and reached when $\chi < 2$. To complete the proof one writes the Euler equation.

To prove the lower bound is reached, one needs the following inequality (Aubin [2], Trüdinger [20], Hempel *et al.* [8]): There exists two constants A and k such that for all $\psi \in H_1$, satisfying $\int_M \psi\,dv = 0$:

$$\int_M e^{\psi}\,dv \leqslant A e^{k\int_M \nabla^\nu\psi\nabla_\nu\psi\,dv}$$

As I have mentioned (Aubin [2]; see also Moser [14]), the best constant k is $1/16\pi$. Thus it fails when $\chi = 2$. For $\chi = 2$, it is well known that Equation (1) has a solution when R' = constant. But (1) has no solution for some functions f.

Nirenberg raised the question: let f be a C^∞ function on the sphere S_2, does there exist $\epsilon > 0$, such that the Equation (1) has a solution with $R' = e^{\eta f}$, for all $\eta \neq 0$, verifying $|\eta| < \epsilon$? The answer is no: Kazdan and Warner [10] showed that the problem of Nirenberg has no solution, when f verifies $\Delta f = \lambda_1 f$, where λ_1 is the first eigenvalue of the laplacian.

2. The theorem of Yamabe

From now on, the riemannian manifolds M_n are C^∞ compact and of dimension $n \geqslant 3$.

THE PROBLEM OF YAMABE. On the riemannian manifold M_n, let us consider the conformal deformation $g' = \varphi^{4/(n-2)}g$ of the metric g, where φ is a C^∞ strictly positive function everywhere. The scalar curvature R' (related to g') expresses itself by means of the scalar curvature R (related to g) according to

$$R'\varphi^{(n+2)/(n-2)} = 4\frac{n-1}{n-2}\Delta\varphi + R\varphi, \tag{2}$$

where $\Delta\varphi = -\nabla^\nu\nabla_\nu\varphi$.

DEFINITION. Let us say a riemannian manifold has Yamabe's property, if there exists a conformal metric g' the scalar curvature R' of which is constant.

CONJECTURE 1. (of Yamabe) Every compact C^∞ riemannian manifolds M_n ($n \geqslant 3$) has Yamabe's property. In other words: there exists a C^∞ function, $\varphi > 0$, and a constant R' verifying Equation (2). In his famous article, Yamabe stated this conjecture as a theorem. Unfortunately his demonstration contains a gap as Trüdinger pointed out. And it is impossible to draw a conclusion.

In the negative or null case, ($\mu \leqslant 0$; for the definition of μ see Theorem 1), it is easy to overcome the mistake. But in the positive case ($\mu > 0$), it is impossible. There exists a counter-example: on the sphere, the functions $\varphi(r)$ of the Corollary 3 are not uniformly bounded. So Yamabe did not prove his conjecture but:

THEOREM 1. (of Yamabe) (See Aubin [1, p. 386], Trüdinger [19], Yamabe [21].) Let M_n be a riemannian C^∞ compact manifold; there exists a conformal metric for which the scalar curvature is either a non positive constant or is everywhere positive.

The following is an outline of the proof. Yamabe considers the expression, for $2 \leqslant q \leqslant N = 2n/(n-2)$:

$$J_q(\varphi) = \frac{4\dfrac{n-1}{n-2}\displaystyle\int_M \nabla^\nu \varphi \nabla_\nu \varphi \, dv + \int_M R\varphi^2 \, dv}{\|\varphi\|_q^2},$$

where φ is a non-negative function belonging to the first Sobolev space $H_1(M_n)$. Let $\mu_q = \inf J_q(\varphi)$ for all $\varphi \in H_1$, $\varphi \geqslant 0$, $\varphi \not\equiv 0$, and $\mu = \mu_N$. First of all Yamabe proves:

THEOREM 2. For $2 \leqslant q < N$, there exists a C^∞ positive function φ_q verifying: $J_q(\varphi_q) = \mu_q$ and

$$4\frac{n-1}{n-2}\Delta\varphi_q + R\varphi_q = \mu_q \varphi_q^{q-1}. \tag{3}$$

And it is possible to prove:

PROPOSITION 1. μ_q is a continuous function of q for $q \in [2, N]$, which is either everywhere positive, or everywhere null, or everywhere negative.

In the positive and null cases, let us make the change of metric $g' = \varphi_{q_0}^{4/(n-2)} g$ for some $q_0 \in]2, N[$. For the metric g', $R' = \mu_{q_0} \varphi_{q_0}^{q_0 - N}$. In the positive case ($\mu > 0$), it is possible to make the scalar curvature everywhere positive. In the case $\mu = 0$, it is possible to make the scalar curvature everywhere null.

Now, let us prove Theorem 1 in the negative case.

PROPOSITION 2. In the negative case ($\mu < 0$), the function φ_q, $q \in [2, N[$ is uniformly bounded.

Let the Green function of the laplacian be such that

$$G(P, Q) \geqslant 0 \quad \text{(see Aubin [3])}.$$

Since φ_q verifies (3):

$$\varphi_q(P) = \int_M \varphi_q(Q)\,\mathrm{d}v(Q) + \frac{n-2}{4(n-1)}\int_M G(P, Q) \times [\mu_q \varphi_q^{q-1}(Q) - R(Q)\varphi_q(Q)]\,\mathrm{d}v(Q) \tag{4}$$

and

$$\varphi_q(P) \leqslant \|\varphi\|_1 + \frac{n-2}{4(n-1)} \sup_{x \in M} |R(x)| \int_M G(P, Q)\varphi_q(Q)\,\mathrm{d}v(Q).$$

Let us iterate r times, $r > n/2$, the preceding inequality. From the properties of $G(P, Q)$, there exists a constant A, such that

$$\varphi_q \leqslant A\|\varphi_q\|_1 \leqslant A\left[\int_M \mathrm{d}v\right]^{1-1/q},$$

since $\|\varphi_q\|_q = 1$. Differentiating (4), one sees that $\varphi_q \in C^1$ uniformly, and it is possible to find a sequence φ_{q_i} with $q_i \to N$, such that φ_{q_i} converges to a C^∞ everywhere positive function φ_N verifying the Equation (2) with $R' = \mu$.

PROPOSITION 3. μ is a conformal invariant.

Let us consider the change of conformal metric $g' = \psi^{4/(n-2)}g$, one verifies that $J(\varphi\psi) = J'(\varphi)$, so that $\mu = \mu'$.

Since $\mu < 0$, if $J(\varphi) < 0$ for one function $\varphi \in H_1$, it occurred to me to make a change of metric (a non conformal one, obviously) so as to be always in this case (Aubin [1, p. 388]). Thus one proves:

THEOREM 3. A compact riemannian manifold M_n of dimension $n \geqslant 3$, carries a metric for which the scalar curvature is a negative constant.

It is possible now to answer a question of Chern: Has the sign of the scalar curvature any topological implications? Aubin [1, p. 402].

COROLLARY 1. The negative case for the scalar curvature has no topological signification, but on the other hand the positive case has topological meaning.

Indeed Hitchin [9] showed that a compact spin manifold with non zero Â-genus of Hirzebruch exists. And according to a theorem of Lichnerowicz [13]: every compact spin manifold, with a non-negative, not identically null, scalar riemannian curvature, has an Â-genus of Hirzebruch equal to zero.

Thus on such a manifold there does not exist a riemannian metric with a positive scalar curvature. As one sees, the null case remains open: Has every manifold a riemannian metric with an identically null scalar curvature?

3. The problem of Yamabe

THEOREM 4. A compact riemannian manifold M_n, $(n \geqslant 3)$, for which $\mu < n(n-1)(\omega_n)^{2/n}$ has the property of Yamabe.

Equation (2) has a C^∞ positive solution φ, with $R' = \mu$ and $\|\varphi\|_N = 1$. ω_n is the measure of the unit sphere $S_n(1)$ and for the meaning of μ see Theorem 1. Yamabe proved this theorem in the case $\mu \leqslant 0$ (Theorem 1). When $\mu \geqslant 0$, we shall employ a method suggested by Trüdinger [19]. First of all let us prove:

PROPOSITION 4. There exists a sequence φ_{q_i}, with $\lim_{i\to\infty} q_i = N$, so that φ_{q_i} converges weakly in H_1 and strongly in L_2 to a non-negative C^∞ function φ_0, verifying:

$$4\frac{n-1}{n-2}\Delta\varphi_0 + R\varphi_0 = \mu\varphi_0^{N-1}. \tag{5}$$

As μ is a conformal invariant, one may suppose that $\int_M dv = 1$. The set of the functions φ_q of Theorem 2 is bounded in H_1 because

$$\|\varphi_q\|_{H_1}^2 \leqslant \frac{n-2}{4(n-1)}\left[\mu_q + \sup_{x\in M}|R(x)|\,\|\varphi_q\|_2^2\right] + \|\varphi_q\|_2^2,$$

$\|\varphi_q\|_2 \leqslant \|\varphi_q\|_q = 1$

and

$$\mu_q \leqslant \int_M R\, dv.$$

The set of functions φ_q is also bounded in L_N, because the inbedding of H_1 in L_N is continuous (Sobolev [18]). Hence the set of functions, $\{\varphi_q^{q-1}\}$ is bounded in $L_{N/(N-1)}$; for $2 \leqslant q < N$, $N/(N-1) < N/(q-1)$. Therefore there exist $\varphi_0 \in H_1$ and a sequence $q_i \to N$ such that: $\varphi_{q_i} \to \varphi_0$ weakly in H_1, strongly in L_2 and almost everywhere, and such that $\varphi_{q_i}^{q_i-1} \to \varphi_0^{N-1}$ weakly in $L_{N/(N-1)}$. Let be $\psi \in H_1$, hence $\psi \in L_N$. As $1/N + (N-1)/N = 1$ and $\lim_{q\to N} \mu_q = \mu$: φ_0 verifies $\varphi_0 \geqslant 0$, $\varphi_0 \in H_1$ and for all $\psi \in H_1$

$$4\frac{n-1}{n-2}\int_M \nabla^\nu \varphi_0 \nabla_\nu \psi\, dv + \int_M R\varphi_0\psi\, dv = \mu \int_M \psi\varphi_0^{N-1}\, dv.$$

And according to a theorem of Trüdinger [19], $\varphi_0 \in C^\infty$, so that φ_0 verifies Equation (5).

By the maximum principle and a method of Nirenberg [16], one shows that two cases can happen: either φ_0 is identically null, or φ_0 is everywhere strictly positive. It is enough now to show that $\|\varphi_0\|_2 \neq 0$, for the proof of Theorem 4. According to Aubin [4], for all $\epsilon > 0$, there exists a constant A, such that for all $\varphi \in H_1$:

$$\|\varphi\|_N^2 \leqslant \left[\frac{4}{n(n-2)}(\omega_n)^{-2/n} + \epsilon\right] \|\nabla\varphi\|_2^2 + A\|\varphi\|_2^2.$$

Since $\|\varphi_q\|_q = 1$, $\|\varphi_q\|_N \geqslant 1$ and

$$1 \leqslant \left[\frac{(\omega_n)^{-2/n}}{n(n-1)} + \frac{(n-2)\epsilon}{4(n-1)}\right]\left[\mu_q - \int_M R\varphi_q^2\, dv\right] + A\,\|\varphi_q\|_2^2$$

when $\mu < n(n-1)(\omega_n)^{2/n}$, there exists $\eta > 0$ such that for $|N-q| < \eta$:

$$\left[\frac{1}{n(n-1)(\omega_n)^{2/n}} + \frac{(n-2)\epsilon}{4(n-1)}\right]\mu_q < 1$$

if one chooses ϵ small enough. For instance if

$$2\mu\epsilon < 1 - \frac{\mu}{n(n-1)(\omega_n)^{2/n}},$$

there exists $C > 0$, such that

$$\underline{\lim}_{q\to N}\|\varphi_q\|_2 \geqslant C. \qquad \text{Hence} \quad \|\varphi_0\|_2 \geqslant C.$$

To complete the proof, we have to show that the lower bound μ of $J_N(\varphi)$ is reached. Let us multiply Equation (5) by φ_0 and integrate it over M; we find

$$J_N(\varphi_0) = \mu\|\varphi_0\|_N^{N-2}, \qquad \text{then} \quad \|\varphi_0\|_N \geq 1.$$

Furthermore the sequence φ_{q_i} converges weakly to φ_0 in L_N, so that

$$\|\varphi_0\|_N \leq \varliminf_{i\to\infty} \|\varphi_{q_i}\|_N = 1$$

because $\mu \leq \mu\|\varphi_q\|_N^2 \leq J_N(\varphi_q)\|\varphi_q\|_N^2 = \mu_q \to \mu$ when $q \to N$. Therefore

$$\|\varphi_0\|_N = 1 \quad \text{and} \quad J_N(\varphi_0) = \mu.$$

Moreover by the theorem of Radon, $\varphi_{q_i} \to \varphi_0$ strongly in L_N and strongly in H_1.

COROLLARY 2. Equation (2) has a positive C^∞ solution with $R' = \mu$ and $\|\varphi\|_N = 1$ if

$$\int_M R \,\mathrm{d}v \leq n(n-1)(\omega_n)^{2/n}\left[\int_M \mathrm{d}v\right]^{(n-2)/n}.$$

In fact, in that case $\mu \leq J_N(1) \leq n(n-1)(\omega_n)^{2/n}$. Either $\mu < n(n-1)(\omega_n)^{2/n}$ and one can apply Theorem 4, or $\mu = J_N(1)$ and $\varphi \equiv 1$ is a solution of Equation (2) with $R' = \mu$.

Now let us apply Theorem 4 to some manifolds. Remember that

$$J(\varphi) = \left[4\frac{n-1}{n-2}\int_M \nabla^\nu\varphi\nabla_\nu\varphi \,\mathrm{d}v + \int_M R\varphi^2 \,\mathrm{d}v\right]\|\varphi\|_N^{-2}, \quad \text{with} \quad N = \frac{2n}{n-2}.$$

Let $B_p(\delta)$ be the ball with centre P and radius $\delta(0 < \delta \leq d$ the radius of injectivity), and compute $J(\psi_k)$ for the sequence of functions ψ_k with support in $\bar{B}_p(\delta)$ defined by

$$\psi_k = \left(\frac{1}{k} + \frac{1-\cos\alpha_r}{\alpha^2}\right)^{1-n/2} - \left(\frac{1}{k} + \frac{1-\cos\alpha_r}{\alpha^2}\right)^{1-n/2},$$

where $r = r(P, Q)$ and $R(P) = n(n-1)\alpha^2$. If $R(P) < 0$, $\cos\alpha r = \operatorname{ch} i\alpha r$

and if $R(P) = 0$,

$$\frac{1-\cos\alpha_r}{\alpha^2} = \frac{r^2}{2}.$$

One assumes also that δ verifies $\delta^2 R(P) \leqslant n(n-1)\pi^2$.

LEMMA 1. For the proof see Aubin [5]. When $k \to \infty$, $J(\psi_k) \sim n(n-1)(\omega_n)^{2/n}$. And a limited expansion gives for $n > 6$:

$$J(\psi_k) = n(n-1)(\omega_n)^{2/n}\left[1 + \frac{1}{9n(n-4)(n-6)k^2} \times \left(4\frac{2n-7}{n-2} T_{ij}T^{ij} - 3S_{ijkl}S^{ijkl}\right) + 0\left(\frac{1}{k^2}\right)\right];$$

and for $n = 6$

$$J(\psi_k) = n(n-1)(\omega_n)^{2/n} + \frac{\log k}{k^2}\frac{\omega_{n-1}}{\omega_n}(\omega_n)^{2/n}\frac{(n-2)(n-1)}{45n(n+2)} \times \left(4\frac{2n-7}{n-2} T_{ij}T^{ij} - 3S_{ijkl}S^{ijkl}\right) + 0\left(\frac{\log k}{k^2}\right).$$

The expressions of the tensors S_{ijkl} and T_{ij} must be taken at P. Let us now prove:

THEOREM 5. A compact non-locally conformally flat riemannian manifold of dimension $n \geqslant 6$, has Yamabe's property.

After Lemma 1, if at a point P,

$$4\frac{2n-7}{n-2} T_{ij}T^{ij} - 3S_{ijkl}S^{ijkl}$$

is negative, μ will be strictly less than $n(n-1)(\omega_n)^{2/n}$, and in that case Theorem 4 can be applied. Now S^j_{ikl} is a conformal invariant, like μ. On the other hand $T_{ij}T^{ij}$ is not.

If the manifold is not locally conformally flat, there exists a point P where $S_{ijkl}S^{ijkl} \neq 0$. In this case take a conformal deformation of the metric such that $T_{ij}T^{ij} = 0$ at P. For instance if f and its gradient vanish at P, let us consider the metric $g'_{ij} = e^f g_{ij}$. We have at P (see Aubin [1, p. 385]):

$$R'_{ij} = R_{ij} - \frac{n-2}{2}\nabla_i\nabla_j f + \frac{\Delta f}{2} g_{ij}$$

and $R' = R + (n-1)\Delta f$. Hence

$$T'_{ij} = T_{ij} - \frac{n-2}{2}\nabla_i\nabla_j f + \frac{n-2}{2n}\Delta f g_{ij}.$$

We take

$$(\Delta f)_P = 0 \quad \text{and} \quad (\nabla_i\nabla_j f)_P = \frac{2}{n-2}(T_{ij})_P,$$

and we apply Lemma 1 in this new metric.

THEOREM 6. A locally conformally flat riemannian manifold $M_n (n \geqslant 3)$ with a finite fundamental group has Yamabe's property.

Let $\tilde{M}_n$ be the universal covering of M_n. According to the theorem of Kuiper [11], there exists a conformal diffeomorphism between $\tilde{M}_n$ and S_n, since $\tilde{M}_n$ is simply connected and locally conformally flat. Thus there exists a C^∞ function $\varphi_0 > 0$ verifying:

$$4\frac{n-1}{n-2}\tilde{\Delta}\varphi_0 + \tilde{R}\varphi_0 = \tilde{\mu}_0\varphi_0^{N-1} \quad \text{and} \quad \int_{\tilde{M}} \varphi_0^N \, d\tilde{v} = 1. \tag{6}$$

Let γ_i, $(i = 1, 2, \ldots, k)$ be the isometries of $\tilde{M}_n$ corresponding to the covering. Let us call $f_i = \varphi_0 \circ \gamma_i$ and $\tilde{f} = \sum_{i=1}^k f_i$. f_i verifies (6) and $\tilde{f}$ is projectable, we can call its projection f. Let us calculate $J(f)$. $J(f) = k^{(2/N)-1}\tilde{J}(\tilde{f})$ and

$$\tilde{J}(\tilde{f}) = \frac{4\frac{n-1}{n-2}\int_{\tilde{M}} \tilde{f}\Delta\tilde{f}\, d\tilde{v} + \int_{\tilde{M}} \tilde{R}(\tilde{f})^2\, d\tilde{v}}{\left[\int_{\tilde{M}} (\tilde{f})^N \, d\tilde{v}\right]^{2/N}} = \tilde{\mu}_0 \frac{\int_{\tilde{M}} \tilde{f}\sum_{j=1}^k f_j^{N-1}\, d\tilde{v}}{\left[\int_{\tilde{M}} (\tilde{f})^N\, d\tilde{v}\right]^{2/N}}.$$

Now we apply

LEMMA 2. If the functions f_j are positive, and if $k > 1$, the following inequality is strict:

$$\int_{\tilde{M}} \left(\sum_{j=1}^k f_j^{N-1}\right)\left(\sum_{j=1}^k f_j\right) d\tilde{v} < \left[\sum_{j=1}^k \int_{\tilde{M}} f_j^N \, d\tilde{v}\right]^{1-(2/N)} \left[\int_{\tilde{M}} \left(\sum_{i=1}^k f_i\right)^N d\tilde{v}\right]^{2/N}$$

As for all j, $\int_{\tilde{M}} f_j^N \, d\tilde{v} = 1$, $J(f) < \tilde{\mu}_0$ and $\mu < \tilde{\mu}_0$. From Proposition 3, $\tilde{\mu}_0 = \tilde{J}(\varphi_0) = \tilde{J}'(1) = n(n-1)(\omega_n)^{2/n}$, $\tilde{g}'$ being the metric of the sphere. Thus μ satisfies $\mu < n(n-1)(\omega_n)^{2/n}$ and it is possible to apply Theorem 4.

Proof of Lemma 2. Let $a_i (i = 1, 2, \ldots, k;\ k > 1)$ and β be positive real numbers. From an inequality of Hardy *et al.* [7], it follows that:

$$\sum_{i=1}^{k} (a_i)^{N\beta+1} < \left(\sum_{i=1}^{k} a_i\right)\left(\sum_{i=1}^{k} a_i^N\right)^{\beta}.$$

Let us take $\beta = (N-2)/N$, $[N = 2n/(n-2) > 2]$, multiply both members of the preceding inequality by $\sum_{i=1}^{k} a_i$, put $a_i = f_i$ and integrate over $\tilde{M}$:

$$\int_{\tilde{M}} \left(\sum_{j=1}^{k} f_j\right)\left(\sum_{j=1}^{k} f_j^{N-1}\right) \mathrm{d}\tilde{v} < \int_{\tilde{M}} \left(\sum_{j=1}^{k} f_j\right)^2 \left(\sum_{j=1}^{k} f_j^N\right)^{(N-2)/N} \mathrm{d}\tilde{v}$$

$$\leqslant \left[\int_{\tilde{M}} \left(\sum_{j=1}^{k} f_j\right)^N \mathrm{d}\tilde{v}\right]^{2/N} \left[\int_{\tilde{M}} \sum_{j=1}^{k} f_j^N \,\mathrm{d}\tilde{v}\right]^{(N-2)/N}.$$

4. Other problems

Instead of making constant the scalar curvature as Yamabe, one might more generally want to make it equal to a given function. That is the purpose of:

THEOREM 7. Let $M_n (n \geqslant 3)$ be a riemannian manifold, for which $\mu < n(n-1)(\omega_n)^{2/n}$, it is possible to find a real number $\varepsilon > 0$, such that for each C^∞ function $f \geqslant 0$ on M_n, verifying $\sup_{x \in M_n} f(x) < \varepsilon$, there exists a conformal metric g' of g, such that scalar curvature $R' = \mu(1-f)$.

The proof is similar to that of the Theorem 4. One considers

$$I_q(\varphi) = \left[4\frac{n-1}{n-2}\int_M \nabla^\nu\varphi\nabla_\nu\varphi \,\mathrm{d}v + \int_M R\varphi^2 \,\mathrm{d}v\right]\left[\int_M (1-f)\varphi^q \,\mathrm{d}v\right]^{-(2/q)}$$

for $2 \leqslant q \leqslant N$, where $\sup_{x \in M_n} f(x) < 1$. If $\inf_{\varphi \in H_1, \varphi \neq 0} I_N(\varphi) < n(n-1)(\omega_n)^{2/n}$,

there exists a C^∞ strictly positive function verifying:

$$4\frac{n-1}{n-2}\Delta\varphi + R\varphi = \mu(1-f)\varphi^{N-1}.$$

And now the main theorem is:

THEOREM 8. The compact C^∞ locally conformally flat riemannian manifolds with finite fundamental group, and the compact C^∞ non-locally conformally flat riemannian manifolds of dimension $n \geq 6$, admit a conformal riemannian metric the scalar curvature R of which is constant and satisfies:

$$R\left[\int_{M_n} dv\right]^{2/n} \leq n(n-1)(\omega_n)^{2/n}$$

with equality only in the case of the sphere. And the first eigenvalue $\lambda_1 > 0$ of the laplacian verifies $\lambda_1 \geq R/(n-1)$.

This theorem is proved partly by Theorems 5 and 6. But there are two results to demonstrate. First of all on the sphere S_n, $\mu = n(n-1)(\omega_n)^{2/n}$. Indeed if $\mu < n(n-1)(\omega_n)^{2/n}$, one could apply Theorem 7 for the sphere. But that is impossible, according to Kazdan and Warner [10]. They showed that there is no solution of (2) on the sphere S_n, with $R' = 1 + \nu\psi_1$ for all real number $\nu \neq 0$ if ψ_1 verifies $\Delta\psi_1 = \lambda_1\psi_1$ where λ_1 is the first eigenvalue of the laplacian.

Secondly in the metric with constant curvature R, as $\mu = J_N(1) = R[\int_M dv]^{2/n}$ for all functions $\psi \in H_1$, with $|\psi| < 1$, $J_N(1+\psi) \geq \mu$:

$$4\frac{n-1}{n-2}\int_M \nabla^\nu\psi\nabla_\nu\psi \, dv + R\int_M (1+\psi)^2 \, dv \geq \mu\left[\int_M (1+\psi)^N \, dv\right]^{2/N}.$$

Terms of the first order in ψ obviously vanish. But the second order gives:

$$\int_M \nabla^\nu\psi\nabla_\nu\psi \, dv \geq \frac{R}{n-1}\left[\int_M \psi^2 \, dv - \left(\int_M \psi \, dv\right)^2\left(\int_M dv\right)^{-1}\right].$$

So that $\lambda_1 \geq R/(n-1)$.

COROLLARY 3. On the sphere S_n, $(n \geq 3)$, $\mu = n(n-1)(\omega_n)^{2/n}$ and Equation (2) has an infinity of non-proportional solutions with $R' =$ constant. All the functions $\varphi(r) = (\beta - \cos \alpha r)^{1-n/2}$, with $1 < \beta$ a real number and $\alpha^2 = R/n(n-1)$, are solutions of (2) with $R' = R(\beta^2 - 1)$, r being the geodesic distance at a given point P of S_n. On the sphere S_n, normed by $\int_{S_n} dv = 1$ all the functions $\varphi \in H_1$ verify:

$$\|\varphi\|^2_{2n/(n-2)} \leq \frac{4}{n(n-2)}(\omega_n)^{-2/n}\|\nabla\varphi\|^2_2 + \|\varphi\|^2_2.$$

The last inequality follows from Theorem 8:

$$\mu = R = n(n-1)(\omega_n)^{2/n}.$$

About the uniqueness. When $\mu \leqslant 0$, two solutions of (2) with $R' =$ constant are proportional (Aubin [1, p. 386]). On the other hand, when $\mu > 0$, there can exist several non-proportional solutions. We have seen already the case of the sphere, but there are other examples.

Let M_n be the manifold ($n \geqslant 6$) product of the sphere $S_{n-2}(1)$ by a torus T_2 product two circles of large radius so that the scalar curvature R, which is constant, verifies $R[\int_M \mathrm{d}v]^{2/n} \geqslant n(n-1)(\omega_n)^{2/n}$. As M_n is non-locally conformally flat, Theorem 4 can be applied. Therefore there exists at least two non-proportional solutions $\varphi_1 \equiv 1$ and the function φ_2, such that $J_N(\varphi_2) = \mu$ because $J(\varphi_1) > \mu$. But here is a theorem of uniqueness:

THEOREM 9. On a compact Einstein-manifold of dimension $n \geqslant 6$, different from the sphere, two solutions of Equation (2) with $R' = R$ are proportional. And the scalar curvature R verifies

$$R\left[\int_M \mathrm{d}v\right]^{2/n} = \mu < n(n-1)(\omega_n)^{2/n}. \tag{7}$$

If $R \leqslant 0$, then $\mu \leqslant 0$ and this result has already been shown. If $R > 0$, according to Obata [17], there is only one solution of (2) with $R' = R$. Furthermore inequality (7) is strict, because Theorem 8 can be applied. In fact, as the Ricci curvature is positive in this case, the fundamental group is finite according to Myers [15]. Finally by Theorem 8, $\lambda_1 \geqslant R/(n-1)$; the well known result of Lichnerowicz [12] is again found, but in a very indirect way.

In conclusion we make:

CONJECTURE 2. Every compact riemannian manifold M_n, $n \geqslant 3$, has his conformal invariant $\mu < n(n-1)(\omega_n)^{2/n}$, except the simply connected and locally conformally flat manifolds (for which $\mu = n(n-1)(\omega_n)^{2/n}$, by Theorem 8).

It should be remembered that for every compact riemannian manifold $M_n (n \geqslant 3)$: $\mu \leqslant n(n-1)(\omega_n)^{2/n}$ by Lemma 1.

When $n \geqslant 6$, we proved Conjecture 2 for almost all manifolds (Theorem 8). Conjecture 2 rests only on the locally conformally flat manifolds with infinite fundamental group.

By Theorem 4 and Kuiper's theorem [11], Conjecture 2 involves:

CONJECTURE 3. On every compact riemannian manifold M_n, $(n \geqslant 3)$, there exists a C^∞ function $\varphi \geqslant 0$, such that $J_N(\varphi) = \mu$.

The conjecture of Yamabe follows from this conjecture. From Conjecture 2 and Theorem 7, we get:

CONJECTURE 4. On a riemannian manifold M_n, $(n \geqslant 3)$, which is not simultaneously simply connected and locally conformally flat, it is possible to find a real number $\varepsilon > 0$, such that for each C^∞ function $f \geqslant 0$ on M_n, verifying $\sup_{x \in M_n} f(x) < \varepsilon$, there exists a conformal metric g' of g, such that the scalar curvature $R' = \mu(1 - f)$.

University Pierre et Marie Curie, Paris

REFERENCES

[1] Aubin, T., 'Metriques riemanniennes et courbure', *J. Diff. Geo.* **4** (4) (1970), 383–424.
[2] Aubin, T., 'Sur la fonction exponentielle', *C.R. Acad. Sc. Paris* **270** (1970), 1514.
[3] Aubin, T., 'Fonction de Green et valeurs propres du laplacien', *J. Math. Pures et appl.* **53** (1974), 347–371.
[4] Aubin, T., 'Problèmes isopérimétriques et espaces de Sobolev', *J. Diff. Geo.*
[5] Aubin, T., 'Equations différentielles non linéaires et problème de Yamabe concernant la courbure scalaire', *J. Math. Pures et appli.*
[6] Berger, M., Riemannian Structures of Prescribed Gaussian Curvature for Compact 2-manifolds', *J. Diff. Geo.* **5** (1971), 325–332.
[7] Hardy, G., Littlewood, J., and Polya, G., *Inequalities.*
[8] Hempel, J. A., Morris, G. R., and Trüdinger, N. S., On the Sharpness of a Limiting Case of the Sobolev Imbedding Theorem', *Bull. Austral. Math. Soc.* **3** (1970), 369–373.
[9] Hitchin, N., 'Compact Four-dimensional Einstein Manifolds', *J. Diff. Geo.* **9** (1974), 435–441.
[10] Kazdan, J. L. and Warner, F. W., 'Scalar Curvature and Conformal Deformation of Riemannian Structure', *J. Diff. Geo.* **10** (1975), 113–134.
[11] Kuiper, N. H., 'On Conformally Flat Spaces in the Large', *Ann. of math.* **50** (1949), 916–924.
[12] Lichnerowicz, A., Géométrie des groupes de tranformations', *Dunod* **58**.
[13] Lichnerowicz, A., 'Spineurs harmoniques', *C.R. Acad. Sc. Paris* **257** (1963), 7.
[14] Moser, J., 'A Sharp Form of an Inequality by N. Trüdinger', *Indiana Un. Math. J.* **20** (11) (1971), 1077–1092.
[15] Myers, S., *Duke Math. J.* **8** (1941), 401–404.
[16] Nirenberg, L., 'A Strong Maximum Principle for Parabolic Equations', *Comm. on pure and applied Math.* **6** (1953), 167–177.
[17] Obata, M., 'The Conjectures on Conformal Transformations of Riemannian Manifolds', *J. Diff. Geo.* **6** (1971), 247–258.

[18] Sobolev, S. L., 'Sur un théorème d'analyse fonctionnelle' *Mat. Sbornik* **45** (1938), 471–496.
[19] Trüdinger, N. S., 'Remarks Concerning the Conformal Deformation of Riemannian Structures on Compact Manifolds', *Ann. Scuola Norm. Sup. Pisa (3)* **22** (1968), 265–274.
[20] Trüdinger, N. S., 'On Imbeddings into Orlicz Spaces and Some Applications', *J. of Math. and Mech.* **17** (5) (1967), 473–483.
[21] Yamabe, H., 'On a Deformation of Riemannian Structures on Compact Manifolds', *Osaka Math. J.* **12** (1960), 21–37.

GENEVIÈVE AVÉROUS and SHOSHICHI KOBAYASHI*

ON AUTOMORPHISMS OF SPACES OF NONPOSITIVE CURVATURE WITH FINITE VOLUME

1. Introduction

The purpose of this paper is to show that the automorphism group of a space of nonpositive curvature with finite total volume is usually finite.

Let M be a Riemannian manifold and $\tilde{M}$ its universal covering manifold with the deck transformation group Γ. Let $I(M)$ and $I(\tilde{M})$ denote the group of isometries of M and $\tilde{M}$ respectively. Let $N = N_{I(\tilde{M})}(\Gamma)$ be the normalizer of Γ in $I(\tilde{M})$. Since every element of $I(M)$ lifts to an element of $I(\tilde{M})$ (unique modulo Γ), it follows that $I(M)$ is naturally isomorphic to N/Γ.

THEOREM 1. Let M be a complete Riemannian manifold of nonpositive sectional curvature with finite volume. If M admits no nonzero parallel vector field (in particular, if the Ricci tensor of M is negative-definite at some point), then $I(M) = N/\Gamma$ is finite.

We should point out that the fact that $I(M)$ is discrete is already in [1] and [5].

COROLLARY. Let $\tilde{M}$ be a Riemannian symmetric space of noncompact type and Γ a discrete subgroup of $I(\tilde{M})$ acting freely on $\tilde{M}$. Let $M = \Gamma\backslash\tilde{M}$. If M has finite volume, then $I(M) = N/\Gamma$ is finite.

A complex analogue of Theorem 1 would be the following:

CONJECTURE. Let M be a complete Kähler manifold of nonpositive holomorphic bisectional curvature with finite volume. If M admits no nonzero parallel vector field, then the group of holomorphic transformations of M is finite.

Although we are unable to verify this conjecture, we can derive a complex analogue of the corollary above from Theorem 1. For a

* Partially supported by NSF Grant GP-42020X.

Cahen & Flato (eds.), Differential Geometry and Relativity, 19–26.

complex manifold M, we denote its group of holomorphic transformations by $A(M)$.

THEOREM 2. *Let $\tilde{M}$ be a hermitian symmetric space of noncompact type and Γ a discrete subgroup of $A(\tilde{M})$ acting freely on $\tilde{M}$. Let $M = \Gamma\backslash\tilde{M}$. If M has finite volume, then $A(M)$ $(= N/\Gamma$ where $N = N_{A(\tilde{M})}(\Gamma))$ is finite.*

This follows immediately from Theorem 1 (or its corollary) and from the fact that the hermitian metric of $\tilde{M}$ is the so-called Bergman metric which is invariant by $A(\tilde{M})$ so that $A(\tilde{M}) \subset I(\tilde{M})$ and $A(M) \subset I(M)$.

In both Corollary to Theorem 1 and Theorem 2 we assumed that Γ acts freely on $\tilde{M}$, i.e., Γ has no element of finite order (other than the identity element) so that $\tilde{M}$ is the unramified universal covering space of a nonsingular manifold M. But it is of some interest to consider also the case where the action of Γ is not necessarily free. This is particularly the case for hermitian symmetric domains.

Let $\tilde{M}$ be a hermitian symmetric space of noncompact type and Γ a discrete subgroup of $A(\tilde{M})$. For each element γ of Γ different from the identity, let F_γ denote the set of points of $\tilde{M}$ left fixed by γ. Since γ is a holomorphic isometry, F_γ is a totally geodesic closed complex submanifold of $\tilde{M}$ and is a hermitian symmetric submanifold of $\tilde{M}$. Let $F = \bigcup_{1\neq\gamma\in\Gamma} F_\gamma$. Then F is invariant by Γ and $\tilde{M} - F$ is an unramified covering manifold of $\pi(\tilde{M} - F) = M - \pi(F)$, where π denotes the projection $\tilde{M} \to M$. If the complex codimension of F_γ is at least 2, then $\tilde{M} - F$ is simply connected and hence every transformation of $M - \pi(F)$ lifts to a transformation of $\tilde{M} - F$. Thus Gottschling [2] shows that $A(M)$ is isomorphic to N/Γ, where $N = N_{A(\tilde{M})}(\Gamma)$. From this we derive the following

THEOREM 3. *Let $\tilde{M}$ be a hermitian symmetric space of noncompact type and Γ a discrete subgroup of $A(\tilde{M})$ such that $M = \Gamma\backslash\tilde{M}$ has finite volume. Assume that each F_γ, $1 \neq \gamma \in \Gamma$, has complex codimension at least 2. Then $A(M) = N/\Gamma$, where $N = N_{A(\tilde{M})}(\Gamma)$, is finite.*

We shall see that all the results stated above, except Theorem 1, can be derived also from a purely group theoretic result of Wang [7].

Although the Riemannian analogue of Theorem 3 may be obtained by imposing the condition that the real codimension of F_γ be at least 3, it is probably of little interest for a lack of applications.

The last section will be devoted to examples including those of Gottschling [2].

2. Proofs of theorems

The proof of Theorem 1 is based on the following result of Bishop and O'Neill [1].

LEMMA 1. If M is a complete Riemannian manifold with finite volume, then all convex functions on M are constant.

Although this lemma is valid for nonsmooth continuous convex functions as well [8], we need only the original, smooth version of Bishop and O'Neill. The following lemma is in [1, 5].

LEMMA 2. If M is a complete Riemannian manifold of nonpositive sectional curvature with finite volume, then the only Killing vector fields of M are the parallel vector fields.

The proof is similar to that of the well known theorem of Bochner on the non-existence of Killing vector fields on a compact Riemannian manifold with negative Ricci tensor and goes as follows. Let X be a Killing vector field on M and let $f = \|X\|^2$. We claim that f is a convex function if the sectional curvature of M is nonpositive. In fact, a simple calculation shows (see, for instance [4, pp. 42–44]) that

$$\nabla^2 f(Y, Y) = \|\nabla_Y X\|^2 - R(X, Y, X, Y)$$

for every tangent vector Y, where $\nabla^2 f$ is the Hessian or the second covariant derivative of f. This formula shows that $\nabla^2 f$ is positive semi-definite, i.e., f is convex. By Lemma 1, f is constant and the formula above implies that $\nabla X = 0$ and $R(X, Y, X, Y) = 0$ for all Y.

LEMMA 3. If M is a complete Riemannian manifold having finite volume, then the group of isometries of M is compact.

As the first step of the proof of this lemma, we imbed the group $I(M)$ of isometries in the bundle $0(M)$ of orthonormal frames of M as follows. Fix a frame $u_0 \in 0(M)$. Since each isometry $\varphi \in I(M)$ induces an automorphism of the bundle $0(M)$, we obtain a mapping

$$\varphi \in I(M) \to \varphi(u_0) \in 0(M).$$

This mapping imbeds $I(M)$ in $0(M)$ as a *closed* submanifold [4, pp. 39–41]. Hence $I(M)$ is compact if and only if the image of $I(M)$ in $0(M)$ lies in $p^{-1}(K)$ for some compact subset K of M, where p denotes the projection $0(M) \to M$. To complete the proof, it suffices to verify that, for a given point x_0 of M, the set $\{\varphi(x_0) : \varphi \in I(M)\}$ is contained in some compact subset K of M. We fix a positive number r and let B be the closed ball of radius r and center x_0 in M. For each positive integer m, let K_m be the closed ball of radius $2mr$ and center x_0. If the set $\{\varphi(x_0) : \varphi \in I(M)\}$ is not contained in any compact subset of M, we can choose a sequence of isometries φ_i together with an increasing sequence of integers m_i such that

$$\varphi_i(x_0) \in K_{m_i} \quad \text{but} \quad \varphi_i(x_0) \notin K_{m_i - 1} \quad \text{for each} \quad i.$$

Then we obtain a sequence of balls $\varphi_i(B)$ which are disjoint. Since each of these balls has the same volume as B, M will have infinite volume. This is absurd.

Theorem 1 follows immediately from Lemmas 2 and 3. Note that the Killing vector fields in Lemma 2 generate a torus group since $I(M)$ is compact by Lemma 3, and this fact is consistent with the structure theorem in [5].

In connection with the conjecture stated in §1, we can ask if the following complex analogue of Lemma 1 is true. If M is a complete Kähler manifold with finite volume, are all plurisubharmonic functions on M constant? If this is true, we obtain the complex analogue of Lemma 2 to the effect that on a complete Kähler manifold of nonpositive holomorphic bisectional curvature with finite volume every holomorphic vector field is parallel.

Before we discuss theorem 3, we point out that Corollary to Theorem 1 and Theorem 2 may be derived from the following lemma of Wang [7] as well as from Theorem 1.

LEMMA 4. Let G be a connected semisimple Lie group without compact factor and Γ a discrete subgroup such that G/Γ has finite volume. Then the normalizer $N_G(\Gamma)$ of Γ in G is discrete and $N_G(\Gamma)/\Gamma$ is finite.

In applying Lemma 4 to the situation in Corollary or Theorem 2, we let G be the identity component of $I(\tilde{M})$. Since $I(\tilde{M})$ has a finite number of components, we can replace $I(\tilde{M})$ by G in the argument.

Finally, we shall indicate how Theorem 3 can be proved. There are two parts to the theorem: (i) $A(M)=N/\Gamma$ and (ii) the finiteness of N/Γ. The main point of Gottschling's proof of $A(M)=N/\Gamma$ is that $\pi(F)$ is precisely the set of singular points of the complex space $M=\Gamma\backslash\tilde{M}$ if every F_γ, $1\neq\gamma\in\Gamma$, has codimension at least 2. For this, see [2] and references therein. Then every transformation $\varphi\in A(M)$ maps $M-\pi(F)$ onto itself. Since $\tilde{M}-F$ is simply connected because of codim $F_\gamma\geqslant 2$, it is a universal covering space of $M-\pi(F)$, and φ lifts to a holomorphic transformation of $\tilde{M}-F$ which extends to a holomorphic transformation $\tilde{\varphi}$ of $\tilde{M}$. It then follows easily that the natural (injective) homomorphism $N/\Gamma\to A(M)$ is surjective. The fact that N/Γ is finite follows from Lemma 4. It can be derived also differential geometrically by extending Lemma 1 to Γ-invariant convex functions on $\tilde{M}$.

3. Examples

Let $\tilde{M}$ be a hermitian symmetric space of noncompact type. Gottschling calls a holomorphic transformation $\gamma\in A(\tilde{M})$ a *reflection* if it has finite order and its fixed point set F_γ is of codimension 1. If $A(\tilde{M})$ contains no reflections, then Theorem 3 applies to $M=\Gamma\backslash\tilde{M}$ for any discrete subgroup Γ such that $\Gamma\backslash\tilde{M}$ has finite volume.

The irreducible hermitian symmetric spaces of noncompact type are classified in four classical types and two exceptional types:

$D_{\mathrm{I}}(m,n)$, $1\leqq m\leqq n$, of rank m and complex dimension mn;

$D_{\mathrm{II}}(n)$, $5\leqq n$, of rank $[n/2]$ and complex dimension $n(n-1)/2$;

$D_{\mathrm{III}}(n)$, $2\leqq n$, of rank n and complex dimension $n(n+1)/2$;

$D_{\mathrm{IV}}(n)$, $5\leqq n$, of rank 2 and complex dimension n;

E_{III} of rank 2 and complex dimension 16;

E_{VII} of rank 3 and complex dimension 27.

For the classical domains, Gottschling [3] determined the reflections completely. His result shows:

THEOREM 4. There are no reflections in the domains

$D_{\mathrm{I}}(m,n)$, $3\leqq m\leqq n$ or $2=m<n$;

$D_{\mathrm{II}}(n)$, $5\leqq n$;

$D_{\mathrm{III}}(n)$, $3\leqq n$;

and reflections do exist in

$$D_{\mathrm{I}}(2,2) \quad \text{or} \quad D_{\mathrm{I}}(1,n),\ 1 \leqq n;$$
$$D_{\mathrm{III}}(2);$$
$$D_{\mathrm{IV}}(n),\ 5 \leqq n.$$

His proof is based on the explicit calculation of the group action of $A(M)$. We assert:

THEOREM 5. There are no reflections in the exceptional domains E_{III} and E_{VII}.

To prove Theorem 5 it suffices to show that there are no totally geodesic complex hypersurfaces in E_{III} or E_{VII}. We shall indicate how this can be done for E_{VII}. We make use of the three facts: (i) rank $(E_{\mathrm{VII}}) = 3$, (ii) dim $(E_{\mathrm{VII}}) = 27$, (iii) dim $(A(E_{\mathrm{VII}})) = 133$. Then we have only to show that there are no hermitian symmetric spaces G/K of noncompact type (irreducible or otherwise) such that (i) rank $(G/K) \leqslant 3$, (*ii*) dim $(G/K) = 26$, (iii) dim $(G) \leqq 133$. But this can be easily established since conditions (i) and (ii) limit the possibility for G/K. The proof for E_{III} is even easier.

The domain $D_{\mathrm{III}}(n)$ is the so-called Siegel's upper half-plane $H(n)$ of degree n consisting of complex symmetric matrices $Z = X + iY$ of order n with positive definite imaginary part Y. The automorphism group of $H(n)$ is $\mathrm{Sp}(n;\mathbf{R})$ and, according to Ramanathan [6], the Siegel modular group $\mathrm{Sp}(n;\mathbf{Z})$ is a maximal discrete subgroup of $\mathrm{Sp}(n;\mathbf{R})$. This fact, together with Theorems 3 and 4 yields the result of Gottschling [2]:

THEOREM 6. The automorphism group of $\mathrm{Sp}(n;\mathbf{Z})\backslash H(n)$ is trivial for $n \geqq 3$.

Next, we consider Hilbert-Siegel modular groups. Let k be a totally real algebraic number field of degree r and let $k^{(1)}\ (=k), k^{(2)}, \ldots, k^{(r)}$ be the conjugate fields. Let o be the ring of integers of k. We let the Hilbert-Siegel modular group $\mathrm{Sp}(n;o)$ act on $H(n)^r = H(n) \times \cdots \times H(n)$ by

$$(A,(Z_1,\ldots,Z_r)) \to (A^{(1)}Z_1,\ldots,A^{(r)}Z_r) \qquad \text{for}$$
$$A \in \mathrm{Sp}(n;o), \quad Z_i \in H(n),$$

where $A^{(2)},\ldots,A^{(r)}$ are the conjugates of A. (The action of $A^{(i)}$ on $H(n)$ is by the usual linear fractional transformation.)

THEOREM 7. If $r \geqq 2$, then $\mathrm{Sp}(n; o)$ acting on $H(n)^r$ contains no reflections.

Proof. Assume $A \in \mathrm{Sp}(n; o)$ leaves a point $(Z_1^0, \ldots, Z_r^0) \in H(n)^r$ fixed. The differential $\mathrm{d}A$ of A defines a linear transformation of the tangent space $T_{Z_1^0}(H(n)) \times \cdots \times T_{Z_r^0}(H(n))$ given by $((\mathrm{d}A^{(1)})_{Z_1^0}, \ldots, (\mathrm{d}A^{(r)})_{Z_r^0})$. If A is a reflection, then all of $(\mathrm{d}A^{(1)})_{Z_1^0}, \ldots, (\mathrm{d}A^{(r)})_{Z_r^0}$ except one are identity transformations. Without loss of generality, assume that $(\mathrm{d}A^{(1)})_{Z_1^0}$ is the identity transformation of the tangent space $T_{Z_1^0}(H(n))$. Then $A^{(1)}$ must be the identity transformation of $H(n)$ (since a holomorphic transformation of a bounded domain $H(n)$, being an isometry, is determined by its first jet at a single point). Hence, all its conjugates $A^{(2)}, \ldots, A^{(r)}$ are identity transformations. Q.E.D.

Actually we proved that, for every element $A \in \mathrm{Sp}(n; o)$, the fixed point set F_A is of codimension at least r.

According to Ramanathan [6, pp. 147, 150] there is a maximal discrete subgroup of $G = \mathrm{Sp}(n; \mathbf{R}) \times \cdots \times \mathrm{Sp}(n; \mathbf{R})$ containing $\mathrm{Sp}(n; o)$ and it contains $\mathrm{Sp}(n; o)$ as a normal subgroup of index 2^l, where l is a nonnegative integer independent of n. This maximal discrete subgroup must be therefore $N_G(\mathrm{Sp}(n; o))$ since $N_G(\mathrm{Sp}(n; o))$ is discrete.

THEOREM 8. Let S_r denote the symmetric group permuting the r factors of $H(n)^r$, and let $\Gamma = \mathrm{Sp}(n; o)$ and $G = \mathrm{Sp}(n; \mathbf{R}) \times \cdots \times \mathrm{Sp}(n; \mathbf{R})$. Then we have a natural exact sequence:

$$1 \to N_G(\Gamma)/\Gamma \to A(\Gamma \backslash H(n)^r) \to S_r \to 1,$$

where $N_G(\Gamma)/\Gamma$ is of order 2^l independent of n and every element except the identity of $N_G(\Gamma)/\Gamma$ is of order 2.

This follows from the exact sequence:

$$1 \to G \to A(H(n)^r) \to S_r \to 1$$

which induces the exact sequence:

$$1 \to N_G(\Gamma) \to N_{A(H(n)^r)}(\Gamma) \to S_r \to 1.$$

The assertion that every element except the identity of $N_G(\Gamma)/\Gamma$ is of order 2 is also due to Ramanathan.

He proves that $N_G(\Gamma) = \Gamma$ if $k = Q(\sqrt{5})$.

Conservatoire National des Arts et Métiers, Paris and University of California, Berkeley

REFERENCES

[1] Bishop, R. and O'Neill, B., 'Manifolds of Negative Curvature', *Trans. Amer. Math. Soc.* **145** (1969), 1–49.
[2] Gottschling, E., 'Invarianten endlicher Gruppen und biholomorphe Abbildungen', *Inventiones math.* **6** (1969), 315–326.
[3] ------, 'Reflections in Bounded Symmetric Domains', *Communications on pure & appl. Math.* **22** (1969), 693–714.
[4] Kobayashi, S., 'Transformation Groups in Differential Geometry', *Ergebnisse* **70** (1972), Springer-Verlag.
[5] Lawson, H. B. and Yau, S.T., 'Compact Manifolds of Nonpositive Curvature', *J. Diff. Geometry* **7** (1972), 211–228.
[6] Ramanathan, K. G., 'Discontinuous Groups II', *Nachr. Akad. Wiss. Göttingen, II. Math-Phys. Klasse*, (1964), 145–164.
[7] Wang, H. C., 'On a Maximality Property of Discrete Subgroups with Fundamental Domain of Finite Measure, *Amer. J. Math.* **89** (1967), 124–132.
[8] Yau, S. T., 'Non-Existence of Continuous Convex Functions on Certain Riemannian Manifolds', *Math. Ann.* **207** (1974), 269–270.

A. AVEZ

HARMONIC FUNCTIONS ON GROUPS*

ABSTRACT. Bounded harmonic functions on groups of non exponential growth are constant.

INTRODUCTION

E. Nelson [8] gave a remarkable proof of Liouville's theorem: Consider a bounded harmonic function f on Euclidean space. Given two points x, y, choose two balls $B(x, r)$, $B(y, r)$ with the given points as centers and of radius r. Since f is harmonic, its value at x (resp. y) is its average over $B(x, r)$ [resp. $B(y, r)$]. Then

$$|f(x)-f(y)| = \frac{1}{|B(x,r)|}\cdot\left|\int_{B(x,r)} f - \int_{B(y,r)} f\right| \leqslant$$

$$\sup|f|\cdot\frac{|B(x,r)\Delta B(y,r)|}{|B(x,r)|} = 0(r^{-1}),$$

where $|\ |$ refers to volume and Δ to symmetric difference. As $r\to+\infty$, we obtain $f(x)=f(y)$.

The key point is the growth of the volume $|B(x, r)|$ as $r\to+\infty$. One expects a generalization to homogeneous spaces the balls of which have 'non exponential growth'.

I. GROWTH

I.1. DEFINITION. A locally compact group G is compactly generated if there exists a compact neighbourhood A of the identity e such that G is the smallest subgroup which contains A. We say that A generates G.

I.2. GROWTH. Let m be a right invariant Haar measure of G. The measure of a Borel subset K of G is denoted by $|K|$. We write K^n for $\{k_1 \dots k_n : k_i \in K\}$.

* A harmonic function on a compact connected riemannian manifold is constant. André Lichnerowicz dignified this result up to the rank of a powerful method of proof. This paper, dedicated to him with gratitude and affection, contains an extension to certain non compact spaces.

Cahen & Flato (eds.), Differential Geometry and Relativity, 27–32.

THEOREM (Y. Guivarc'h [6] and J. Milnor [7]). Let V be a compact neighbourhood of e which generates G. Then $\lim_{n=\infty} |V^n|^{1/n} = C(V)$ and $C(V) \geq 1$.

If $C(V) = 1$ we say that G has non exponential growth. This property is independent of a particular choice of V. Indeed, if W is another compact neighbourhood of e which generates G, $W^k \subset V$ for some positive integer k, and $C(W) \leq [C(V)]^k$. Therefore $C(V) = 1$ implies $C(W) = 1$.

The growth of a finitely generated group was introduced by V. Arnold (private communication, 1965) and Svarc [9] and Milnor [7]. Guivarc'h [6] handled the general case.

Finite extensions of finitely generated nilpotent groups have non exponential growth [10]. For connected LIE groups Guivarc'h [6] proved the equivalence of the following properties:

(a) G has non exponential growth,
(b) G is unimodular and amenable,
(c) G is of type R, i.e. weights of its adjoint representation have all no real part.

2. Random walks

From now on G is a locally compact group, compactly generated.

II.2. DEFINITION. A random walk on G is a probability measure μ on G. We shall assume:

(1) μ has compact support S,
(2) μ is aperiodic: S generates G,
(3) μ has a density $p \in L^\infty(G, m)$ with respect to the right invariant Haar measure m.

Example. G is discrete and μ has a finite support which generates G.

Here is a more sophisticated (and fundamental) example:

II.2. LAPLACE OPERATOR ON LIE GROUPS. Assume G is an n-dimensional connected Lie group, carrying a right invariant riemannian metric. The riemannian volume element dV defines a right invariant Haar measure. Let $B(x,r)$ [resp. $S(x,r)$] denote the riemannian ball [resp. sphere] centered at x and of radius r. The area element of $S(x,r)$ will be $\mathrm{d}S$. Finally, Δ is the Laplace operator [it could be replaced by any right invariant differential operator satisfying the maximum principle].

Fix a positive number r. If f is harmonic, i.e. $\Delta f = 0$, then

$$f(x) = \int_{S(x,s)} N(x,y)\cdot f(y)\cdot \mathrm{d}S(y) \qquad \text{for any} \quad x \in G, s \in [0,r], \tag{1}$$

where N is the Poisson kernel of Δ.

Since $N \geqslant 0$ and $N(x,y) = 0(|x,y|^{-n+1})$, where $|\ ,\ |$ denotes the riemannian distance, there exists a C^∞ function $u:[0,r]\to R^+$ such that $p: g \rightsquigarrow N(e,g)\cdot u(|e,g|)$ is C^∞ on $B(e,r)$, non negative and of compact support $\overline{B(e,r)}$.

Since $\int_{S(x,s)} N(x,y)\cdot \mathrm{d}S(y) = 1$ for any $s \in [0,r]$, we may even assume

$$\int_G p(g)\cdot \mathrm{d}V(g) = \int_0^r \left[\int_{S(e,s)} N(e,g)\cdot \mathrm{d}S(g)\right] u(s)\cdot \mathrm{d}s$$

$$= \int_0^r u(s)\,\mathrm{d}s = 1.$$

Therefore $\mathrm{d}\mu = p\cdot \mathrm{d}V$ is a probability measure on G, satisfying the assumptions of II.1.

Moreover Equation (1) and the right invariance of N and $\mathrm{d}V$ imply

$$f(x) = \int_{B(x,r)} N(x,y)\cdot u(|x,y|)\cdot f(y)\cdot \mathrm{d}V(y)$$

$$= \int_{B(e,r)} p(g)\cdot f(gx)\cdot \mathrm{d}V(g). \tag{2}$$

We proceed to generalize this situation in III.

III. Liouville's theorem

III.1. μ-HARMONIC FUNCTIONS. Under the assumptions of II.1, we say that a bounded Borel function $f: G\to R$ is μ-harmonic if $\mu * f = f$, i.e.

$$f(x) = \int_G p(g)\cdot f(gx)\cdot \mathrm{d}m(g).$$

Then, for any positive integer n, we have

$$f(x)=\int_G p^{(n)}(g)\cdot f(gx)\cdot \mathrm{d}m(g)=\int_G p^{(n)}(gx^{-1})\cdot f(g)\cdot \mathrm{d}m(g), \tag{3}$$

where $p^{(n)}$ is defined inductively: $p^{(1)}=p$,

$$p^{(n+1)}(g)=\int_G p^{(n)}(gh^{-1})\cdot p(h)\cdot \mathrm{d}m(h). \tag{4}$$

Constants are bounded μ-harmonic functions. The converse is true if G is finitely generated and Abelian (Choquet-Deny [3]), nilpotent (Dynkin-Maliutov [5]) or finite extension of a nilpotent group (Azencott [2] and Guivarc'h [6]). *A posteriori* these groups have all non exponential growth. The following theorem gives the reason. It is a generalization of [1] and, even for finitely generated groups, the proof is simpler.

III.2. THE MAIN RESULT. Let G be a locally compact group, compactly generated, of non exponential growth, carrying a right invariant Haar measure m. Assume $\mu=p\cdot m$, $p\in L^\infty(G,m)$, is a probability measure, aperiodic and of compact support S. Then, any bounded μ-harmonic function is constant.

III.3. PRELIMINARY LEMMA. Let $z(t)=-t\cdot\log t$ if $t>0$ and $z(0)=0$. Then,

$$\limsup_{n=\infty} n^{-1}\int_G z[p^{(n)}(g)]\cdot \mathrm{d}m(g)\leqslant 0.$$

Proof. The support of $p^{(n)}$ is S^n and

$$\int_G p^{(n)}(g)\cdot \mathrm{d}m(g)=1. \tag{5}$$

Since z is a concave function, Jensen's inequality gives

$$n^{-1}\cdot\int_G z[p^{(n)}(g)]\,\mathrm{d}m(g)\leqslant n^{-1}\cdot|S^n|\cdot z\left[|S^n|^{-1}\cdot\int_{S^n} p^{(n)}(g)\,\mathrm{d}m(g)\right]=$$

$$n^{-1}\cdot|S^n|\cdot z(|S^n|^{-1})=\log|S^n|^{1/n}\to 0 \qquad \text{as } n\to+\infty.$$

III.4. PROOF OF III.2. We may assume $0<f\leqslant M=\sup f$. Let $\psi(t)=t\cdot\log t-t+1$ if $t>0$ and $\psi(0)=1$. Consider $I=$

$\int_G \psi[f(u)/f(e)] \cdot p(u) \cdot \mathrm{d}m(u)$, and use relation (3) to write

$$\frac{f(u)}{f(e)} = \int_{S^{n+1}} \frac{p^{(n)}(gu^{-1})}{p^{(n+1)}(g)} \cdot \frac{p^{(n+1)}(g) \cdot f(g)}{f(e)} \cdot \mathrm{d}m(g).$$

Since ψ is a convex function and

$$\frac{p^{(n+1)}(g) \cdot f(g)}{f(e)} \geq 0,$$

$$\int_{S^{n+1}} \frac{p^{(n+1)}(g) \cdot f(g) \cdot \mathrm{d}m(g)}{f(e)} = 1,$$

Jensen's inequality implies

$$\psi\left(\frac{f(u)}{f(e)}\right) \leq \int_{S^{n+1}} \frac{p^{(n+1)}(g) \cdot f(g)}{f(e)} \cdot \psi\left[\frac{p^{(n)}(gu^{-1})}{p^{(n+1)}(g)}\right] \mathrm{d}m(g)$$

and

$$I \leq \frac{M}{f(e)} \int_S \int_{S^{n+1}} p(u) \cdot p^{(n+1)}(g) \cdot \psi\left[\frac{p^{(n)}(gu^{-1})}{p^{(n+1)}(g)}\right] \mathrm{d}m(g) \cdot \mathrm{d}m(u).$$

Expand $\psi(\ )$ and use Fubini's theorem and the right invariance of m. Relations (4) and (5) yield

$$I \leq \frac{M}{f(e)} \left[\int_G z[p^{(n+1)}(g)] \cdot \mathrm{d}m(g) - \int_G z[p^{(n)}(g)] \cdot \mathrm{d}m(g)\right].$$

Telescope and use the lemma:

$$I \cdot f(e) \leq M \cdot \limsup_{n=\infty} n^{-1} \cdot \int_G z[p^{(n)}(g)] \cdot \mathrm{d}m(g) \leq 0.$$

Since $\psi(t) > 0$ if $t \neq 1$, we obtain $f(u) = f(e)$ for $u \in S$. Now $F: g \rightsquigarrow f(ga)$ is still a bounded μ-harmonic function for any a. Therefore $F(u) = F(e)$, i.e. $f(ua) = f(a)$ for $u \in S$. Take $a = u^{-1}: f(e) = f(u) = f(u^{-1})$ and, consequently, $f(b) = f(ub) = f(u^{-1}b)$ for any $u \in S$, $b \in G$. Since S generates G, $f =$ constant.

III.5. COROLLARY. A bounded harmonic function on a homogeneous space of a Lie group of type R is constant.

III.6. REMARK. One may wonder if one can dispense with homogeneity in III.5.

Non exponential growth still makes sense for arbitrary riemannian manifold: let M be a riemannian manifold. Denote by $|B(x, r)|$ the volume of a riemannian ball centered at x and of radius r. It is easy to prove that $\lim_{r\to\infty}|B(x,r)|^{1/r}=1$ implies the same relations $\lim_{r\to\infty}|B(y,r)|^{1/r}=1$ for any $y\in M$. In such a case we say that M has non exponential growth.

Do bounded-harmonic-functions reduce to constants on riemannian manifolds of non exponential growth? The following is a counter-example (known of O. Chung): M is the cylinder $\{(x,y): x\in R,\ y \text{ (mod. 1)}\}$ equipped with the riemannian metric $dx^2+(1+x^2)^2\cdot dy^2$. Then M has non exponential growth, since

$$|B(0,r)|=\int_{B(0,r)}(1+x^2)\,dx\cdot dy\leqslant\int_{-r}^{r}(1+x^2)\,dx<r^3+r,$$

where 0 has coordinates $(0,0)$. However $\arctan(x)$ is a non constant bounded harmonic function. Note that the curvature $-2\cdot(1+x^2)^{-1}$ is negative.

Debiard *et al.* [4] proved recently that bounded harmonic functions on riemannian manifolds of non negative Ricci curvature are constant. Better, under the same hypothesis, Yau [11] proved that *positive* harmonic functions are constant.

University Pierre et Marie Curie, Paris

REFERENCES

[1] Avez, A., 'Théorème de CHOQUET-DENY pour les groupes à croissance non exponentielle', *C.R.A.S. Paris* **279** (1974), 25–28.
[2] Azencott, R., 'Espaces de POISSON des groupes localement compacts', *Lectures Notes*. Springer (1970).
[3] Choquet, G. and Deny, J., 'Sur l'équation de convolution $\mu=\mu*\sigma$', *C.R.A.S. Paris* **250** (1960), 799–801.
[4] Debiard, A., Gaveau, B., and Mazet, E., Théorèmes de comparaison en géométrie (to appear).
[5] Dynkin, E. B. and Maliutov, M. B., 'Random Walk on Groups with a Finite Number of Generators', *Dokl. Akad. Nauk. S.S.S.R.* **137** (1961), 1042–1045.
[6] Guivarc'h, Y., 'Croissance polynomiale et périodes des fonctions harmoniques', *Bull. Soc. Math. France* **101** (1973), 333–379.
[7] Milnor, J., 'A Note on Curvature and Fundamental Group', *J. Diff. Geometry* **2** (1968), 1–7.
[8] Nelson, E., 'A Proof of Liouville's Theorem', *Proc. A.M.S.* **12** (1969), 995.
[9] Svarc, A. S., 'A Volume Invariant of Coverings', *Dokl. Akad. Nauk. S.S.S.R.* **105** (1955), 32–34.
[10] Wolf, J., 'Growth of Finitely Generated Groups and Curvature of Riemannian Manifolds', *J. Diff. Geom.* **2** (1968), 421–446.
[11] Yau, S. T., Harmonic Functions on Complete Riemannian Manifolds', *Comm. Pure and Appl. Math.* **27** (1975), 201–228.

M. BERGER

SOME RELATIONS BETWEEN VOLUME, INJECTIVITY RADIUS, AND CONVEXITY RADIUS IN RIEMANNIAN MANIFOLDS

ABSTRACT. One knows how to attach to a compact riemannian manifold (M, g) three real numbers: its volume $v(g)$, its injectivity radius $i(g)$ and its convexity radius $c(g)$. The present article studies the following problems: do there exist universal constants $\lambda(n)$, $\mu(n)$ such that $v(g) \geqslant \lambda(n) i^n(g)$, $v(g) \geqslant \mu(n) c^n(g)$ for every riemannian manifold of dimension n? An affirmative answer is given to the second problem for any n but with an unsharp constant; an affirmative answer is given to the first problem only when $n = 2$ but with the sharp bound $\lambda(2) = 4/\pi$.

In Section 1 we will recall the classical definitions of the volume $v(g)$, the injectivity radius $i(g)$ and the convexity radius $c(g)$ of a compact riemannian manifold (M, g). It is tempting to expect a universal inequality involving $v(g)$ and $i(g)$, the best possible answer would be

$$v(g) \geqslant \frac{\alpha(n)}{\pi^n} i^n(g)$$

(where $\alpha(n)$ denotes the volume of the n-dimensional euclidean sphere), that inequality being valid for any riemannian manifold of dimension n and equality occurring if and only if (M, g) is a standard sphere. Existence of such an inequality would be basic in riemannian geometry; for example, it would immediately solve the generalized Blaschke conjecture in even dimension, thanks to Weinstein result ([11, remark 2 on p. 514]).

To our knowledge the preceding question is for the moment completely open when $n \geqslant 3$, one does not even know if there exists a $\lambda(n) > 0$ with $v(g) \geqslant \lambda(n) i^n(g)$ for any riemannian manifold of dimension n. In this article we only show that the question is easily solved when $n = 2$ (Proposition 2.1).

For the convexity radius the situation is a little better, since we can prove that there exists for any n a strictly positive $\mu(n)$ with $v(g) \geqslant \mu(n) c^n(g)$ for any (M, g) of dimension n (Proposition 3.1). But the bound is not the best possible; it seems reasonable to expect the bound will be $\alpha(n)/(\pi/2)^n$, with equality characterizing standard spheres; see 3.2.

Cahen & Flato (eds.), Differential Geometry and Relativity, 33–42.

NOTE ADDED IN PROOF. In an article to appear in *Osaka Math. J.*, a non-best bound is obtained when $n = 3$.

1. Definitions

1.1. Throughout this article by manifold will be meant a *compact*, C^∞ and without boundary, its dimension will always be denoted by n and be $\geqslant 2$; all riemannian structures are also C^∞. By a *standard sphere* we mean a riemannian manifold (M, g) where $M = S^n$ is the n-dimensional sphere and g is a riemannian structure of constant sectional curvature on S^n.

To any (M, g) we attach first its *volume* $v(g)$ (*area* $a(g)$ when $n = 2$) defined as $v(g) = \int_M v_g$, the total mass of (M, g) for its canonical measure v_g.

1.2. The injectivity radius $i(g, m)$ of (M, g) at the point $m \in M$ is the supremum of real numbers ε such that the exponential map at m, $\exp_m : T_mM \to M$, is a diffeomorphism onto its image when restricted to the open ball of radius ε in the tangent space T_mM of M at m. The *injectivity radius* $i(g)$ of (M, g) is

$$i(g) = \inf\{i(g, m) : m \in M\};$$

we have $i(g) > 0$ thanks to the compactness of M.

1.3. The notion of convexity radius is a little more subtle, and moreover changes according to authors. We define first a *segment* from m to n in (M, g) to be a geodesic from m to n whose length is equal to the distance $d(m, n)$; then a subset A of M is said to be *convex* if any two points m, n in A are connected by a unique segment in M, this segment moreover having to lie in A. *Denote by* $B(m, \varepsilon)$ the open ball in the metric space (M, g) of center m and radius ε. Then the convexity radius $c(g, m)$ of (M, g) at $m \in M$ is the supremum of the real numbers ε such that all the balls $B(m, \eta)$ with $\eta \leqslant \varepsilon$ are convex. Finally the *convexity radius* of (M, g) is

$$c(g) = \inf\{c(g, m) : m \in M\},$$

with again $c(g) > 0$ thanks to the compactness of M.

1.4. For a standard sphere (S^n, g), we set $\alpha(n) = v(g)$ (for these classical values, if necessary, see [2, p. 209]). We have $i(g) = \pi$ and $c(g) = \pi/2$.

1.5. *Denote by* $\mathrm{l}(g)$ the length of the smallest nonconstant periodic geodesic of (M, g); then a result Klingenberg (see for example [1, p. 245]) asserts that: $i(g) = \mathrm{l}(g)/2$ or there exists a segment γ of length $i(g)$ with ends m, n such that m, n are conjugate points on γ. In particular $i(g) = \mathrm{l}(g)/2$ as soon as (M, g) is of nonpositive sectional curvature.

1.6. REMARK. One always has $i(g) \geqslant 2c(g)$.

We prove it by contradiction: suppose $i(g) < 2c(g)$. Let m, n be points with $d(m, n) = i(g)$, let γ be a segment from m to n and denote by p the middle point of γ; by assumption $m, n \in B(p, c(g))$ which is a convex ball. This implies γ is the unique segment from m to n and now 1.5 implies that m, n are conjugate on γ. By a classical result (see [1], pp. 237–238), if one parametrizes γ with $\gamma(0) = m$, $\gamma(t) = n$ and extends it beyond t for every $t' > t$, then there exists a segment $\sigma_{t'}$ from m to $\gamma(t') = n'$ of length $d(m, n') < t'$ and, moreover, the uniqueness of segments from m to n implies that $\sigma_{t'}$ converges to γ when t' tends to t; in particular we can choose $\sigma_{t'}$ as close to the nonsegment geodesic $\gamma|_{[0, t']}$ as we want. Consider now the two balls $B(m, t'/2)$, $B(n', t'/2)$; they are convex as soon as $t' \leqslant 2c(g)$, hence by the study of Section 11.8 of [1] they have no common point besides $\gamma(t'/2)$ in a neighborhood of $\gamma(t'/2)$. But this contradicts the fact that the middle point of m, n' on $\sigma_{t'}$ belongs to these balls.

1.7. We will need the following property (see for example [1, p. 247]): let γ a geodesic in (M, g), starting from $m = \gamma(0)$ and Y a Jacobi field along with initial value $Y(0) = 0$. Then, if $t \leqslant c(g)$, one has $\langle Y(t), Y'(t)\rangle \geqslant 0$.

2. Area and injectivity radius.

2.1. PROPOSITION. *For any manifold M of dimension* 2 *and for any riemannian structure g on M one has*

$$a(g) \geqslant \frac{4}{\pi} i^2(g).$$

Moreover $a(g) = (4/\pi) i^2(g)$ *if and only if* (M, g) *is a standard sphere.*

In case $M = S^2$ the inequality follows directly from 1.5 and [7, Theorem 4.2, p. 296]. Moreover equality holds if and only if we have a standard sphere. To take care of the case $M \neq S^2$ we need the following:

2.2. LEMMA. *For any (M, g) of dimension 2 and any $m \in M$ one has*

$$a(B(m, i(g)/2)) \geq i^2(g)/2.$$

Denote, for $t \leq i(g)/2$, by $\mathrm{l}(C(t))$ the length of the circle $C(t) = \{p \in M : d(p, m) = t\}$. We have

$$a(B(m, i(g)/2)) = \int_0^{i(g)/2} \mathrm{l}(C(t))\, \mathrm{d}t$$

(apply [5, 3.2.11, p. 248] for example). Pick now any geodesic γ through m with $\gamma(0) = m$; then the circle $C(t)$ is divided by the points $\gamma(t)$, $\gamma(-t)$ into two curves $C'(t)$, $C''(t)$ from $\gamma(t)$ to $\gamma(-t)$. By the very definition of $i(g)$ we then have for their lengths: $\mathrm{l}(C'(t))$, $\mathrm{l}(C''(t)) \geq d(\gamma(t), \gamma(-t)) = 2t$, hence $\mathrm{l}(C(t)) \geq 4t$ and $a(B(m, i(g)/2)) \geq i^2(g)/2$.

To resume the proof of 2.1 assume now $M \neq S^2$; we apply Klingenberg's trick: pick any $m, n \in M$ with $d(m, n) \geq i(g)$. Thus, we know that if $d(p, m) < i(g)$ or $d(p, n) < i(g)$ for every $p \in M$, then necessarily M is homeomorphic to S^2: see [6, p. 260–261]. We are sure now that there exists $p \in M$ with $d(p, m) \geq i(g)$ and $d(p, n) \geq i(g)$, which implies that the three balls $B(m, i(g)/2)$, $B(n, i(g)/2)$, $B(p, i(g)/2)$ have no common point. Using 2.2 three times one gets $a(M) \geq \frac{3}{2} i^2(g) > (4/\pi) i^2(g)$; the proof of 2.1 is now complete.

After 2.1, a natural question which arises is the following: for a given compact two-dimensional manifold M, what is the best constant $k(M)$ such that $a(g) \geq k(M) i^2(g)$ for every g on M and what implies equality? A partial answer is given by the following

2.4. PROPOSITION. *If $M = P^2$ the two-dimensional projective space, then*

$$a(g) \geq \frac{8}{\pi} i^2(g) \quad \textit{for every } g \textit{ on } P^2,$$

with the equality if and only if g is of constant curvature.

If $M = T^2$ the two-dimensional torus, then

$$a(g) \geq 2\sqrt{3}\, i^2(g) \quad \textit{for every } g \textit{ on } T^2,$$

with the equality if and only if (T^2, g) is a flat equilateral torus.

If $M = K$ the Klein bottle, then

$$a(g) \geq 4i^2(g) \qquad \textit{for every } g \textit{ on } K,$$

with the equality if and only if (K, g) is a flat square Klein bottle.

For T^2 and P^2 the proposition follows directly from 1.5 and Pu's and Loewner's theorems, see for convenience [2, p. 303 up]. For K it follows from 1.5 again and

2.5. PROPOSITION. *For any riemannian structure g on K one has $a(g) \geq l'^2(g)$, where $l'(g)$ denotes the length of the smallest periodic geodesic non-homotopic to zero, with equality if and only if (K, g) is a flat square Klein bottle.*

This result is an easy modification of Loewner's one; since to our knowledge it has not appeared in the literature we give the proof here. Let $(T^2, \tilde{g}) \to (K, g)$ be a riemannian covering of (K, g) by the torus: this covering induces an involutive isometry φ of $(T^2, \tilde{g})$; by the conformal representation theorem there exists a flat riemannian structure $\bar{g}$ on T^2 which is conformally related to $\tilde{g}$. Hence φ is a conformal automorphism of $(T^2, \bar{g})$ hence an isometry (elementary: the automorphisms of $\mathbb{C}$ are linear). Then we can go back down to K to get a flat riemannian structure g_0 on K conformally related to g. And then the proof of Loewner (see [2, p. 303 up]) applies; we get first

$$\frac{a(g)}{l'^2(g)} \geq \frac{a(g_0)}{l'^2(g_0)}.$$

The conclusion follows now from the structure of flat Klein bottles as described in [12, 2.5.9, p. 78].

2.6. REMARKS. The only remaining case is that of a two-dimensional compact surface, orientable or not, of genus $\gamma \geq 2$. At the moment we know only

2.7 $\qquad a(g) \geq \frac{3}{2} i^2(g)$ for every g on M.

To determine the best possible constant $k(M)$ is an open problem. But we would like to mention an amusing connection between 2.7 and

Blatter's theorem; that result ([4, Satz 4, p. 156]) asserts that for every integer γ there exists a real number $k(\gamma)>0$ with $a(g)\geq k(\gamma)l'^2(g)$ for every compact orientable two-dimensional manifold of genus γ and for every riemannian structure g on M (where $l'(g)$ is again the length of the smallest periodic geodesic non-homotopic to zero).

The number $k(\gamma)$ is the inverse of the Hermite constant in the geometry of numbers; its exact value is only known for $\gamma = 1, 2, 3, 4$: $k(1)=\sqrt{3}/2$, $k(2)=1/\sqrt{2}$, $k(3)=\sqrt[3]{6}/3$, $k(4)=1/2$; and one knows that $k(\gamma)\to 0$ as $\gamma\to\infty$ (see [10, (20) p. 299 and (27) p. 318]). But suppose now our riemannian structure g on our M of genus γ is of non-positive curvature; then $i(g)=l(g)/2$ by 1.5 and $l(g)=l'(g)$ (for example by the Gauss-Bonnet formula). Finally for any such non-positive curvature g one has

2.8 $$a(g)\geq \tfrac{3}{8}l'^2(g).$$

We cannot say much more about it but this fact seems to suggest that the constant $k(\gamma)$ in Blatter's theorem might be very weak. It would be fun if 2.8 were holding for any g on any $M\neq S^2$.

3. Volume and convexity radius

3.1. PROPOSITION. *For any n, for any riemannian manifold* (M, g), *the following inequality holds*

$$v(g)\geq \frac{\alpha(n-1)}{n2^{n-2}}c^n(g).$$

3.2. REMARK. When $n=2$, the best bound

$$\frac{4\pi}{(\pi/2)^2}=\frac{16}{\pi}$$

is obtained by 1.6 and 2.1, with equality characterizing standard spheres.

3.3. To prove 3.1 we introduce the following function f on $U(M)\times \mathbf{R}_+$, where $U(M)$ is the *unit tangent bundle* to M; *if* $x\in U(m)$ the *unit sphere at* $m\in M$, let $\{x, u_2, \ldots, u_n\}$ be an orthonormal basis of T_mM, let γ be the geodesic $t\mapsto \exp_m(tx)$ and let $Y_i (i=2,\ldots,n)$ be the Jacobi field along γ with initial conditions

$Y_i(0) = 0$ and $Y'_i(0) = u_i$. Then we set

$$f(x, t) = \|Y_2(t) \wedge \cdots \wedge Y_n(t)\|.$$

It is classical that the volume $v(B(m, \varepsilon))$ of the ball of center m and radius ε, for $\varepsilon \leqslant i(g, m)$, is given by

$$v(B(m, \varepsilon)) = \int_{x \in U(m)} \left(\int_0^s f(x, t)\, dt \right) dx,$$

where dx is the canonical measure on the unit sphere $U(m)$ in the euclidean space T_mM (see [1, pp. 253–256]). Then 3.1 will follow from the

3.4. LEMMA. *For any $x \in U(M)$ and any $t \leqslant c(g)$: $f(x, t) \geqslant (t/2)^{n-1}$.*
Pick any $m, n \in M$ with $d(m, n) \geqslant i(g)$. By 1.6 we have $B(m, c(g)) \cap B(n, c(g)) = \emptyset$, hence

$$v(g) \geqslant 2\alpha(n-1) \int_0^{c(g)} \left(\frac{t}{2}\right)^{n-1} dt = \frac{\alpha(n-1)c^n(g)}{n2^{n-2}}.$$

We think it will be easier for the reader to grasp the idea behind the proof if he sees it first when $n = 2$ (even after the remark 3.2!).

3.5. When $n = 2$, Jacobi fields along γ are identified themselves with numerical functions, and moreover are positive ones between conjugate points; for $t \leqslant c(g)$ let y (resp. z) be the Jacobi field along γ with initial conditions $y(0) = 0$ and $y'(0) = 1$ (resp. $z(t) = 0$ and $z'(t) = -1$). By 1.7 we see that y is a nondecreasing function; and, reversing the time and starting from the time t, we see that z is a nonincreasing function; hence there exists an $s \in]0, t[$ with $y(s) = z(s)$. Since y, z are solutions of the same differential equation $y'' + \sigma y = 0$, $z'' + \sigma z = 0$, a classical computation in Sturm-Liouville theory yields

$$z(u) = y(t)y(u) \int_u^t \frac{dv}{y^2(v)} \quad \text{for every } u \in]0, t].$$

Set $k = y(t)$; then

$$\int_s^t \frac{dv}{y^2(v)} = k^{-1},$$

but y is nondecreasing so

$$\int_s^t \frac{dv}{y^2(v)} \geqslant \frac{t-s}{k^2},$$

i.e. $t-s \leqslant k$. If we play the same game with z instead of y we get $s \leqslant k$. Finally $t \leqslant 2k$, hence $y(t) \geqslant t/2$, which is 3.4 when $n=2$.

3.6. Now let n be any integer; here we use [8, pp. 31–32]. With the notations of 3.3 *define* matrices $A(t)$ by

$$A(t)(u_2, \ldots, u_n) = (Y_2(t), \ldots, Y_n(t)).$$

We have $f(x,t) = \det(A(t))$ for $t \leqslant c(g)$; moreover by 1.7 and [1, p. 254] we know that $f(x,t)$ is nondecreasing. Following [8], the map $t \mapsto A(t)$ is the solution of a differential equation $A'' + KA = 0$; let us now fix some $t \leqslant c(g)$ and define B to be the solution of $B'' + KB = 0$ with condition $B(t) = 0$ and $B'(t) = -I$ (the identity matrix). By the same token as in 3.5 we know that $v \mapsto \det(B(v))$ is nonincreasing on $[0, t]$; so that again there exists an s with $\det(A(s)) = \det(B(s))$ in $]0, t[$. By [8]:

$$B(u) = A(u)\left(\int_u^t A^{-1}(v)[A^{-1}(v)]^* \, dv\right) A(t)^*, \quad \forall u \in]0, t],$$

where the * denotes transposition. Setting $k = \det(A(s)) = \det(B(s))$, we have

$$\det\left(\int_s^t A^{-1}(v)[A^{-1}(v)]^* \, dv\right) = k^{-1}.$$

We need now two classical lemmas from analysis:

3.7. LEMMA (see for example [3, 11.8.9.5]). *On the set of positive definite symmetrical matrices the function $X \mapsto (\det(X))^{-1/2}$ is strictly convex.*

3.8. LEMMA (see [9, 204, pp. 150–151]). *Let $\varphi : [a, b] \to E$ a continuous map with values in a finite dimensional real vector space E and $\xi : E \to \mathbb{R}$ a convex map defined on a convex subset C of E. Then*

$$\xi\left(\frac{\int_a^b \varphi(u)\, du}{b-a}\right) \leqslant \frac{\int_a^b \xi(\varphi(u))\, du}{b-a}.$$

To complete the proof of 3.4 we apply 3.8 with

$$\varphi : [s, t] \ni v \mapsto A^{-1}(v)[A^{-1}(v)]^* \quad \text{and} \quad \xi : X \mapsto (\det(X))^{-1/2}.$$

Yielding:

$$\left[\det\left(\frac{\int_s^t A^{-1}(v)[A^{-1}(v)]^* \, \mathrm{d}v}{t-s}\right)\right]^{-1/2} \leqslant \frac{\int_s^t [\det(A^{-1}(v)[A^{-1}(v)]^*)]^{-1/2} \, \mathrm{d}v}{t-s}$$

But

$$\left[\det\left(\frac{\int_s^t A^{-1}(v)[A^{-1}(v)]^* \, \mathrm{d}v}{t-s}\right)\right]^{-1/2} = (t-s)^{(n-1)/2} k^{1/2},$$

and

$$\frac{\int_s^t [\det(A^{-1}(v)[A^{-1}(v)]^*)]^{-1/2} \, \mathrm{d}v}{t-s} = \frac{\int_s^t \det(A(v)) \, \mathrm{d}v}{t-s} \leqslant \frac{k(t-s)}{t-s} = k$$

because $\det(A(t))$ is nondecreasing. Finally: $t - s \leqslant k^{1/(n-1)}$; and reversing the time as in 3.5: $s \leqslant k^{1/(n-1)}$. Hence $k^{1/(n-1)} \geqslant t/2$ or, equivalently: $k \geqslant (t/2)^{n-1}$.

3.9. REMARK. It is clear that the equality in 3.1 can never be obtained; for we should then have $\det(A(t)) = (t/2)^{n-1}$, contradicting $A'(0) = I$.

ACKNOWLEDGMENTS

The author is happy to have the opportunity to thank the Japan Society for Promotion of Science and the mathematics department of Osaka University for their hospitality.

Centre National de la Recherche Scientifique et Université Paris VII

REFERENCES

[1] Bishop, R. L. and Crittenden, R. J., *Geometry of Manifolds*, Academic Press.
[2] Berger, M., 'Lectures on Geodesics in Riemannian Geometry', Tata Institute, Bombay.

[3] Berger, M., Cours de Géométrie, Éditions CEDIC.
[4] Blatter, C., 'Über Extremallängen auf geschlossenen Flächen', *Comment. Math. Helv.* **35** (1961), 153–168.
[5] Federer, H., *Geometric Measure Theory*, Springer.
[6] Gromoll, D., Klingenberg, W., and Meyer, W., 'Riemannsche Geometrie im Grossen', *Springer Lectures Notes in Mathematics* no. 55.
[7] Green, L. W., 'Auf Wiedersehenflächen', *Ann. of Math.* **78** (1963), 289-299.
[8] Green, L. W., 'A theorem of E. Hopf', *Michigan Math. J.* **5**, (1958), 31–34.
[9] Hardy, G. H., Littlewood, J. E., and Polya, G., *Inequalities*, Cambridge University Press.
[10] Lekerkerker, C. G., *Geometry of Numbers*, Wolters-Noordhoff.
[11] Weinstein, A., "On the Volume of Manifolds All of Whose Geodesics Are Closed', *J. of Differential Geometry* **9** (1974), 513–517.
[12] Wolf, J. A., *Spaces of Constant Curvature*, 2nd edition, J. Wolf.

ROBERT S. CAHN, PETER B. GILKEY†, and JOSEPH A. WOLF‡

HEAT EQUATION, PROPORTIONALITY PRINCIPLE, AND VOLUME OF FUNDAMENTAL DOMAINS

1. INTRODUCTION

In this note, we extend the Hirzebruch proportionality principle to the coefficients in the asymptotic expansions for the Laplacians on differential forms with values in homogeneous vector bundles over symmetric spaces. The case zero-forms and the trivial line bundle is a proportionality principle for the trace of the heat kernel. For $2m$-dimensional manifolds, the case of m-th order terms of some asymptotic expansions is a proportionality principle for the indices of certain elliptic complexes. These results have implications for the volumes of fundamental domains of discrete subgroups, and in refining these implications we also develop a proportionality principle for equivariant characteristic classes.

Hirzebruch's original work [9] on proportionality studied ordinary characteristic classes on hermitian symmetric spaces, and Serre [10] studied the Euler class in a general setting. N. Wallach tells us that his student Miatelo obtained a result similar to Corollary 2.8 below, by methods of harmonic analysis on semisimple Lie groups. And of course many mathematicians, starting with C. L. Siegel [11], looked for lower bounds on the volume of fundamental domains for discrete groups.

Our result on the volumes of fundamental domains is a qualitative improvement on previous work. The latter was a matter of positive lower bounds for the volume, while we show that the volume is an integral multiple of a certain number.

2. ASYMPTOTIC EXPANSIONS

Let G/K be a symmetric space of noncompact type, and let $M' = G'/K$ denote the compact dual symmetric space. Here we take G and G' to be analytic subgroups of a complex group G_C and $K = G \cap G'$. Let Γ

† Research partially supported by NSF Grant MPS72-04357.
‡ Research partially supported by NSF Grant MPS74-01477.

Cahen & Flato (eds.), Differential Geometry and Relativity, 43–54.

denote a discrete, co-compact, torsion-free subgroup of G, and denote $M = \Gamma \backslash G/K$ the compact locally symmetric space with fundamental group Γ and universal cover G/K. We view M and M' as riemannian manifolds with invariant metric arising from the Cartan-Killing form on the complexification of the Lie algebra of G.

Let $\pi : K \to GL(V)$ be a (finite dimensional, unitary) representation of K. We consider the associated hermitian vector bundles.

$$E'_\pi : G' \times_K V \to M' \quad \text{and} \quad E_\pi : (\Gamma \backslash G) \times_K V \to M \tag{2.1}$$

with typical fibre V. The corresponding second order elliptic operators

$$\Delta'_{\pi,p} : \text{Laplacian on } \Lambda^p(M') \otimes E'_\pi \quad \text{and} \tag{2.2}$$

$$\Delta_{\pi,p} : \text{Laplacian on } \Lambda^p(M) \otimes E_\pi$$

define zeta functions (trace heat kernel)

$$\zeta_{M',\pi,p}(t) = \sum e^{-\lambda' t} \quad \text{and} \quad \zeta_{M,\pi,p}(t) = \sum e^{-\lambda t}, \tag{2.3}$$

where the summations run over the eigenvalues (with multiplicities) of $\Delta'_{\pi,p}$ and $\Delta_{\pi,p}$. If $d = \dim M' = \dim M$, then these zeta functions have asymptotic expansions at $t = 0$ (see Gilkey [8]),

$$\zeta_{M',\pi,p}(t) \sim (4\pi t)^{-d/2}(a'_{0,p,\pi} + a'_{1,p,\pi}t + \cdots + a'_{n,p,\pi}t^n + 0(t^{n+1})) \tag{2.4a}$$

and

$$\zeta_{M,\pi,p}(t) \sim (4\pi t)^{-d/2}(a_{0,p,\pi} + a_{1,p,\pi}t + \cdots + a_{n,p,\pi}t^n + 0(t^{n+1})). \tag{2.4b}$$

Our main result is the observation.

2.5. THEOREM. $a_{k,\pi,p} = (-1)^k \dfrac{\text{Vol}(M)}{\text{Vol}(M')} a'_{k,\pi,p}$.

Proof. $a_{k,\pi,p}$ is a local invariant of M, that is $a_{k,\pi,p} = \int_M P_{k,\pi,p}(x)\,dx$, where dx is the volume element and $P_{k,\pi,p}(x)$ is a polynomial in the curvatures of M and E at x and their covariant derivatives. All those covariant derivatives vanish here, so $P_{k,\pi,p}(x)$ is homogeneous of degree k in the curvatures. Since M is locally homogeneous, now $a_{k,\pi,p} = P_{k,\pi,p}(x) \cdot \text{Vol}(M)$. If $y \in M'$, the same considerations hold, and the curvatures are the negatives of those at $x \in M$, so $a'_{k,\pi,p} = P_{k,\pi,p}(y) \cdot \text{Vol}(M') = (-1)^k P_{k,\pi,p}(x) \cdot \text{Vol}(M')$. Q.E.D.

Calculation of the $a'_{k,\pi,p}$ is an algebraic problem with the Peter-Weyl Theorem for G', Cartan's highest weight theory for the representations of G', Frobenius' Reciprocity Theorem, and the decomposition of a representation of G' under restriction to K. See Cahn-Wolf [5, §1]. Theorem 2.5 carries the result from the compact symmetric space M' to the compact locally symmetric space M.

In case π is trivial and $p = 0$, the operators (2.2) reduce to

(2.6) $\quad \Delta'$: Laplacian on $L_2(M')$ and Δ: Laplacian on $L_2(M)$.

Then (2.3) reduce to the ordinary zeta functions $\zeta_{M'}$ and ζ_M, and (2.4) becomes

(2.7a) $$\zeta_{M'}(t) \sim (4\pi t)^{-d/2}(a'_0 + a'_1 t + \cdots + a'_n t^n + 0(t^{n+1})) \quad \text{as } t \downarrow 0$$

and

(2.7b) $$\zeta_M(t) \sim (4\pi t)^{-d/2}(a_0 + a_1 t + \cdots + a_n t^n + 0(t^{n+1})) \quad \text{as } t \downarrow 0.$$

Theorem 2.5 now specializes to

2.8. COROLLARY. $a_k = (-1)^k \dfrac{\mathrm{Vol}(M)}{\mathrm{Vol}(M')} a'_k.$

In this case, the a'_k can be calculated explicitly (see Cahn-Wolf [5, 6]) and that gives the a_k.

In the hermitian symmetric case, Theorem 2.5 and Corollary 2.8 hold for the complex Laplacians. If M and M' are spin manifolds in a consistent way, then they hold for the Dirac Laplacians (see Wolf [12, 13]).

3. Elliptic complexes

We now assume that M' and M have even dimension $d = 2m$. Let $\mathcal{S}'_\pi$ and $\mathcal{S}_\pi$ denote their respective twisted signature complexes (Gilkey [8]) using forms with values in E'_π and E_π. Then

(3.1) $$\mathrm{Index}(\mathcal{S}'_\pi) = \sum (-1)^p a'_{m,\pi,p} \quad \text{and}$$
$$\mathrm{Index}(\mathcal{S}_\pi) = \sum (-1)^p a_{m,\pi,p}.$$

Using this, Theorem 2.5 gives us

(3.2) $$\mathrm{Index}(\mathcal{S}_\pi)\cdot \mathrm{Vol}(M') = (-1)^m\,\mathrm{Index}(\mathcal{S}'_\pi)\cdot \mathrm{Vol}(M).$$

More generally, if $\mathscr{C}'$ is any natural elliptic complex over M', and $\mathscr{C}$ is the corresponding complex over M, then our considerations give

3.3. THEOREM. $\operatorname{Index}(\mathscr{C}) \cdot \operatorname{Vol}(M') = (-1)^m \operatorname{Index}(\mathscr{C}') \cdot \operatorname{Vol}(M)$.

Of course Theorem 3.3 can also be derived from the Atiyah-Singer Index Theorem.

The point of Theorem 3.3 is that $\operatorname{Index}(\mathscr{C}')$ can often be calculated directly. Here are a few examples.

de Rham Complex. The index of the de Rham complex is the Euler-Poincaré characteristic χ. If $\operatorname{rank} K < \operatorname{rank} G$ then $\chi(M') = 0$. If $\operatorname{rank} K = \operatorname{rank} G$ then $\chi(M') = |W(G')|/|W(K)|$, quotient of the orders of the Weyl groups. Now Theorem 3.3 gives

(3.4)
$$\chi(M) = 0 \textit{ if } \operatorname{rank} K < \operatorname{rank} G, \textit{ and}$$
$$\chi(M) = (-1)^m \frac{\operatorname{Vol}(M)}{\operatorname{Vol}(M')} \frac{|W(G')|}{|W(K)|} \textit{ if } \operatorname{rank} K = \operatorname{rank} G.$$

Dolbeault complex. The index of the Dolbeault complex is the arithmetic genus A. In the hermitian case, $A(M') = 1$, e.g. by the Bott-Borel-Weil Theorem [3]. So Theorem 3.3 gives

(3.5) $$A(M) = (-1)^m \operatorname{Vol}(M)/\operatorname{Vol}(M') \quad \textit{in the hermitian case.}$$

This is equivalent to Hirzebruch's [9, Satz 4].

Signature complex. When the dimension $d = 2m$ is divisible by 4, the signature τ is defined to be the signature of the symmetric bilinear form $H^m \times H^m \to H^{2m}$ (cup product). It is the index of the ordinary (trivial line bundle) signature complex. Evidently $\tau(M') = \pm 1$ if $\dim H^m(M'; R) = 1$, so Theorem 3.3 and the fact [14, Theorem 6] that $\tau(M) \geqq 0$ give

(3.6) $$\tau(M) = \frac{\operatorname{Vol}(M)}{\operatorname{Vol}(M')} \quad \textit{for } m \textit{ even and} \quad \dim H^m(M'; R) = 1.$$

That applies, for example, to the even quaternion projective spaces $\operatorname{Sp}(2l+1)/\operatorname{Sp}(2l) \times \operatorname{Sp}(1)$.

At this point we tabulate some information for the $M' = G'/K$ where $\operatorname{rank} K = \operatorname{rank} G'$ and G' is an exceptional group. Note the curious fact: if $\dim H^m(M'; R) \neq 0$ then $\dim H^m(M'; R)$ divides $\chi(M')$. This is also the case for many classical M', but not all, e.g. not $\operatorname{Sp}(8)/\operatorname{SU}(8)$.

(3.7)	$M' = G'/K$	$2m$	$\chi(M')$	$\dim H^m(M'; R)$	$\tau(M')$
(i)	$G_2/SO(4)$	8	3	1	1
(ii)	$F_4/\mathrm{Spin}(9)$	16	3	1	1
(iii)	F_4/A_1C_3	28	12	0	0
(iv)	E_6/T_1D_5	32	27	3	3
(v)	E_6/A_1A_5	40	36	4	4
(vi)	E_7/T_1E_6	54	56	0	0
(vii)	E_7/A_1D_6	64	63	7	7
(viii)	E_7/A_7	70	72	0	0
(ix)	E_8/A_1E_7	112	120	8	8
(x)	E_8/D_8	128	135	9	7

So (3.6) holds directly for $G_2/SO(4)$ and the Cayley plane $F_4/\mathrm{Spin}(9)$, and holds after possible division by a small number for E_6/T_1D_5, E_6/A_1A_5, E_7/A_1D_6, E_8/A_1E_7, and E_8/D_8.

The information in (3.7) comes out of the Hirsch Formula [0, §26], the primitive invariants for the various simple groups (Borel-Chevalley [2]), and the recent results of Shaw Mong [14].

4. Volume of fundamental domain

In the appropriate normalization of Haar measures, Theorem 3.3 says

4.1. THEOREM. Let $\mathscr{C}'$ be a natural elliptic complex over M'. If $\mathrm{Index}(\mathscr{C}') \neq 0$ then

$$\mathrm{Vol}(\Gamma\backslash G) = |\mathrm{Index}(\mathscr{C})| \cdot \mathrm{Vol}(G')/|\mathrm{Index}(\mathscr{C}')|.$$

{See Cahn [4] for $\mathrm{Vol}(G')$ here, and note that $\mathscr{C}'$ exists with $\mathrm{Index}(\mathscr{C}') \neq 0$ precisely when rank K = rank G.}

We interpret Theorem 4.1 as saying: if rank K = rank G then there is a number $\mathrm{Vol}(G')/|\mathrm{Index}(\mathscr{C}')| > 0$ such that $\mathrm{Vol}(\Gamma\backslash G)$ is a positive integral multiple of that number. At worst, one can use the de Rham complex and conclude

$$(4.2) \qquad \mathrm{Vol}(\Gamma\backslash G) = n \cdot \mathrm{Vol}(G') \cdot |W(K)|/|W(G')|, \qquad n \text{ positive integer.}$$

But that could be derived directly from Chern's Gauss-Bonnet Theorem [7]; see Serre [10, §3]. Evidently, the information contained in Theorem 4.1, under this interpretation, is optimized by minimizing $|\mathrm{Index}(\mathscr{C}')|$. Here are a few examples.

Dolbeault Complex. If $\mathscr{C}'$ is the Dolbeault complex then $\mathrm{Index}(\mathscr{C}') = A(M') = 1$, so

$$(4.3) \qquad \mathrm{Vol}(\Gamma\backslash G) = n \cdot \mathrm{Vol}(G'), \qquad n \text{ positive integer, in the hermitian case.}$$

That applies whenever G is a connected linear group locally isomorphic to a product whose factors are of the form $SU(p,q)$, $SO(2,l)$, $SO^*(2l)$, $Sp(l;R)$, E_{6,T_1D_5} and E_{7,T_1E_6}.

Signature complex. If the dimension $d = 2m$ is divisible by 4, then Theorem 4.1 says

$$(4.4) \qquad \mathrm{Vol}(\Gamma\backslash G) = n \cdot \mathrm{Vol}(G')/\tau(M'), \quad n \text{ integer, if } \tau(M') \neq 0.$$

As noted in (3.6) and (3.7), this applies to

$Sp(2l,1)$, G_{2,A_1A_1} and F_{4,B_4} with $\tau(M') = 1$;
E_{6,T_1D_5} with $\tau = 3$,
E_{7,A_1D_6} and E_{8,D_8} with $\tau = 7$;
E_{6,A_1A_5} with $\tau = 4$, E_{8,A_1E_7} with $\tau = 8$.

5. K-Characteristic classes

Recall the definition of the K-characteristic ring of M'. If B_K is the classifying space for K and $\sigma: M' \to B_K$ induces the principal K-bundle $G \to M'$, then $\sigma^*H^*(B_K; Z)$ is the K-characteristic subring of $H^*(M'; Z)$. Its elements, the K-characteristic classes, are obtained modulo torsion as follows. Let $\mathfrak{K}$ be the Lie algebra of K, $\mathscr{C}(\mathfrak{K})$ the graded associative algebra of Ad_K-invariant polynomials $\mathfrak{K} \to C$. Fix a K-connection on $G' \to M'$ and let Ω' denote its curvature form. If $c \in \mathscr{C}(\mathfrak{K})$, then the de Rham class $[c(\Omega')] \in H^*(M'; C)$ is independent of the connection. We say that c is integral if, in this way, it gives an integral class on B_K. The coefficient homomorphism $H^*(M'; Z) \to H^*(M'; C)$ maps the K-characteristic ring onto $\{[c(\Omega')] : c \in \mathscr{C}(\mathfrak{K})$ and c is integral.$\}$ Now we have a mild variation on Hirzebruch Proportionality [9]:

5.1. THEOREM. Suppose that M and M' have even dimension $d = 2m$ and let $c \in \mathscr{C}(\mathfrak{K})$ be integral and of degree m. Then the K-characteristic numbers

$$c[M'] = \int_{M'} c(\Omega') \quad \text{and} \quad c[M] = \int_M c(\Omega)$$

are integers, and

$$c[M'] \cdot \mathrm{Vol}(M) = (-1)^m c[M] \cdot \mathrm{Vol}(M').$$

{As in the proof of Theorem 2.5, this is a matter of local homogeneity and $\Omega = -\Omega'$.}

5.2. COROLLARY. If rank K = rank G, so $d = 2m$ and $\sigma^* H^{2m}(B_K; Z)$ has finite index r in $H^{2m}(M'; Z)$, then $\mathrm{Vol}(\Gamma \backslash G)$ is an integral multiple of $\mathrm{Vol}(G')/r$.

{For then we have $c \in \mathscr{C}(\mathfrak{K})$ of degree m with $c[M'] = r$.}

As an application of Corollary 5.2, suppose that

(5.3a) M' is a product of complex and quaternionic grassmannians.

In other words, suppose that

(5.3b) G is covered by a group $\Pi\, SU(p_i, q_i) \times \Pi\, Sp(l_j, m_j)$.

Taking G_C simply connected,

$$G' = \Pi\, SU(p_i + q_i) \times \Pi\, Sp(l_j + m_j)$$

and

$$K = \Pi S(U(p_i) \times U(q_i)) \times \Pi\, Sp(l_j) \times Sp(m_j),$$

so

$$H^*(G'; Z) \quad \text{and} \quad H^*(K; Z) \quad \text{are torsion free.}$$

Since it is now known that a compact connected group modulo a maximal torus is torsion free in cohomology, a result of Borel [0; Prop. 30.2] says $H^*(M'; Z) = \sigma^* H^*(B_K; Z)$, and in particular $r = 1$ in Corollary 5.2. In summary,

(5.4) $$\operatorname{Vol}(\Gamma\backslash G) = n \cdot \operatorname{Vol}(G'), \quad n \textit{ positive integer, in the case } (5.3)$$

Borel's result [0; Prop. 30.2] in fact shows that the number r in Corollary 5.2 is a product of powers of primes p for which $H^*(G'; Z)$ or $H^*(K; Z)$ has p-torsion. In view of (3.7), this sharpens (4.2) in the cases where M' is E_7/A_1D_6, E_8/D_8 or a real grassmannian of even dimension, telling us

(5.5a) if $G = E_{7,A_1D_6}$ then $\operatorname{Vol}(\Gamma\backslash G) = n \cdot \operatorname{Vol}(G')$, some integer n;

(5.5b) if $G = E_{8,D_8}$ then $\operatorname{Vol}(\Gamma\backslash G) = n \cdot \operatorname{Vol}(G')$, some integer n;

(5.6) $$\begin{cases} \text{if } G \text{ is locally isomorphic to } \mathrm{SO}(2u, 2v) \text{ or } \mathrm{SO}(2u, 2v+1) \\ \text{and if } 2^c \text{ is the highest power of 2 that divides } 2\binom{u+v}{v}, \\ \text{then } \operatorname{Vol}(\Gamma\backslash G) = n \cdot \operatorname{Vol}(G')/2^c \text{ for some integer } n. \end{cases}$$

6. Real grassmannians

We now make some calculations to improve (5.6). The result is

6.1. THEOREM. The even dimensional oriented real Grassman manifolds $M' = G'/K = \mathrm{SO}(2u+l)/\mathrm{SO}(2u) \times \mathrm{SO}(l)$ have 2 as a K-characteristic number.

As an immediate consequence we will have

(6.2) $$\begin{cases} \text{if } G \text{ is locally isomorphic to } \mathrm{SO}(2u, 2v) \text{ or } \mathrm{SO}(2u, 2v+1) \\ \text{then } \operatorname{Vol}(\Gamma\backslash G) = n \cdot \operatorname{Vol}(G')/2 \text{ for some positive integer } n. \end{cases}$$

6.3. LEMMA. Express the non-oriented real Grassmann manifold as $M'' = G''/K'' = O(k+l)/O(k) \times O(l)$. Then $H^*(M''; Z)$ is equal to its K''-characteristic subring, that is, is the image of the cohomology $H^*(B_{K''}; Z)$ of the classifying space.

Proof. Here $M' = \mathrm{SO}(k+l)/\mathrm{SO}(k) \times \mathrm{SO}(l) = G'/K$ and $\xi : M' \to M''$ denotes the standard 2-sheeted covering. Consider the diagram

$$\begin{array}{ccc} M' & \xrightarrow{\alpha} & B_K \\ \Big\downarrow{\scriptstyle \xi} & & \Big\downarrow{\scriptstyle \eta} \\ M'' & \xrightarrow{\beta} & B_{K''} \end{array}$$

where the horizontal maps to classifying spaces induce the principal bundles $G' \to M'$ and $G'' \to M''$, and where η is induced by $K \subset K''$. We must prove β^* surjective on integral cohomology.

Borel [1; Theorem 11.1] proved that η^* is bijective and β^* is surjective for Z_2-cohomology.

Let p be an odd prime. The coverings ξ and η are of order 2 and 4, and the spaces in the diagram have no cohomology p-torsion, so ξ^* and η^* are injective for Z_p-cohomology. Also, from the Stieffel-Whitney classes, $B_{SO(k)} \to B_{O(k)}$ and $B_{SO(l)} \to B_{O(l)}$ induce surjective maps for Z_p-cohomology, so η^* is surjective as well. Borel's result [0; Prop. 30.2] shows α^* surjective for Z_p-cohomology. We conclude that β^* is surjective.

Now $\beta^* : H^*(B_{K''}; Z_p) \to H^*(M''; Z_p)$ is surjective for all p, and that proves the lemma. Q.E.D.

6.4. LEMMA. Let $k = 2u$ and $l = 2v$, even, so that M'' has a G'-invariant orientation. Give M'' the riemannian metric such that $M' \to M''$ is a local isometry and let Ω'' be the curvature form. Then there is an integral, Ad (K'')-invariant, polynomial c on $\mathfrak{K}$ such that

$$\int_{M''} c(\Omega'') = 1 \quad \text{and} \quad \int_{M'} c(\Omega') = 2.$$

{This proves Theorem 6.1 for l even.}

Proof. Lemma 6.3 gives us c, as required, with $\int_{M''} c(\Omega'') = 1$. Since $\xi : M' \to M''$ is riemannian, $\Omega' = \xi^*\Omega$, and so $\int_{M'} c(\Omega') = 2 \int_{M''} c(\Omega'') = 2$ because ξ is 2-sheeted. Q.E.D.

Proof of Theorem. Write $K = K_1 \times K_2$ where $K_1 = SO(2u)$ and $K_2 = SO(l)$, and let $l = 2v$ or $2v + 1$.

The usual (vector) representation τ_1 of K_1 has weight system $\{a_1, \ldots, a_u, -a_1, \ldots, -a_u\}$ and has Pfaffian $\mathrm{Pf}(\tau_1) = a_1 a_2 \ldots a_u \in \mathscr{C}(\mathfrak{K}_1)$. The vector representation τ_2 of K_2 has weight system

$\{b_1, \ldots, b_v, -b_1, \ldots, -b_v, ((0))\}$, where (( )) indicates that the term occurs just when $l = 2v+1$. Now the representation of K on the tangent space of M' is $\tau_1 \otimes \tau_2$, which has weight system $\{\pm(a_j + b_k), \pm(a_j - b_k), ((\pm a_j))\}$, $1 \leq j \leq u$ and $1 \leq k \leq v$. That has Pfaffian

$$E = ((a_1 a_2 \ldots a_u)) \cdot \prod_{j,k} (a_j^2 - b_k^2) \in \mathscr{C}(\mathfrak{K}). \tag{6.5}$$

and $E(\Omega')$ is the Euler class on M'.

We apply Lemma 6.4 to the case $l = 2v$. The fact that c is integral and K''-invariant, says that, on the Lie algebra of the maximal torus, c is an integral polynomial in the elementary symmetric functions $\sigma_s(a_1^2, \ldots, a_u^2)$ and the $\sigma_t(b_1^2, \ldots, b_v^2)$. Since $c(\Omega')[M'] = 2$ and

$$E(\Omega')[M'] = \chi(M') = 2\binom{u+v}{v}, \quad \text{(6.5) tells us that}$$

$$\prod_{j,k} (a_j^2 - b_k^2) = \binom{u+v}{v} c + \sum p_i q_i, \tag{6.6}$$

where the p_i are integral polynomials in the $\sigma_s(a_1^2, \ldots, a_u^2)$ and the $\sigma_t(b_1^2, \ldots, b_v^2)$, and the q_i are integral polynomials in the $\sigma_w(a_1^2, \ldots, b_v^2)$.

Now let $l = 2v+1$. Multiplying (6.6) by $a_1 a_2 \ldots a_u$, we see from (6.5) that

$$E = \binom{u+v}{v} c' + \sum p'_i q_i, \tag{6.7}$$

where $c' = (a_1 a_2 \ldots a_u)c$ and the $p'_i = (a_1 a_2 \ldots a_u) p_i$ are integral polynomials in the

$$\sigma_s(a_1^2, \ldots, a_u^2), \quad \mathrm{Pf}(\tau_1) \quad \text{and} \quad \sigma_t(b_1^2, \ldots, b_v^2),$$

and where the q_i are integral polynomials in the $\sigma_w(a_1^2, \ldots, b_v^2)$. Now c', p_i and $q_i \in \mathscr{C}(\mathfrak{K})$, all integral, with the q_i invariant by conjugation from G'. Thus (6.7) says that

$$c'(\Omega')[M'] = \binom{u+v}{v}^{-1} \cdot E(\Omega')[M'] = 2$$

is a K-characteristic number of M'. Q.E.D.

7. Summary for Fundamental Domains

We proved a number of integrality statements for volumes of fundamental domains, and here we assemble them for the case of simple groups. Remember, that the 2-sphere is hermitian and that the 4-sphere is the quaternion projective line.

7.1. THEOREM. Let G be a connected linear simple Lie group with rank K = rank G. If Γ is a torsion free discrete subgroup with $\Gamma\backslash G$ compact, then

(7.2) $$\mathrm{Vol}\,(\Gamma\backslash G) = n \cdot \mathrm{Vol}\,(G')/r \quad \text{for some integer } n > 0$$

where r depends only on G, and where $r = 1$ except possibly in the cases

(7.3)	G (to local isomorphism)	r is a divisor of
	$SO(2u, l)$, $(1, 1) \neq (u, l) \neq (2, 1)$	2
	F_{4,A_1C_3}	12
	E_{6,A_1A_5}	4
	E_{7,A_7}	72
	E_{8,A_1E_7}	8

It seems clear that a closer look at the exceptional groups will improve the result on the value of r, especially for E_{7,A_7}.

University of Miami, Coral Gables, Florida

Princeton University, Princeton, New Jersey

The Hebrew University of Jerusalem and
The University of California at Berkeley

REFERENCES

[0] Borel, A., 'Sur la cohomologie des espaces fibres et des espaces homogènes de groupes de Lie compacts', *Ann. of Math.* **57** (1953), 115–207.
[1] Borel, A., 'La cohomolgie mod 2 de certains espaces homogènes', *Comment. Math. Helv.* **27** (1953), 165–197.
[2] Borel, A. and Chevalley, C., 'The Betti Numbers of the Exceptional Groups', *Memoirs Amer. Math. Soc.* **14** (1955), 1–9.
[3] Bott, R., 'Homogeneous Vector Bundles', *Ann. of Math.* **66** (1957), 203–248.
[4] Cahn, R. S., 'Lattice points and Lie Groups III', *Proc. Amer. Math. Soc.* **46** (1974), 247–249.
[5] Cahn, R. S. and Wolf, J. A., 'Zeta Functions and Their Asymptotic Expansions for Compact Symmetric Spaces of Rank One', *Comm. Math. Helv.* **51** (1976), 1–21.
[6] Cahn, R. S. and Wolf, J. A., 'Zeta Functions and Their Asymptotic Expansions for Compact Locally Symmetric Spaces of Negative Curvature', *Bull. Amer. Math. Soc.* **81** (1975), 1086.
[7] Chern, S.-S., 'A Simple Intrinsic Proof of the Gauss-Bonnet Formula for Closed Riemannian Manifolds', *Ann. of Math.* **45** (1944), 747–752.
[8] Gilkey, P. B., *The Index Theorem and the Heat Equation*, Publish or Perish Inc., Boston, 1974.
[9] Hirzebruch, F., 'Automorphe Formen und der Satz von Riemann-Roch', *Symposium Internacional de Topologia Algebraica*, pp. 129–144, Mexico City, 1958.
[10] Serre, J. P., Cohomolgie des groupes discrets, 'Prospects in Mathematics', *Ann. of Math. Studies* **70** (1971), 77–169.
[11] Siegel, C. L., 'Some Remarks on Discontinuous Groups', *Ann. of Math.* **46** (1945), 708–718.
[12] Wolf, J. A., 'Essential Self Adjointness for the Dirac Operator and Its Square', *Indiana Univ. Math. J.* **22** (1973), 611–640.
[13] Wolf, J. A., 'Partially Harmonic Spinors and Representations of Reductive Lie Groups', *J. Functional Analysis* **15** (1974), 117–154.
[14] Mong, S., 'The Signature of Compact Homogeneous Spaces' (to appear).

E. COMBET and C. MORENO

SOME REMARKS ON THE FUNDAMENTAL KERNELS OF A PSEUDO-RIEMANNIAN MANIFOLD

INTRODUCTION

We consider in this paper the elementary solutions of the wave operator, or dalembertian, on a pseudo-Riemannian manifold of normal-hyperbolic type. We review some results and problems concerning these solutions; in particular their relations with the geometry and analysis on the manifold (G and G_1 kernels) are considered. Professor Lichnerowicz's work has greatly inspired this paper. We wish to present it to him as a mark of our heartiest thanks.

It is known that the ordinary Laplace differential operator can be generalized to functions, differential forms, tensor and spinor fields on a Riemannian (elliptic) manifold [13–15] leading to the now classical relations between the geometry and the topology of the manifold. In the same way, the d'Alembert differential operator can be generalized to the functions, tensors and spinors on a pseudo-Riemannian manifold of normal-hyperbolic type (or Lorentzian manifold): [16–18] and it plays a central role in the theory of fields in General Relativity.

Let (M, g) be a Lorentzian manifold: M is a smooth differentiable manifold and g a smooth metric of signature $(+,-,-,\ldots,-)$ on M. The scalar dalembertian operator is the second order differential operator of local expression:

$$\Delta u = -\frac{1}{\sqrt{|g|}}\,\partial_\alpha(\sqrt{|g|}\,g^{\alpha\beta}\partial_\beta u),$$

where $|g| = |\det(g_{\alpha\beta})|$. For every fixed point x' of M, it exists locally (see [4, 5] for global results) on M two distributions $E^{\pm}_{x'}$ with supports in the future and past half cones with vertex x'. These distributions are the elementary solutions of Δ relatively to x':

$$\Delta E^{\pm}_{x'} = \delta_{x'}.$$

They are unique and they define two kernels, called fundamental kernels, in a neighbourhood of the diagonal of $M \times M$. Other kernels

Cahen & Flato (eds.), Differential Geometry and Relativity, 55–62.

relative to Δ exist naturally on M: the propagator $G = E^+ - E^-$ and a kernel G_1 that we shall construct in §2; they generalize respectively the D and D_1 kernels of the Special Relativity [2, 3, 6, 7, 16–18].

1. Fundamental kernels

(a) The singularities of the elementary solutions $E^{\pm}_{x'}$ are asymptotically developed through some distributions deduced from the square power P of the geodesic length on (M, g). This result has been established by Hadamard [11].

In a normal geodesic coordinate system (x^α) relative to x', this function P is written:

$$P(x, x') = g_{\alpha\beta}(x')x^\alpha x^\beta. \tag{1.1}$$

The Hadamard method can be easily adapted to the tensorial and spinorial cases by means of the transport tensor of the appropriate connexion and gives then analogous expansion [7].

(b) On the canonical Mikowskian space $(\mathbf{R}^n, g_0)$ it is generally considered a complex quadratic form:

$$\mathscr{P} = P + i\alpha P', \tag{1.2}$$

where P is given by (1.1), P' is definite positive and α is real >0. Then when α tends to 0 in the $\mathscr{P}^\lambda$ (λ complex) distributions the D and D_1 kernels considered in the introduction are simultaneously obtained.

Can this method be extended to a Lorentzian manifold (M, g)?

On this manifold M, a field $g + i\alpha g'$ of complexes quadratic forms can be considered, g' being of elliptic type. The foregoing method can be extended if a normal geodesic coordinate system common to g and g' exists at each point; but then the Euclidian connexions associated to g and g' are the same ones, and the holonomy group at every point x' keeps invariant a one dimensional subspace of the tangent space $T_{x'}M$. Therefore [13, §§69–70] the manifold (M, g) is locally reducible:

$$ds^2 = dx^1 \otimes dx^1 - \sum_{i,j=2}^{n} g_{ij}(x^2, \ldots, x^n)\, dx^i \otimes dx^j,$$

$n = \dim M$. Compare with §2 below.

(c) The singular part of the elementary solutions are therefore asymptotically developed through certain powers P^{λ_n} (or $\log P \ldots$)

restricted to the past and future half-cones of M. Up to a numerical coefficient, the coefficients of these developments are kernels u_k which can be calculated by an iterative process. In a normal geodesic coordinate system at x', we have:

$$\left.\begin{aligned} u_0(x,x') &= |g(x)|^{-1/4} \\ u_k(x,x') &= \frac{u_0(x,x')}{s^k(x)}\int_0^{s(x)} \frac{t^{k-1}\Delta_x u_{k-1}(x,x')}{u_0(x,x')}\,dt \end{aligned}\right\} \tag{1.3}$$

where the integration, in the calculus of u_k, is taken along the geodesic joining x' to x.

We can recognize in (1.3) the coefficients of the powers of t in the asymptotic development of the fundamental solution of the heat equation on a compact (elliptic) Riemannian manifold. In this case (elliptic compact), the canonical measure v of (M, g), and the following a_k numbers are introduced:

$$a_k = \int u_k(x,x)v(x). \tag{1.4}$$

Following M. Berger these numbers play a central role in the 'spectral geometry' of the Riemannian manifold (M, g) (this problem started with a physical problem: to search the relations between the spectrum and the form of a membrane).

In the case of a Lorentzian manifold (M, g) with a compact basis M, we can still introduce the sequence (1.4) and state in the same way the problem of the relations between this sequence and the geometry of (M, g) (the physical background is now the link of the u_k with the 'Huygen's principle' of the Lorentzian manifold (M, g)[2, 10, 21]. The results of a recent paper [19] of Gilkey should be taken into account in the calculus of the terms of the u_k sequence.

(d) These calculi are more easily carried out on some particular spaces and specially on the harmonic spaces [19, 20]; in these spaces we have the equality:

$$\Delta_x P(x,x') = -2n + P(x,x')H(P(x,x')),$$

where $n = \dim M$ and H, the characteristic function of (M, g), is a C^∞ function on a neighbourhood of the origin of $\mathbf{R}$. Now the functions

(1.3) may be written as $u_k(P(x, x'))$ and they satisfy the system:

$$\begin{cases} u_0'(z) = -\frac{1}{8} H(z) u_0(z); \; u_0(0) = 1 \\ 2z\, u_k'(z) + \left(\frac{z}{4} H(z) + k\right) u_k(z) = 4z\, u_{k-1}''(z) + (2n + zH(z)) u_{k-1}'(z). \end{cases}$$

In this way we can find out, for these particular spaces, but in a normal hyperbolic metric, some analogous results to those of the elliptic 'spectral geometry'.

It is probable that when the 'descent method' of Hadamard is applied to the solutions of the dalembertian, on these particular spaces, it will give some interesting 'integral relations' between the fundamental kernels associated to the different dimensions (compare with [1]).

2. THE G_1 KERNEL

(a) For simplicity, let us consider the scalar kernel G_1, corresponding to the Klein-Gordon scalar equation:

$$(\Delta - m^2) u = 0, \tag{2.1}$$

with a mass term $m > 0$.

When M is the Minkowski space of the Special Relativity we can define through the Fourier transform, the positive u^+ and negative u^- frequencies parts of a real solution u of (2.1). This splitting can also be carried out if we introduce on $M \times M$ a G_1 kernel having the following properties:

(i) G_1 is real, symmetric, and a solution of (2.1),

(ii) let G be the propagator of (2.1); the following Lichnerowicz's equality is satisfied [18]:

$$G(x, x') = \int_\sigma \{G_1(x, y) \,.\, \partial_\lambda G_1(x', y) - G_1(x', y) \,.\, \partial_\lambda G_1(x, y)\} \,\mathrm{d}\sigma^\lambda(y), \tag{2.2}$$

where this Volterra product is taken over a spacelike hypersurface σ of M,

(iii) the G_1 kernel defines a correspondence $J : u \to u_1$, between the real solutions of (2.1) in such a way that if we write:

$$u_1(x') = \int_\sigma \{u(y) \,.\, \partial_\lambda G_1(x', y) - G_1(x', y) \,.\, \partial_\lambda u(y)\} \, d\sigma^\lambda(y)$$

and:

$$b(u, v) = \int_\sigma \{u(y) \,.\, \partial_\lambda v(y) - \partial_\lambda u(y) \,.\, v(y)\} \, d\sigma^\lambda(y)$$

for two solutions of (2.1), we have:

$$b(u, u_1) \geqslant 0.$$

With such a kernel G_1, the frequencies parts of u are then the functions:

$$u^\pm = \frac{1}{2}(u \pm iu_1). \tag{2.3}$$

(b) As remarked by Lichnerowicz [18] this construction of $u^\pm$ can be formulated on a Lorentzian manifold M and it is reduced to the construction of a G_1 kernel satisfying (i), (ii) and (iii) of (a). G. Rideau [25] has deduced of these conditions a uniqueness theorem for G_1 when M is the flat Minkowski space-time.

In the following we suppose that (M, g) is a stationary Lorentzian manifold (see [20]) and $M = V \times \mathbb{R}_{x^0}$ with M closed and:

$$ds^2 = \sum_{\mu,\nu=0,1,\ldots,n} g_{\mu\nu}(x^1, \ldots, x^n) \, dx^\mu \, dx^\nu.$$

A calculus of G_1 is possible when adapting to this case some Seeley's results on powers of differential operators [26]. A detailed description of the following can be found in [24]. See also Chevalier [2], Combet [7′] and Duistermaat-Hörmander [8].

Let us consider the following metric conform to the quotient metric of M:

$$\gamma_{ij} = g_{00}^{1/(n-2)} \tilde{\gamma}_{ij}; \qquad \tilde{\gamma}_{ij} = g_{ij} - \frac{g_{0i} g_{0j}}{g_{00}},$$

$i, j = 1, \ldots, n$, and let us write:

$$\sigma^2 = g_{00}^{-[1/(n-2)]}, \qquad W^\mu = \sigma^2 g^{0\mu}, \qquad \gamma = \det(\gamma_{ij}).$$

Introduce now the following differential operators:

$$N = \frac{1}{v^0}(2L_W - \operatorname{div}(W)),$$

$$M = \frac{1}{v^0}\left(-\overset{*}{\Delta} + \frac{m^2}{\sigma^2}\right),$$

where L_W is the Lie derivative with respect to the vector $W = (W^1, \ldots, W^n)$, and $\overset{*}{\Delta}$ is the Laplacian on V relative to γ_{ij}. With these operators Equation (2.1) can be written:

$$\partial_0^2 u + N\partial_0 u + Mu = 0,$$

and also, in the matricial form (here we set $x^0 = t$):

$$\frac{\mathrm{d}}{\mathrm{d}t}\begin{pmatrix} u^1 \\ u^2 \end{pmatrix} = \begin{pmatrix} 0 & I \\ -M & -N \end{pmatrix}\begin{pmatrix} u^1 \\ u^2 \end{pmatrix}, \tag{2.4}$$

where $u^1 = u$, $u^2 = u'_t$. This leads to the following definitions of the operators:

$$T^{-1} = \begin{pmatrix} 0 & I \\ -M & -N \end{pmatrix}, \qquad T = \begin{pmatrix} -M^{-1}N \\ I \end{pmatrix}\begin{pmatrix} -M^{-1} \\ 0 \end{pmatrix}.$$

(c) In order to study the operators in (b) we need to introduce the Sobolev spaces H^s on the closed manifold V. Notice that H^0 coincides with the completed space of the C^∞ space relatively to:

$$(u, v)_0 = \int_V u\bar{v}v^0\eta(\gamma)$$

where

$$\eta(\gamma) = \sqrt{\gamma}\,\mathrm{d}x^1 \ldots \mathrm{d}x^n$$

and M is a strictly positive elliptic differential operator for $(,)_0$. This justifies the presence of M^{-1} in the foregoing expression of T. The complex powers M^l of M exist and the properties of T and T^{-1} are easily obtained if we put the following scalar product on $H^s \times H^{s-1}$:

$$([u^1, u^2], [v^1, v^2])_1 = (M^{s/2}u^1, M^{s/2}v^1_0) + (M^{(s-1)/2}u^2, M^{(s-1)/2}v^2)_0.$$

In fact we can obtain an appropriate bound for the resolvent of T^{-1}: $H^{s+1} \times H^s \to H^s \times H^{s-1}$ and then adapt the Seeley's theory to construct the complex powers $(T^{-1})^l$. The T^{-1} operator is skew adjoint on $H^s \times H^{s-1}$ relatively to the scalar product:

$$(u, v)_{s,s-1} = (T^{-s+1}u, T^{-s+1}v)_s,$$

where now, u and v are elements of $H^s \times H^{s-1}$. Equation (2.4) can now be written:

$$\frac{1}{i}\frac{\mathrm{d}}{\mathrm{d}t}u(t) = Hu(t), \qquad H = i^{-1}T^{-1}, \tag{2.5}$$

and the Cauchy's problem for this equation is solved by means of the one parameter group U of unitary operators generated by T^{-1} on $(H^s \times H^{s-1}; (,)_{s,s-1})$. The unitary operator J:

$$J = T(TT^*)^{-1/2}, \qquad T^* = -T$$

satisfies: $J^2 = -I$ and defines a complex structure on the solutions space $H^s \times H^{s-1}$. J commutes with T and it is the unique complex structure $\tilde{J}$ to do it and such that $(\tilde{J}u, Ju)_{s,s-1} \geqslant 0$. Therefore we see that the solutions space $H^s \times H^{s-1}$ of Equation (2.4) has $((,)_{s,s-1}; J)$ as Hermitian structure and $(H^s \times H^{s-1}; \sigma_s)$ with $\sigma_s(u, v) = (Ju, v)_{s,s-1}$ as strongly symplectic structure, the presence of these structures being in agreement with I. E. Segal's theory [27].

(d) We can now obtain the propagator G and the kernel G_1 of (2.1); we write:

$$U_1(t) = U(t)J = JU(t)$$

or, in matricial form:

$$U_1(t) = \begin{pmatrix} U^1_{11}(t) & U^1_{12}(t) \\ U^1_{21}(t) & U^1_{22}(t) \end{pmatrix}.$$

$U(t)$ determines in the same way the $U^0_{ij}(t)$ operators. Let now G^0_{12}, G^1_{12} be respectively the associated kernels to the $U^0_{12}(t)$ and $U^1_{12}(t)$ operators; then the G, G_1 kernels are defined on $M \times M$ as:

$$G = \frac{1}{g^{00}} \otimes \frac{1}{g^{00}} G^0_{12},$$

$$G_1 = \frac{1}{g^{00}} \otimes \frac{1}{g^{00}} G^1_{12}.$$

It can be shown the G, G_1 satisfy (i), (ii) and (iii) of (2.a), that the $\pm$ operators (2.3) are two supplementary projections on $H^s \times H^{s-1}$ and it is possible to relate the splitting in frequencies parts to the spectrum of the H operator (2.5) on $H^s \times H^{s-1}$.

See [24] for further details and for the extension of this method to the spin 1 field with a mass term on a globally reducible Lorentzian manifold with closed space sections.

Université Lyon and Collège de France

REFERENCES

[1] Burridge–Papanicolaou–McLaughin, 'The Solution by Descente of Certain Partial Differential Equations', *Comm. Pure and Appl. Math.* **XXVI** (1973), 105–129.

[2] Chevalier, M., 'Sur le Noyau de Diffusion de l'Opérateur Laplacien', *CRAS* **264-A** (1967), 380.

[3] Chevalier, M. 'Opérateurs de Création et annihilation sur un espacetemps Stationnaire', *J. Math. Pures et Appl.* **53** (1974), No. 3.

[4] Choquet-Bruhat, Y., 'Solutions Élémentaires d'Équations du second Ordre', Colloque CNRS, Nancy, 1956.

[5] Choquet-Bruhat, Y., 'Partial Differential Equations on a Manifold', *Battelle Rencontres 1967*, Benjamin, New York, 1968.

[6] Colleau, J., 'Etude des Propagateurs Tensoriels et Spinoriels', *Annales Inst. H. Poincaré* **3** (1965), 195.

[7] Combet, E., 'Solutions Élémentaires des Dalembertiens Généralisés', *Memorial des Sc. Mathématiques* **CLX** (1965).

[7'] Combet, E., Séminaire de Physique mathématique, Collège de France, 1965.

[8] Duistermaat, J. J. and Hörmander, L., 'Fourier Integral Operators II', *Acta Mathematica* **128** (1972), 183–269.

[9] Gilkey, P. B., 'Local Invariants of a Pseudo-Riemannian Manifold,' *Math. Scand.* **36** (1975), 109–130.

[10] Günther, P. and Wünsch, V., 'Maxwellsche Gleichungen und Huyghensches Prinzip I', *Mathematische Nachrichten* **63** (1974), 97–121.

[11] Hadamard, J., *Lectures on Cauchy's problem*, Dover New York 1952.

[12] Helgason, S., 'Differential Operators on Homogeneous Spaces', *Acta Mathematica* **102** (1959), 239–299.

[13] Lichnerowicz, A., *Théorie globale des connexions et des groupes d'holonomie*, Cremonese, Roma, 1955.

[14] Lichnerowicz, A., *Géométrie des groupes de transformations*, Dunod, Paris, 1958.

[15] Lichnerowicz, A., 'Spineurs Harmoniques', *CRAS Paris* **257** (1963), 7–9.

[16] Lichnerowicz, A. 'Propagateurs et Commutateurs en Relativité Générale', *I.H.E.S. Publications mathématiques* **10** (1961).

[17] Lichnerowicz, A., 'Champs Spinoriels et Propagateurs en Relativité Générale', *Bull. Soc. Math. de France* **92** (1964), 11–100.

[18] Lichnerowicz, A., 'Propagateurs et Quantification en Relativité Générale', *Cahiers de Physique* **18** (1964), 83–92.

[19] Lichnerowicz, A., *Equations de Laplace et espaces harmoniques*, Colloque équations aux dérivées partielles, Louvain, 1953.

[20] Lichnerowicz, A., *Théories relativistes de la gravitation et de l'électromagnétisme*, Masson, Paris, 1955.

[21] McLenaghan, R. G., 'On the Validity of Huygen's Principle', *Ann. Inst. H. Poincaré (A)* **20** (1974), 153–188.

[22] Moreno, C., 'Sur la Quantification dans un Espace-Temps Courbe', *CRAS Paris* **273** (1971), 830–833.

[23] Moreno, C. 'Commutateurs et Opérateurs de Création-Annihilation', *CRAS Paris* **274** (1972), 281–284.

[24] Moreno, C., 'Commutateurs et Opérateurs de Création-Annihilation', Thèse, Université Lyon 1.

[25] Rideau, G. *CRAS Paris* **260** (1967), 2719–2721.

[26] Seeley, R. T., 'Complex Powers of an Elliptic Operator, In *Proceedings of Symp. pure math.* **X**, AM.S (1967).

[27] Segal, I. E., 'Symplectic Structure and the Quantization Problem for Wave Equation', Conference on Symplectic geometry and Mathematical Physics, Roma, 1973.

ALFRED GRAY

GEODESIC BALLS IN RIEMANNIAN PRODUCT MANIFOLDS

Let M be a Riemannian manifold and $m \in M$. We denote by $A(r)$ and $V(r)$ the surface area and volume of a geodesic ball of radius r centered at m. In [1] a power series development for $A(r)$ and $V(r)$ has been given. The coefficients are invariant polynomials in the curvature operator and its covariant derivatives.

In this note we consider the case when M is a Riemannian product manifold, $M = M_1 \times M_2$. Let $A_1(r)$, $A_2(r)$, $V_1(r)$, $V_2(r)$ be the corresponding area and volume functions at $m_1 \in M_1$ and $m_2 \in M_2$. If $M_1 \times M_2$ are flat then

$$V(r) = \binom{\frac{1}{2}n}{\frac{1}{2}k}^{-1} V_1(r)\,V_2(r), \tag{1}$$

where $n = \dim M$, $k = \dim M_1$ and $(\frac{1}{2}n)! = \Gamma(\frac{1}{2}n + 1)$. Equation (1) follows easily from the formula

$$V(r) = \left(\frac{n}{2}!\right)^{-1} (\pi r^2)^{n/2} \qquad (n = \dim M).$$

However, if M_1 and M_2 are not flat, the situation is considerably more complicated. Instead of a product formula, we show that an integral transform product formula holds.

Let $f : (0, \infty) \to \mathbb{R}$. Provided the integral exists, we define

$$\hat{f}(s) - \int_0^\infty e^{-s^2t^2} f(t)\,\mathrm{d}t.$$

More generally if f is analytic, the power series $e^{-s^2t^2} f(t)$ can be integrated term by term. Thus $\hat{f}$ can always be formally defined, even though it is not a function.

Our basic theorem is the following.

THEOREM. Suppose M_1 and M_2 are complete Riemannian manifolds $M = M_1 \times M_2$. Then at any point in M we have

$$\hat{A}(s) = \hat{A}_1(s)\hat{A}_2(s). \tag{2}$$

Cahen & Flato (eds.), Differential Geometry and Relativity, 63–66.

Before proving this theorem we must clarify the definitions of $A(r)$ and $V(r)$. According to [1] we have

$$A(r)=\int_{S^{n-1}(1)} r^{n-1}\omega_{1\ldots n}(\exp_m (ru))\,du, \tag{3}$$

where $S^{n-1}(1)$ is the unit sphere in the tangent space M_m and $\omega_{1\ldots n}^2=\det(g_{ij})$ in normal coordinates. If r is less than the distance from m to the first conjugate point, then $A(r)$ is the surface area of the geodesic ball of radius r centered at m. However, the right hand side of (3) makes sense for all values of r, provided we use the power series expansion for $\omega_{1\ldots n}$ given in [1]:

$$\omega_{1\ldots n}=1-\tfrac{1}{6}\sum_{i,j=1}^{n} R_{ij}(m)x_ix_j-\tfrac{1}{12}\sum_{i,j,k=1}^{n}\nabla_iR_{jk}(m)x_ix_jx_k\cdots. \tag{4}$$

Therefore we define $A(r)$ by (3) for all values of r; note that it is sometimes negative.

Proof of the theorem. We choose normal coordinates $(x_1,\ldots,x_n)$ on M at $m=(m_1,m_2)$ so that $(x_1,\ldots,x_k)$ are normal coordinates on M_1 and $(x_{k+1},\ldots,x_n)$ are normal coordinates on M_2. Write $r_1^2=x_1^2+\cdots+x_k^2$, $r_2^2=x_{k+1}^2+\cdots+x_n^2$, $r^2=r_1^2+r_2^2$. Then we have $\omega_{1\ldots n}(p,q)=\omega_{1\ldots k}(p)\omega_{k+1\ldots n}(q)$. Therefore

$$\begin{aligned}
\hat{A}(s)&=\int_0^\infty e^{-s^2r^2}A(r)\,dr\\
&=\int_0^\infty\int_{s^{n-1}(1)} e^{-s^2r^2}r^{n-1}\omega_{1\ldots n}(\exp_m(ru))\,du\,dr\\
&=\int_{-\infty}^{\infty}\cdots\int_{-\infty}^{\infty} e^{-s^2(x_1^2+\cdots+x_n^2)}\omega_{1\ldots k}(\exp_{m_1}(r_1u_1))\\
&\quad\cdot\omega_{k+1\ldots n}(\exp_{m_2}(r_2u_2))\,dx_1\ldots dx_n\\
&=\left\{\int_{-\infty}^{\infty}\cdots\int_{-\infty}^{\infty} e^{-s^2(x_1^2+\cdots+x_k^2)}\omega_{1\ldots k}(\exp(r_1u_1))\,dx_1\ldots dx_k\right\}\\
&\quad\cdot\left\{\int_{-\infty}^{\infty}\cdots\int_{-\infty}^{\infty} e^{-s^2(x_{k+1}^2+\cdots+x_n^2)}\omega_{k+1\ldots n}(\exp(r_2u_2))\,dx_{k+1}\ldots dx_n\right\}\\
&=\hat{A}_1(s)\hat{A}_2(s).
\end{aligned}$$

We now use (2) to determine the relations between $A(r)$ and $A_1(r)$, $A_2(r)$, and also between $V(r)$ and $V_1(r)$, $V_2(r)$. We expand $V(r)$, $V_1(r)$,

$V_2(r)$ in power series

$$\left.\begin{aligned} V(r) &= \pi^{n/2}\left(\left(\frac{n}{2}\right)!\right)^{-1}\sum_{p=0}^{\infty} a_{2p}r^{2p+n} \\ V_1(r) &= \pi^{k/2}\left(\left(\frac{k}{2}\right)!\right)^{-1}\sum_{p=0}^{\infty} b_{2p}r^{2p+k} \\ V_2(r) &= \pi^{l/2}\left(\left(\frac{l}{2}\right)!\right)^{-1}\sum_{p=0}^{\infty} c_{2p}r^{2p+l} \end{aligned}\right\}, \tag{5}$$

where $n = k + l$. A method is given in [1] for computing the coefficients in (5). For example,

$$a_0 = 1,\; a_2 = -\frac{1}{6(n+2)}\tau(R),$$

$$a_4 = \frac{1}{360(n+2)(n+4)}\{-3\|R\|^2 + 8\|\rho(R)\|^2 + 5\tau(R)^2 - 18\Delta R\},$$

where

$$\tau(R) = \sum_i R_{ii},$$

$$\|\rho(R)\|^2 = \sum_{i,j} R_{ij}^2,$$

$$\|R\|^2 = \sum_{i,j,k,l} R_{ijkl}^2,$$

and

$$\Delta R = \sum_i \nabla_{ii}^2\tau(R).$$

(In [1] it is shown that $a_{2p+1} = 0$ for all p). From the formula

$$A(r) = \frac{\mathrm{d}}{\mathrm{d}r}V(r) = \pi^{n/2}\left\{\left(\frac{n}{2}\right)!\right\}^{-1}\sum_{p=0}^{\infty}(2p+n)a_{2p}r^{2p+n-1} \tag{6}$$

and (2) we can compute the a_{2p} in terms of the b_{2p} and c_{2p}. Using Laplace transforms we see that

$$\begin{aligned} \hat{A}(s) &= \sum_{p=0}^{\infty}\left\{\left(\frac{n}{2}\right)!\right\}^{-1}(2p+n)\pi^{n/2}a_{2p}\int_0^{\infty} e^{-s^2r^2}\, r^{2p+n-1}\,\mathrm{d}r \\ &= \sum_{p=0}^{\infty}\left\{\left(\frac{n}{2}\right)!\right\}^{-1}\left(p+\frac{n}{2}\right)!\,\pi^{n/2}\, s^{-(2p+n)}a_{2p}, \end{aligned} \tag{7}$$

and similarly for $\hat{A}_1(s)$ and $\hat{A}_2(s)$. Therefore from (2) and (7) we obtain

$$\sum_{p=0}^{\infty}\left(\frac{n}{2}+1\right)\cdots\left(\frac{n}{2}+p\right)a_{2p}s^{-2p}=\left\{\sum_{p=0}^{\infty}\left(\frac{k}{2}+1\right)\cdots\left(\frac{k}{2}+p\right)b_{2p}s^{-2p}\right\}$$
$$\cdot\left\{\sum_{p=0}^{\infty}\left(\frac{l}{2}+1\right)\cdots\left(\frac{l}{2}+p\right)c_{2p}s^{-2p}\right\}. \tag{8}$$

By equating powers of s, we can find a_p in terms of the b_q and c_r where $q, r \leqslant p$. We have

$$a_0 = b_0 = c_0 = 1$$

and

$$(n+2)(n+4)\ldots(n+2p)a_{2p}$$
$$=\sum_{q=0}^{p}(k+2)(k+4)\ldots(k+2q)(n-k+2)$$
$$\cdot(n-k+4)\ldots(n-k+2p-2q)b_{2q}c_{2p-2q}.$$

Thus

$$a_2=\left\{\frac{k+2}{n+2}\right\}b_2+\left\{\frac{n-k+2}{n+2}\right\}c_2$$

$$a_4=\left\{\frac{(k+2)(k+4)}{(n+2)(n+4)}\right\}b_4+\left\{\frac{(k+2)(n-k+2)}{(n+2)(n+4)}\right\}b_2c_2$$
$$+\left\{\frac{(n-k+2)(n-k+4)}{(n+2)(n+4)}\right\}c_4,$$

and so forth.

University of Maryland and Universität Zürich

REFERENCE

[1] Gray, A., 'The Volume of a Small Geodesic Ball of a Riemannian Manifold', *Michigan Math. J.* **20** (1973), 329–344.

S. HALPERIN and D. LEHMANN

TWISTED EXOTISM

1. Introduction

Recall that for any C^∞ vector bundle $E \to M$, equipped with a connection ∇ of curvature R, one defines (cf. Koszul [6]) an exterior differential $d_\nabla : A^p(M, E) \to A^{p+1}(M, E)$ on the space $A^*(M, E)$ of differential forms on M with coefficients in E (i.e. sections of $\Lambda T^*(M) \otimes E$) by putting:

$$(d_\nabla \alpha)(x_0, \ldots, x_p) = \sum_{i=0}^{p} (-1)^i \nabla_{x_i}(\alpha(x_0, \ldots, \hat{x}_i, \ldots, x_p))$$

$$+ \sum_{i<j} (-1^{i+j}\alpha([x_i, x_j], x_0, \ldots, \hat{x}_i, \ldots, \hat{x}_j, \ldots, x_p).$$

This operator verifies: $d_\nabla d_\nabla \alpha = R \wedge \alpha$ (Ricci identity). The graded vector space

$$H^*_\nabla(M, E) = \frac{\ker d_\nabla}{\operatorname{Ker} d_\nabla \cap \operatorname{Im} d_\nabla}$$

has been introduced in another way by I. Vaisman [9], under the name of 'curvature groups' of (E, ∇). In the sequel, we will call it '*twisted*' *cohomology* of M with coefficients in E, with respect to ∇ (see [3] for a detailed description). 'Scalar' cohomology will mean de Rham cohomology (it corresponds to (E, ∇) trivial).

Let $P \to M$ be a C^∞ principal G-bundle, ω a connection on P and V a finite dimensional G-module. Assume that $E = V \times_G P$ is the vector bundle associated to (P, V), and ∇ the connection on E corresponding to ω. Then (cf. [3]) the Chern-Weil homomorphism generalizes to a map $[\lambda^V_\omega] : [S\mathscr{G}^* \otimes V]^{\mathscr{G}} \to H^*_\nabla(M, E)$, induced by a map $\lambda^V_\omega : [S\mathscr{G}^* \otimes V]^{\mathscr{G}} \to \operatorname{Ker} d_\nabla$, where $\mathscr{G}$ denotes the Lie algebra of G, and $[S\mathscr{G}^* \otimes V]^{\mathscr{G}}$ the space of those polynomials on $\mathscr{G}$ with coefficients in V which are $\mathscr{G}$-equivariant when regarded as maps from $S\mathscr{G}$ (with the adjoint representation) into V. If such a polynomial has degree k, its

Cahen & Flato (eds.), Differential Geometry and Relativity, 67–73.

image under λ_ω^V has dimension $2k$. We call the elements of $\mathrm{Im}\,[\lambda_\omega^V]$ '*twisted* (*primary*) *characteristic classes*'.

On the other hand, '*scalar exotic* (*or secondary*) *characteristic classes*' have been defined, first for foliations (Godbillon-Vey [2], Bott-Haefliger [1], Kamber-Tondeur [4], etc...), then generalized in different ways (see for instance [8]). In particular, assume $\mathscr{G}$ to be reductive, and for any connection ω on a principal G-bundle $P \to M$, let $\lambda_\omega : I^k(\mathscr{G}) \to A^{2k}(M)$ be the Chern-Weil map (at the cochain level), and $\Delta_{\omega,\omega'} : I^k(\mathscr{G}) \to A^{2k-1}(M)$ the Chern map such that $\mathrm{d} \cdot \Delta_{\omega,\omega'} = \lambda_{\omega'} - \lambda_\omega$. Assume, now on, that the curvature Ω of ω satisfies $\Omega^{r+1} = 0$ ($2r+2$-form on P with coefficients in $\overbrace{\mathscr{G} \otimes \mathscr{G} \ldots \otimes \mathscr{G}}^{r+1}$), while ω' is flat ($\Omega' = 0$). Then, one defines a DGA-map

$$\rho_{\omega,\omega'} : \Lambda P_{\mathscr{G}} \otimes I(\mathscr{G})/_{(>r)} \to A^*(M),$$

where $\Lambda P_{\mathscr{G}}$ denotes the exterior algebra over the space $P_{\mathscr{G}}$ of primitive elements of $\Lambda \mathscr{G}^*$. ($\Lambda P_{\mathscr{G}}$ is equal to the invariant cochains on $\mathscr{G}$, and to $H^*(\mathscr{G}, \mathbf{R})$, and $(>r)$ stands for the homogeneous ideal of $I(\mathscr{G})$ (resp. of $S\mathscr{G}^*$) of polynomials of degree $>r$). Let $\tau : P_{\mathscr{G}} \to I(\mathscr{G})$ denote some fixed transgression. For any $x \in I(\mathscr{G})$, denote by $\bar{x}^r$ its equivalence class modulo $(>r)$. Then, the differential d on $\Lambda P_{\mathscr{G}} \otimes I(\mathscr{G})/_{(>r)}$ and the map $\rho_{\omega,\omega'}$ are given, for any $h \in P_{\mathscr{G}}$ and $x \in I(\mathscr{G})$, by:

$$\begin{cases} \mathrm{d}h = \overline{\tau h}^r \\ \mathrm{d}\bar{x}^r = 0 \end{cases} \qquad \begin{cases} \rho_{\omega,\omega'}(h) = \Delta_{\omega,\omega'}(\tau h) \\ \rho_{\omega,\omega'}(\bar{x}^r) = \lambda_\omega(x) \end{cases}$$

(obvious convention: $\dim \bar{x}^r = 2 \times$ degree of x).

The scalar exotic classes are the elements of $H^*(M, \mathbf{R})$ which belong to the image of the map $[\rho_{\omega,\omega'}]$ induced by $\rho_{\omega,\omega'}$ in cohomology. (*Remark*: there is an inclusion)

$$\Lambda P_{\mathscr{G}} \otimes I(\mathscr{G})/_{(>r)} \to \Lambda \mathscr{G}^* \otimes S\mathscr{G}^*/_{(>r)}$$

into the truncated Weil algebra which induces an isomorphism in cohomology).

In the next section we shall define some '*twisted secondary* (*or exotic*) *classes*', in the twisted cohomology, which will be new in the following sense: they cannot be always written as products of twisted primary classes and scalar secondary ones [for the natural product $H^p(M, \mathbf{R}) \times H^q_\nabla(M, E) \to H^{p+q}_\nabla(M, E)$ induced by the exterior product

$$A^p(M) \times A^q(M, E) \to A^{p+q}(M, E)$$

corresponding to the scalar multiplication $\mathbf{R} \times V \to V$].

2. Construction of twisted exotism

Let again ω be a connection on the G-bundle $P \to M$, such that $\Omega^{r+1} = 0$, and ω' be another connection on P which is flat ($\Omega' = 0$). Let V be a G-module, and $E = V \times_G P$ the associated vector bundle, with ∇ corresponding to ω.

Recall that a pseudo-complex [9] is a sequence (C, d) of linear maps $\cdots \to C^p \xrightarrow{\mathrm{d}} C^{p+1} \xrightarrow{\mathrm{d}} \cdots$ and its cohomology the graded vector space

$$H^*(C) = \frac{\operatorname{Ker} \mathrm{d}}{\operatorname{Ker} \mathrm{d} \cap \operatorname{Im} \mathrm{d}}.$$

Call map of pseudo-complex $f:(C, \mathrm{d}) \to (C', \mathrm{d}')$ any family $(f_p : C^p \to C'^p)$ such that $\mathrm{d}' \cdot f_p = f_{p+1} \cdot \mathrm{d}$: such an f induces $[f]: H^*(C) \to H^*(C')$ in cohomology.

Recall (B. Kostant [5]) that there exists a graded $\mathscr{G}$-submodule $L = \bigoplus_{\alpha \geq 0} L^\alpha$ of $S\mathscr{G}^*$ such that multiplication in $S\mathscr{G}^*$ induces an isomorphism $I(\mathscr{G}) \otimes L \xrightarrow{\cong} S(\mathscr{G}^*)$ ($L^0 = \mathbf{R}$). This isomorphism induces an isomorphism

$$\sum_{\alpha=0}^{r} \Lambda P_{\mathscr{G}} \otimes I(\mathscr{G})/_{(>r-\alpha)} \otimes [L^\alpha \otimes V]^{\mathscr{G}} \xrightarrow{\cong} \Lambda P_{\mathscr{G}} \otimes [S(\mathscr{G}^*)/_{(>r)} \otimes V]^{\mathscr{G}}$$

Now for each α $(0 \leq \alpha \leq r)$ define a pseudo-complex structure D on $\Lambda P_{\mathscr{G}} \otimes I(\mathscr{G})/_{(>r-\alpha)} \otimes [L^\alpha \otimes V]^{\mathscr{G}}$ by

$$D(h_1 \Lambda \ldots \Lambda h_s \otimes \bar{x}^{(r-\alpha)} \otimes \psi)$$
$$= \sum_{a=1}^{s} (-1)^{a-1} h_1 \Lambda \ldots \hat{h}_a \ldots \Lambda h_s \otimes \overline{\tau(h_a) \cdot x}^{(r-\alpha)} \otimes \psi,$$

$$h_a \in P_{\mathscr{G}} \quad (a = 1, \ldots, s),\ x \in I(\mathscr{G}),\ \psi \in [L^\alpha \otimes V]^{\mathscr{G}}.$$

(Note that $D^2 = 0$).

Define now a map of pseudo-complexes

$$\rho^V_{\omega,\omega'}: (\Lambda P_{\mathscr{G}} \otimes I(\mathscr{G})/_{(>r-\alpha)} \otimes [L^\alpha \otimes V]^{\mathscr{G}}, D) \to (A^*(M, E), \mathrm{d}_\nabla)$$

by setting

$$\rho^V_{\omega,\omega'}(h_1 \Lambda \ldots \Lambda h_s \otimes \bar{x}^{(r-\alpha)} \otimes \psi)$$
$$= \Delta_{\omega,\omega'}(\tau h_1) \Lambda \ldots \Lambda \Delta_{\omega,\omega'}(\tau h_s) \Lambda \lambda_\omega(x) \Lambda \lambda^V_\omega(\psi).$$

(Because $\lambda_\omega(x)\Lambda\lambda_\omega^V(\psi)=\lambda_\omega^V(x\cdot\psi)$, degree of $(x\cdot\psi)$ = degree of $x+\alpha$, and λ_ω^V vanishes for polynomials of degree $>r$, the given definition of $\rho_{\omega,\omega'}^V$ does not depend upon x, but only on $\bar{x}^{(r-\alpha)}$. It is clear that $d_\nabla\cdot\rho_{\omega,\omega'}^V=\rho_{\omega,\omega'}^V\cdot D$. Therefore $\rho_{\omega,\omega'}^V$ induces a map $[\rho_{\omega,\omega'}]$ from $\bigoplus_{\alpha=0}^r H^*(\Lambda P_{\mathscr{G}}\otimes I(\mathscr{G})/_{(>r-\alpha)}\otimes[L^\alpha\otimes V]^{\mathscr{G}}, D)$ into $H_\nabla^*(M,E)$.

DEFINITION. The *twisted exotism* of ω,ω' is the image of $[\rho_{\omega,\omega'}^V]$.

The twisted exotism is (for each α) a module over the scalar exotic classes, and when $\alpha=0$ this is generated by the twisted primary classes $V^{\mathscr{G}}$. In particular, we get 0 for $\alpha=0$ if $V^{\mathscr{G}}=0$. Twisted exotism becomes more interesting for $\alpha>0$.

3. Examples

(1) Let $G=SO(4)$, $r=2$. Then $I(so(4))=\mathbf{R}[\chi,C_2]$, where χ (resp. C_2) are the polynomials corresponding to the Euler class (resp. first pontryagin class) of an oriented vector bundle of rank 4. Let h_χ and h_2 be the corresponding primitive elements of dimension 3; then we may choose τ so that $\tau(h_\chi)=\chi$ and $\tau(h_2)=C_2$.

For the universal scalar secondary cohomology we have $H^p(\Lambda(h_\chi,h_2)\otimes\mathbf{R}(\chi,C_2)/_{(>2)}=0$, $1\leqslant p\leqslant 6$ while the class $[h_\chi\otimes\chi]$ in dimension 7 is non-zero.

Now take $\alpha=1$. Then $I(so(4))/_{(>1)}=\mathbf{R}$ and $[L^1\otimes so(4)]^{so(4)}=[so(4)^*\otimes so(4)]^{so(4)}$ contains in particular the identity $Id_{so(4)}$ in $so(4)$. (In fact, $\dim[L^1\otimes so(4)]^{so(4)}=4$.) Moreover the differential D is zero in $\Lambda(h_\chi,h_2)\otimes[L^1\otimes so(4)]^{so(4)}$ and so $[h_\chi\otimes 1\otimes I\,d_{so(4)}]$ is a non-trivial universal twisted secondary class. We shall give an example, now of a vector bundle $V\times_G P$ with 2 connections ω,ω' (as in 2) such that $[\rho_{\omega,\omega'}^V]([h_\chi\otimes 1\otimes Id_{so(4)}])$ is non-zero.

In fact, let $M=SO(5)$ and let F be the foliation of M of codimension 4 induced from the fibration $M\to SO(5)/SO(4)=S^4$. The normal bundle ν has a natural bundle like metric, and an orthonormal framing. Let ω be the unique torsion free connection in ν respecting the metric and let ω' be the unique flat connection respecting both metric and framing. Then according to Lazarov and Pasternak [7] the 7-dimensional class $[\rho_{\omega,\omega'}(h_\chi\otimes\chi)]$ is non-zero in $H^7(M;\mathbf{R})$.

Now let $V=so(4)$ (with the adjoint representation). Then the associated bundle $V\times_G P=E$ has fibre at a point p the skew symmetric transformations of the fibre ν_p. Let ∇ be the connection in E corresponding to ω, and denote its curvature by $R \in A^2(M, E)$. Then $d_\nabla R = 0$ (Bianchi identity). Thus we have the class $[R] \in H^2_\nabla(M, E)$.

Now regard χ as a linear map so (4)$\otimes$so (4)$\to \mathbf{R}$. It induces a map $\chi_* : H_\nabla(M, E \otimes E) \to H(M, \mathbf{R})$ and so we have the sequence of maps

$$\gamma : H^5_\nabla(M, E) \otimes H^2_\nabla(M, E) \longrightarrow H^7_\nabla(M, E \otimes E) \xrightarrow{\chi_*} H^7(M; \mathbf{R}).$$

A simple calculation shows that

$$[\rho_{\omega,\omega'}(h_\chi \otimes \chi)] = \gamma\{[\rho^{\mathrm{so}(4)}_{\omega,\omega'}(h_\chi \otimes 1 \otimes I\,\mathrm{d}_{\mathrm{so}(4)})] \otimes [R]\}.$$

Thus $[\rho^{\mathrm{so}(4)}_{\omega,\omega'}(h_\chi \otimes 1 \otimes I\mathrm{d}_{\mathrm{so}(4)})]$ and $[R]$ are both non-zero.

Remarks

1. This 5-dimensional class can *not* be obtained by taking products of scalar exotism (which does not live below dimension 7) with twisted primary classes (which live only in even dimension).

2. Note that $[R] \neq 0$, while E is trivial. Another example of this fact appears in [3].

3. For foliations of codimension 4, with bundle-like metric and orthonormal framing of the transverse bundle, scalar and twisted exotism will both vanish if (but not only if) there exists an orthonormal framing $X_1, \ldots, X_4$ by projectable vector fields X_i, and such that brackets $[X_i, X_j]$ all vanish ($\omega = \omega'$).

4. In the example above we used the fact that the scalar exotism did not vanish to prove that the twisted class was also non-trivial. In the next example all the scalar exotism is zero, but the twisted exotism is non-zero. This example, however, does not arise from a foliation.

(2) Suppose $G = \mathrm{SL}(2, \mathbf{R})$ and $r = 2$. Then $I(\mathrm{sl}(2, \mathbf{R}) = \mathbf{R}[C]$ where C is the determinant:

$$C\begin{pmatrix} h & x \\ y & -h \end{pmatrix} = -(h^2 + xy).$$

Denote by u the corresponding primitive element (dimension 3).

The d.g.a. $\Lambda(u) \otimes \mathbf{R}[C]/_{(>2)}$ has no cohomology below dimension 6, and therefore all scalar exotism vanishes if the base M has dimension 5: we will assume this from now on. This also implies that any connection ω on a principal G-bundle $P \to M$ satisfies $\Omega^3 = 0$. For $\alpha = 1$, $I(\mathrm{sl}(2,\mathbf{R}))/_{(>1)} = \mathbf{R}$, and exactly as in the previous example

$$[L^1 \otimes \mathrm{sl}(2,\mathbf{R})]^{\mathrm{sl}(2,\mathbf{R})} = [(\mathrm{sl}(2,\mathbf{R}))^* \otimes \mathrm{sl}(2,\mathbf{R})]^{\mathrm{sl}(2,\mathbf{R})}$$

contains $Id_{\mathrm{sl}(2,\mathbf{R})}$, and the element $u \otimes 1 \otimes Id_{\mathrm{sl}(2,\mathbf{R})}$ represents a non-zero class $[u \otimes 1 \otimes Id_{\mathrm{sl}(2,\mathbf{R})}]$ in

$$H^5\left(\Lambda(u) \otimes [L^1 \otimes \mathrm{sl}(2,\mathbf{R})]^{\mathrm{sl}(2,\mathbf{R})}\right).$$

We want to prove that the *corresponding twisted exotic class* in $H^5_\nabla(M, E)$

$$[\Delta_{\omega,\omega'}(u) \Lambda R]$$

may be non-zero (*although all scalar exotism vanishes*), for suitable M, E, ω and ω'.

In fact, let $E = M \times \mathrm{sl}(2,\mathbf{R})$, be the trivial bundle associated to $P = M \times \mathrm{SL}(2,\mathbf{R})$: via the adjoint representation of $\mathrm{SL}(2,\mathbf{R})$. Let ω' be the standard (flat) connection and set

$$\omega = \omega' + \pi^* \theta \otimes H,$$

where

$$\pi : P \to M, \quad \theta \in A^1(M) \quad \text{and} \quad H = \begin{pmatrix} 1 & 0 \\ 0 & -1 \end{pmatrix} \in \mathrm{sl}(2,\mathbf{R}).$$

Then an easy computation gives $\Delta_{\omega,\omega'}(u) = -2\theta \Lambda \, \mathrm{d}\theta$ and $R = \mathrm{d}\theta \otimes H$ (R the curvature of ω). It follows that

$$\begin{aligned}[\rho^{\mathrm{sl}(2,\mathbf{R})}_{\omega,\omega'}(u \otimes 1 \otimes I\, \mathrm{d}_{\mathrm{sl}(2,\mathbf{R})})] &= [\Delta_{\omega,\omega'}(u) \Lambda R] \\ &= [-2(\theta \Lambda (\mathrm{d}\theta)^2) \otimes H].\end{aligned}$$

On the other hand let

$$\Phi = \alpha X + \beta Y + \gamma H \in A^4(M, E) = A^4(M) \otimes \mathrm{sl}(2,\mathbf{R}),$$

where

$$\alpha, \beta, \gamma \in A^4(M) \quad \text{and} \quad X = \begin{pmatrix} 0 & 1 \\ 0 & 0 \end{pmatrix}, \quad Y = \begin{pmatrix} 0 & 0 \\ 1 & 0 \end{pmatrix}.$$

Then, denoting by ∇ the linear connection associated to ω, we have

$$\mathrm{d}_\nabla \Phi = (\mathrm{d}\alpha + \theta \Lambda \alpha) X + (\mathrm{d}\beta - \theta \Lambda \beta) Y + \mathrm{d}\gamma H.$$

Thus the twisted class above vanishes only if $\theta\Lambda(d\theta)^2$ is a coboundary.

In particular, suppose M oriented and compact and let $(u_1, \ldots, u_5)$ be a positive system of local coordinates on M with domain $U \subset M$. Let $h \in C^\infty(M)$ be a positive non-zero function with support in U, and set

$$\theta = h(u_1\Lambda\ du_2 + u_3\Lambda\ du_4 + du_5).$$

Then $\theta\Lambda(d\theta^2 = 2h^3(du_1\Lambda\ du_2\Lambda\ du_3\Lambda\ du_4\Lambda\ du_5)$ and so

$$\int_M \theta\Lambda(d\theta)^2 > 0.$$

Hence in this case $[\rho^{sl(2,R)}_{\omega,\omega'}(u \otimes 1 \otimes Id_{sl(2,R)})] \neq 0.$

Université des Sciences et Techniques de Lille

REFERENCES

[1] Bott-Haefliger, R., 'On Characteristic Classes of Γ-Foliations', *Bulletin of Am. Math. Soc.* (1972), and others.
[2] Godbillon, C. and Vey, J., 'Un invariant des feuilletages de codimension 1', *Comptes rendus Acad Sc. Paris* (June 1971).
[3] Halperin, S. and Lehmann, D., 'Cohomologie et classes caractéristiques des choux de Bruxelles', (à paraître dans les Comptes Rendus des journées de Dijon (1974) *Lectures Notes in Mathematics*, Springer.
[4] Kamber, F. and Tondeur, P., 'Characteristic Invariants of Foliated Bundles', *Bulletin de l'AMS* (1973), and others.
[5] Kostant, B., 'Lie Group Representations on Polynomial Rings,' *Am. J. of Maths* **85** (1963), 327–404.
[6] Koszul, J. L., *Lectures Notes in Differential Geometry*, Tata Institute, 1960.
[7] Lazarov, C. and Pasternak, J., 'Secondary Characteristic Classes for Riemannian Foliations' (to appear).
[8] Lehmann, D., 'Classes exotiques de Chern-Simons et $\mathcal{J}$-connexité des espaces de connexions', *Comptes Rendus Ac. Sc. Paris* (October 1972), and others.
[9] Vaisman, I., 'Les pseudo-complexes de cochaînes ...', *Analele stiinfice ale Universitatii din Iasi* **14** (1) (1968), 105–136.

YVETTE KOSMANN

ON LIE TRANSFORMATION GROUPS AND THE COVARIANCE OF DIFFERENTIAL OPERATORS

In an attempt to understand the notions of 'invariance' and 'covariance' of a differential operator, which are of great importance in mathematical physics, we were led to study the actions of a group of transformations of a manifold M (the space-time in general relativity) on the sections of a vector bundle over M (the tensor or spinor fields of a given type). Several equivalent characterizations of these actions are given. A similar study is made for a Lie algebra of vector fields.

The invariance of scalar differential operators has been extensively studied [10]. It is clear that the invariance of the operator is not a necessary condition for the space of solutions to be invariant. That idea, suitably formulated for a differential operator on an arbitrary vector bundle, leads to our definition of the covariance.

1. Algebraic preliminaries

1.1. *Semi-linear representations*

Let K be the field of real or complex numbers. Let $\mathscr{F}$ be an associative K-algebra and let Γ be an $\mathscr{F}$-module.

DEFINITION 1. A K-endomorphism E of Γ is said to be semi-linear if there exists a K-endormorphism $E_{\mathscr{F}}$ of $\mathscr{F}$ such that for $f \in \mathscr{F}$ and $\psi \in \Gamma$,

$$E(f\psi) = (E_{\mathscr{F}}f)(E\psi).$$

If $E(\Gamma)$ is not included in the torsion submodule of Γ, then $E_{\mathscr{F}}$ is uniquely determined and $E_{\mathscr{F}}$ is a ring endomorphism.

Let the ring $\mathscr{F}$ be commutative and let Θ denote the tensor algebra of the $\mathscr{F}$-module Γ.

PROPOSITION 1. Assume Γ has an element which is not a torsion element. The set of automorphisms of the graded K-algebra Θ is isomorphic to the set of semi-linear K-automorphisms of Γ.

Cahen & Flato (eds.), Differential Geometry and Relativity, 75–89.

In fact, a semi-linear automorphism of Γ defines an automorphism of $\mathcal{F}$ and extends uniquely to an automorphism of Θ.

If G is a group, we define the *semi-linear representations* of G into Γ to be the representations of G by semi-linear automorphisms of Γ.

The quotient of two semi-linear representations (with respect to the same representation of G into the K-algebra $\mathcal{F}$) will be called a *multiplier* (see [5, chap. 9]). More precisely, if $\mathcal{R}_0$ is a semi-linear representation of G into Γ, an $\mathcal{R}_0$-multiplier of G into Γ is a mapping $\mathcal{T}$ from G into the set of $\mathcal{F}$-module automorphisms of Γ such that for $\mu \in G$ and $\nu \in G$,

$$\mathcal{T}(\mu\nu) = \mathcal{T}\mu \circ \mathcal{R}_0\mu \circ \mathcal{T}\nu \circ \mathcal{R}_0(\mu^{-1}),$$
$$\mathcal{T}(1) = 1,$$

where 1 denotes the unit element of G and the identity transformation of Γ. Note that the definition of a multiplier in [14] is different.

1.2. *Derivative representations*

Similarly we define

DEFINITION 2. A K-endomorphism e of Γ is said to be derivative if there exists a K-endomorphism $e_{\mathcal{F}}$ of $\mathcal{F}$ such that for $f \in \mathcal{F}$ and $\psi \in \Gamma$,

$$e(f\psi) = f(e\psi) + (e_{\mathcal{F}}f)\psi.$$

If there exists in Γ an element which is not a torsion element, then $e_{\mathcal{F}}$ is uniquely determined and $e_{\mathcal{F}}$ is a derivation of $\mathcal{F}$.

Using a theorem of Bourbaki [3, III, §10, no. 9, prop. 14] we have

PROPOSITION 2. Assume Γ has an element which is not a torsion element. The set of derivations of the graded K-algebra Θ is isomorphic to the set of derivative K-endomorphisms of Γ.

If $\mathcal{G}$ is a Lie algebra, we define the *derivative representations* of $\mathcal{G}$ into Γ to be the representations of $\mathcal{G}$ by derivative endomorphisms of Γ.

The difference of two derivative representations (with respect to the same representation of $\mathcal{G}$ by derivations of the K-algebra $\mathcal{F}$) will be called an *adder*. More precisely, if $\mathcal{M}_0$ is a derivative representation of $\mathcal{G}$ into Γ, an $\mathcal{M}_0$-adder of $\mathcal{G}$ into Γ is a linear map $\mathcal{N}$ from $\mathcal{G}$ into the space of $\mathscr{F}$-module endomorphisms of Γ such that for $X \in \mathscr{G}$ and $Y \in \mathscr{G}$,

$$\mathcal{N}[X, Y] = [\mathcal{N}X, \mathcal{N}Y] + [\mathcal{M}_0 X, \mathcal{N}Y] + [\mathcal{N}X, \mathcal{M}_0 Y].$$

2. G-VECTOR BUNDLES, INDUCED REPRESENTATIONS AND FIRST-ORDER DIFFERENTIAL OPERATORS

All manifolds and mappings are assumed to be C^∞.

Let M be a manifold and let $\mathsf{K}(M)$ be the K-algebra of C^∞ functions on M with values in K.

2.1. *Semi-linear representations and actions on vector bundles*

Let ξ be a K-vector bundle over the manifold M.

Let G be a Lie group. Suppose that ξ is a G-vector bundle. Then M is a G-manifold (the action of G on ξ projects onto an action of G on M), and we have a representation of G into $\mathsf{K}(M)$ defined by

$$\mu f = f \circ \mu^{-1}$$

for $\mu \in G$ and $f \in \mathsf{K}(M)$.

Let $\Gamma(\xi)$ denote the $\mathsf{K}(M)$-module of all C^∞ sections of ξ. We have a semi-linear representation of G into $\Gamma(\xi)$ defined by

$$(\mu\psi)(x) = \mu(\psi(\mu^{-1}x))$$

for $\mu \in G$, $\psi \in \Gamma(\xi)$ and $x \in M$.

For example, if ξ is a tensor bundle over M and M is a G-manifold, we can consider the usual semi-linear representation of G into $\Gamma(\xi)$, G acting on ξ by tensor products of the differentials of the diffeomorphisms $\mu \in G$. If M is a spin manifold and G is a subgroup of the connected component of the conformal group, for any bundle ξ of tensor-spinors, there is still a semi-linear representation of G into $\Gamma(\xi)$ which we shall call the *usual representation* [6].

Let $\Theta(\xi)$ be the tensor algebra of the $\mathsf{K}(M)$-module $\Gamma(\xi)$. From proposition 1 we have

PROPOSITION 3. The set of representations of G into the graded K-algebra $\Theta(\xi)$ is isomorphic to the set of semi-linear representations of G into $\Gamma(\xi)$.

We shall say that a representation of G into $\Gamma(\xi)$ is C^∞ if, for all

$\psi \in \Gamma(\xi)$, the mapping $(\mu, x) \to (\mu\psi)(x)$ from $G \times M$ into ξ is C^∞. We have the following proposition:

PROPOSITION 4. Suppose that M is second countable. There is a one-to-one correspondence between actions of G on ξ and C^∞ semi-linear representations of G into $\Gamma(\xi)$.

Proof. If ξ is a G-vector bundle, we have seen that there is a semi-linear representation of G into $\Gamma(\xi)$. It is clearly C^∞.

Conversely, a C^∞ semi-linear representation of G into $\Gamma(\xi)$ defines for each μ in G an automorphism of $\mathsf{K}(M)$. It was shown by R. Bkouche [1] that, if M is second countable, every automorphism of $\mathsf{K}(M)$ corresponds to a unique diffeomorphism of M. Let us denote by $(\mu, x) \to \mu x$ the action of G on M thus defined. The given C^∞ semi-linear representation arises from a unique action of G on ξ defined by

$$\mu u = (\mu\psi)(\mu x)$$

where u is an element of the fiber of ξ over x, $\mu \in G$, $x \in M$ and $\psi \in \Gamma(\xi)$ is such that $\psi(x) = u$. (It is easy to show that μu is independent of the choice of ψ and that the action of G on ξ is C^∞.)

We have also proved the following result [7]:

Let G act on M in a fixed way; the set of actions of G on ξ over the given action of G on M is isomorphic to the set of C^∞ representations of G into $\Gamma(\xi)$ which are semi-linear with respect to the representation of G into $\mathsf{K}(M)$ defined by the action of G on M.

2.2. *Semi-linear representations and induced representations*

Let H be a closed subgroup of the Lie group G. Let us fix a representation of H into a finite-dimensional K-vector space F. The induced bundle is the G-vector bundle over G/H with fiber F associated with the principal bundle $G \to G/H$. The induced representation [2] is the semi-linear representation of G into the vector space of sections of the induced bundle defined by the action of G on the induced bundle.

Let us denote by $R(H)$ the category associated with the ring obtained from the semi-ring of equivalence classes of finite-dimensional H-modules. Let us denote by $K_G(G/H)$ the category associated with the ring obtained from the semi-ring of equivalence classes of G-vector bundles over G/H. We have the following result [12]:

PROPOSITION 5. The restriction and the induction of representations define an equivalence of categories from $R(H)$ into $K_G(G/H)$.

Let G act on M by a fixed *transitive* action and let H be the isotropy subgroup of a point x_0 in M. Then M is isomorphic to G/H. If ξ is a G-vector bundle over M, we have a representation of H into the fiber F of ξ at x_0, ξ is isomorphic to the induced bundle and the semi-linear representation of G into $\Gamma(\xi)$ is equivalent to the induced representation.

As a consequence, we see that for a vector bundle ξ over M, there exists an action of G on ξ over a given transitive action of G on M if and only if ξ is isomorphic to an induced bundle.

2.3. *Derivative representations and first-order differential operators*

Let ξ be a vector bundle over the manifold M. We first state that, given a Lie group G and its Lie algebra $\mathscr{G}$, the differential of a semi-linear representation of G is a derivative representation of $\mathscr{G}$. More precisely, we define the differential of a C^∞ semi-linear representation of G into $\Gamma(\xi)$ to be the representation of $\mathscr{G}$ into $\Gamma(\xi)$ defined by

$$(X\psi)(x) = \lim_{t\to 0} \frac{1}{t}\,[(\exp(tX)\psi)(x) - \psi(x)]$$

where $X \in \mathscr{G}$, $\psi \in \Gamma(\xi)$, $x \in M$, $t \in \mathbb{R}$.

If we denote by $\bar{X}$ the Killing vector fields associated with the action of G on M, we have

$$Xf = -\bar{X}f$$

for every $f \in \mathsf{K}(M)$. More generally, for any bundle ξ of tensor-spinors over M, the differential of the usual semi-linear representation of G into $\Gamma(\xi)$ (which is clearly C^∞) is equal, up to a change in sign, to the Lie derivation with respect to the Killing vector fields.

PROPOSITION 6. A C^∞ representation of a connected Lie group G into $\Gamma(\xi)$ is semi-linear if and only if its differential is derivative.

The differential of a semi-linear representation is clearly derivative. Conversely, suppose the differential is derivative. Setting for $X \in \mathscr{G}$,

$\mu_t = \exp(tX)$, we have, for $\psi \in \Gamma(\xi)$,

$$\frac{\mathrm{d}}{\mathrm{d}t}(\mu_t\psi) = X(\mu_t\psi);$$

hence, for $f \in \mathsf{K}(M)$,

$$\begin{aligned}\frac{\mathrm{d}}{\mathrm{d}t}&(\mu_t(f\psi) - (\mu_t f)(\mu_t\psi)) \\ &= X(\mu_t(f\psi)) - (X(\mu_t f))(\mu_t\psi) - (\mu_t f)(X(\mu_t\psi)) \\ &= X(\mu_t(f\psi) - (\mu_t f)(\mu_t\psi)),\end{aligned}$$

since X is a derivative endomorphism of $\Gamma(\xi)$. The conclusion follows from the uniqueness of the solution of an ordinary linear differential equation with initial value 0 for $t = 0$.

From proposition 2, we have

PROPOSITION 7. The set of representations of $\mathcal{G}$ by derivations of the graded K-algebra $\Theta(\xi)$ is isomorphic to the set of derivative representations of $\mathcal{G}$ into $\Gamma(\xi)$.

We now show that the derivative representations of $\mathcal{G}$ into $\Gamma(\xi)$ are the appropriate generalization to arbitrary vector bundles over M of the representations by first-order scalar differential operators, corresponding to the case when ξ is the trivial bundle $M \times \mathsf{K}$.

We shall say that a first-order differential operator on the vector bundle ξ has a *scalar symbol* if the local coefficients of its first-order terms are functions (considered as particular endomorphisms of the fibers of ξ). The first-order differential operators on ξ with scalar symbols form a Lie algebra, whereas this is not true in general of the set of all first-order differential operators.

PROPOSITION 8. The derivative representations of $\mathcal{G}$ into $\Gamma(\xi)$ are in a one-to-one correspondence with the homomorphisms of $\mathcal{G}$ into the Lie algebra of first-order differential operators on ξ with scalar symbols.

Proof. If $X \to \mathcal{M}X$ is a derivative representation of $\mathcal{G}$ into $\Gamma(\xi)$, then for each $X \in \mathcal{G}$, $\mathcal{M}X$ is a first-order differential operator on ξ since, for all $\psi \in \Gamma(\xi)$, the mapping $f \to (\mathcal{M}X)(f\psi) - f((\mathcal{M}X)\psi)$ is $\mathsf{K}(M)$-linear, and the symbol of that first-order differential operator is clearly scalar.

Conversely, if $X \to \mathcal{M}X$ is a representation of $\mathcal{G}$ by first-order differential operators on ξ with scalar symbols $\sigma(\mathcal{M}X)$, then $X \to \sigma(\mathcal{M}X) \circ \mathrm{d}$ is a representation of $\mathcal{G}$ into the Lie algebra of derivations of $\mathsf{K}(M)$ (that is to say an *action of* $\mathcal{G}$ on M), and with respect to that representation, $\mathcal{M}$ is derivative.

We can also give the following characterization when ξ is a *tensor bundle*:

PROPOSITION 9. Every derivative representation of $\mathcal{G}$ into $\Gamma(\xi)$ is of the form $X \to -\mathcal{L}\bar{X} + \mathcal{N}X$ where $X \to \bar{X}$ is an action of $\mathcal{G}$ on M, $\mathcal{L}\bar{X}$ is the Lie derivative with respect to the vector field $\bar{X}$ and for all $X \in \mathcal{G}$, $\mathcal{N}X$ is $\mathsf{K}(M)$-linear.

In fact, the associated representation of $\mathcal{G}$ into $\mathsf{K}(M)$ is necessarily of the form $X \to \bar{X}$. The Lie derivation with respect to $-\bar{X}$ is a derivative representation of $\mathcal{G}$ into $\Gamma(\xi)$ and we know that the difference of two derivative representations is necessarily $\mathsf{K}(M)$-linear.

3. Cohomological interpretation

Let us give the cohomological interpretation of the multipliers defined in 1.1. and of the adders defined in 1.2. We consider the case where Γ is the $\mathsf{K}(M)$-module $\Gamma(\xi)$ of sections of a vector bundle ξ over M. The vector space of $\mathsf{K}(M)$-endomorphisms of $\Gamma(\xi)$ can be identified with the vector space of vector bundle endomorphisms of ξ over the identity of M and also with the vector space of sections of the vector bundle $\xi \otimes \xi^*$. It will be denoted by $E(\xi)$. The group of $\mathsf{K}(M)$-automorphisms of $\Gamma(\xi)$ will be denoted by $A(\xi)$.

3.1. *Multipliers as group cocycles*

Let $\mathcal{R}_0$ be a C^∞ semi-linear representation of a Lie group G into $\Gamma(\xi)$. $A(\xi)$ is a G-group under the action

$$\mu \cdot a = \mathcal{R}_0\mu \circ a \circ \mathcal{R}_0\mu^{-1} \qquad (\mu \in G, a \in A(\xi)).$$

A multiplier is the quotient $\mathcal{T}$ of two C^∞ semi-linear representations $\mathcal{R}$ and $\mathcal{R}_0$ (with respect to the same representation of G into $\mathsf{K}(M)$) and it is defined by

$$\mathcal{T}\mu = \mathcal{R}\mu \circ \mathcal{R}_0\mu^{-1} \qquad (\mu \in G).$$

Let us consider the C^∞-differentiable cohomology of G with values in the G-group $A(\xi)$ in dimensions 0 and 1. It is in general a non-abelian cohomology [13]. A 0-cochain is an element of $A(\xi)$ and the coboundary of a 0-cochain a is defined by

$$\mathrm{d}a(\mu) = a^{-1} \circ (\mu \cdot a) \qquad (\mu \in G).$$

The 0-cocycles are those elements a of $A(\xi)$ such that $\mathrm{d}a = 1$, namely those elements a of $A(\xi)$ which commute with $\mathcal{R}_0\mu$ for all $\mu \in G$. They form a subgroup of the group $A(\xi)$.

A 1-cochain is a map C from G to $A(\xi)$ such that the map $(\mu, x) \in G \times M \to (C\mu)_x \in \xi \otimes \xi^*$ is C^∞. The coboundary of a 1-cochain C is defined by

$$\mathrm{d}C(\mu, \nu) = (\mu \cdot C\nu) \circ [C(\mu\nu)]^{-1} \circ C\mu \qquad (\mu \in G, \nu \in G).$$

The 1-cocycles are the 1-cochains C satisfying $\mathrm{d}C = 1$. They do not form a group if $A(\xi)$ is not abelian. It is clear that

PROPOSITION 10. The multipliers of G into ξ are the 1-cocycles of G with values in $A(\xi)$.

The 1-coboundaries are the mappings of G into $A(\xi)$ of the form $\mu \to a^{-1} \circ (\mu \cdot a)$ where a is a fixed element of $A(\xi)$. If $\mathcal{T}$ is a 1-coboundary, then the representations $\mathcal{R}_0$ and $\mathcal{R}$ are related by $\mathcal{R}\mu = a^{-1} \circ \mathcal{R}_0\mu \circ a$ $(\mu \in G, a \in A(\xi))$ and they are therefore associated with *equivalent actions* of G on ξ. Two such representations will be called *equivalent representations.*

It is easily checked that every 1-coboundary is a 1-cocycle and that an equivalence relation is defined on the set of 1-cochains in the following way: two 1-cochains C and C' are equivalent if and only if there exists $a \in A(\xi)$ such that for all $\mu \in G$,

$$C'\mu = a^{-1} \circ C\mu \circ a \circ \mathrm{d}a(\mu) = a^{-1} \circ C\mu \circ (\mu \cdot a).$$

The quotient of the set of 1-cocycles by this equivalence relation will be denoted by $H^1(G, A(\xi))$. It is the set of equivalence classes of actions of G on ξ over the given action of G on M.

3.2. *Adders and Lie algebra cocycles*

Let $\mathcal{M}_0$ be a derivative representation of $\mathcal{G}$ into $\Gamma(\xi)$. The vector space $E(\xi)$ is a $\mathcal{G}$-module under the action

$$X \cdot u = (\mathcal{M}_0 X) \circ u - u \circ (\mathcal{M}_0 X) = [\mathcal{M}_0 X, u] \qquad (X \in \mathcal{G}, u \in E(\xi)).$$

An adder is the difference $\mathcal{N} = \mathcal{M} - \mathcal{M}_0$ of two derivative representations (with respect to the same representation of G into $\mathsf{K}(M)$).

Let us consider the cohomology of $\mathcal{G}$ with values in the $\mathcal{G}$-module $E(\xi)$. A 0-cochain is an element of $E(\xi)$ and the coboundary of a 0-cochain u is defined by

$$\mathrm{d}u(X) = X \cdot u \qquad (X \in \mathcal{G}).$$

The 0-cocycles are those elements u of $E(\xi)$ such that $\mathrm{d}u = 0$, namely those elements u of $E(\xi)$ which commute with $\mathcal{M}_0 X$ for all $X \in \mathcal{G}$.

A 1-cochain is a linear map c from $\mathcal{G}$ to $E(\xi)$. The coboundary of a 1-cochain c is defined by

$$\mathrm{d}c(X, Y) = X \cdot cY - Y \cdot cX - c[X, Y] \qquad (X \in \mathcal{G}, Y \in \mathcal{G}).$$

It is clear that the adders of $\mathcal{G}$ into ξ are the 1-cochains $\mathcal{N}$ such that $\mathrm{d}\mathcal{N} + [\mathcal{N}, \mathcal{N}] = 0$.

The 1-coboundaries are the linear maps of $\mathcal{G}$ into $E(\xi)$ of the form $X \to X \cdot u$ where u is a fixed element of $E(\xi)$. If $\mathcal{N}$ is a 1-coboundary with $\mathcal{N} = \mathrm{d}u (u \in E(\xi))$, then $\mathcal{M}X = \mathcal{M}_0 X + X \cdot u$ for all $X \in \mathcal{G}$. Two such representations $\mathcal{M}_0$ and $\mathcal{M}$ will be called *equivalent representations.*

If in particular the fiber of ξ is one-dimensional, then $E(\xi)$ is an abelian Lie algebra. In this case, the adders of $\mathcal{G}$ into ξ are exactly the 1-cocycles of $\mathcal{G}$ with values in $E(\xi)$.

More generally let us call every 1-cocycle of $\mathcal{G}$ into $\mathsf{K}(M)$ a *scalar adder* of $\mathcal{G}$ into ξ, considering that a function over M acts on the sections of ξ by pointwise multiplication. For example, on a Riemannian or pseudo-Riemannian manifold, the mapping $X \to \delta X$, where δX denotes the divergence of the vector field X, is a scalar adder.

We shall state a theorem of Palais and Stewart which is proved in [11, §9]:

Let M be second countable. If G is a compact and semi-simple Lie group, for any tensor bundle ξ over M,

$$H^1(\mathcal{G}, E(\xi)) = 0.$$

As a corollary to this theorem, we have

PROPOSITION 11. Let M be second countable and let G be a compact semi-simple Lie group acting on M. Every scalar adder of $\mathcal{G}$ into a vector bundle ξ over M is of the form $X \to X \cdot u$, where u is a fixed function.

In fact, the scalar adders of $\mathcal{G}$ into ξ are the 1-cocycles of $\mathcal{G}$ into $\mathbb{K}(M)$ and, under the hypotheses, every such 1-cocycle is a coboundary.

4. Trivial bundles and the Mackey-Varadarajan cocycles

We now consider the case where ξ is the trivial bundle $M \times F$ over the manifold M, with fiber a finite dimensional $\mathbb{K}$-vector space F.

4.1. *Global viewpoint*

Let G be a Lie group acting on M. There is a natural action of G on ξ, namely $\mu(x, v) = (\mu x, v)$ for $\mu \in G$, $(x, v) \in M \times F$.

A $\mathbb{K}(M)$-automorphism of ξ can be identified with a section of $M \times \text{Aut}\, F$, where $\text{Aut}\, F$ denotes the group of automorphisms of the vector space F.

A multiplier with respect to that natural action of G on ξ is a map $\mathcal{T}$ from G into $A(\xi) = \Gamma(M \times \text{Aut}\, F)$ and can be identified with a map from $G \times M$ into $\text{Aut}\, F$. If we set

$$S(\mu, x) = \mathcal{T}(\mu, \mu x)$$

we can verify that the mappings S are exactly those introduced by M. Flato, J. Simon and D. Sternheimer in [4], that is to say that they are the mappings S of $G \times M$ into $\text{Aut}\, F$ such that

(1) $S(1, x) = 1$ for $x \in M$ (where 1 denotes the unit element of G and the identity transformation of F),
(2) $S(\mu\nu, x) = S(\mu, \nu x) \circ S(\nu, x)$ for $\mu \in G$, $\nu \in G$, $x \in M$,
(3) for all $\mu \in G$, the mapping $x \to S(\mu, x)$ from M into $\text{Aut}\, F$ is C^∞.

If we compare this with the definition of the $(G, M, \text{Aut}\, F)$-cocycles in [14, VIII, 5], we can state:

PROPOSITION 12. The mappings S are the strict differentiable $(G, M, \text{Aut}\, F)$-cocycles in the sense of Mackey-Varadarajan.

Of course the mappings S are simpler to handle than the multipliers; once a basis is chosen in F, they are matrix-valued mappings. For instance if M is a vector space and F is the space of tensors of a given type, say (p, q) tensors, then the mapping S associated with the usual action on $\xi = M \times F$ of the conformal group of M (for a given Euclidian or pseudo-Euclidian structure on M) satisfies

$$|\mathrm{Det}\, S(\mu, x)| = ((k\mu)(x))^{-p+q}$$

where $(k\mu)(x)$ denotes the scale factor of the conformal transformation μ at the point x. If we introduce the reduced action of the conformal group on ξ, defined by multiplying the usual action by the scale factor $(k\mu)^{p-q}$, the resulting mappings S' have determinants of modulus 1.

We have the following analogue of Theorem 8.27 of [14]:

PROPOSITION 13. If the Lie group G acts transitively on the manifold M and if the principal bundle $G \to M$ admits a section, the set of equivalence classes of strict differentiable $(G, M, \mathrm{Aut}\, F)$-cocycles is isomorphic to the set of equivalence classes of differentiable representations of H into F, where H is the isotropy subgroup of a point in M.

In fact, given a cocycle, it clearly defines a representation of H and equivalent representations are associated with equivalent cocycles.

Conversely, given a section of $G \to M$, one has a trivialization of $M \times F$ and via this trivialization the natural action of G on $M \times F$ gives rise to an action of G on $M \times F$, which coincides with the given action of H on F. The quotient of this action by the natural action is the required cocycle. By a change of sections, one gets equivalent cocycles.

Thus, provided that the principal bundle $G \to G/H$ admits a section, there is a canonical isomorphism from the set of equivalence classes of actions of G on $G/H \times F$ to the set of equivalence classes of representations of H into F.

4.2. *Infinitesimal viewpoint*

Let $\mathcal{G}$ be a Lie algebra. For a trivial vector bundle $\xi = M \times F$, we can identify a vector bundle endomorphism of ξ over the identity of M with a mapping from M into $\mathrm{End}\, F$, where $\mathrm{End}\, F$ denotes the vector space of endomorphisms of F. Therefore, we can replace an adder $\mathscr{N}$ of $\mathscr{G}$ into ξ by a map s from $\mathscr{G} \times M$ into End F by setting

$$(\mathscr{N}X)_x = s(X, x).$$

But the characteristic relation satisfied by the mappings s is not so simple as in the global case. The set of adders of $\mathscr{G}$ into $\xi = M \times F$ is isomorphic to the set of mappings s from $\mathscr{G} \times M$ into End F which are C^∞ as functions of $x \in M$, linear in $X \in \mathscr{G}$ and which satisfy for all $X \in \mathscr{G}$, $Y \in \mathscr{G}$, $x \in M$, $u \in F$,

$$\begin{aligned}\{s([X, Y], x) - [s(X, x), s(Y, x)]\}u &= (X\phi_2)(x) - s(Y, x)((X\phi)(x)) \\ &\quad + s(X, x)((Y\phi)(x)) - (Y\phi_1)(x)\end{aligned}$$

where $\phi(x) = u$, $\phi_1(x) = s(X, x)(\phi(x))$ and $\phi_2(x) = s(Y, x)(\phi(x))$.

Using Proposition 6, we can show that the differential of a multiplier is an adder, and hence the differential of a mapping S is a mapping s.

Summing up the case of a trivial bundle $\xi = M \times F$, over a second countable manifold M, and a Lie group G with Lie algebra $\mathscr{G}$, we have the following commutative diagram where the horizontal arrows are isomorphisms and each vertical arrow is the differential (which is not an isomorphism!).

Mappings S from $G \times M$ into Aut F	$\to$	Multipliers of G into ξ	$\to$	Actions of G on ξ	$\to$	C^∞ semi-linear representations of G into $\Gamma(\xi)$
$\downarrow$		$\downarrow$		$\downarrow$		$\downarrow$
Mappings s from $\mathscr{G} \times M$ into End F	$\to$	Adders of $\mathscr{G}$ into ξ	$\to$	Actions of $\mathscr{G}$ on ξ	$\to$	Derivative representations of $\mathscr{G}$ into $\Gamma(\xi)$

According to Proposition 8, an *action of* $\mathscr{G}$ *on* ξ is defined to be a homomorphism of $\mathscr{G}$ into the Lie algebra of first-order differential operators on ξ with scalar symbols. We are thus led to consider this Lie algebra as the Lie algebra of the 'infinite dimensional Lie group' of all automorphisms of the vector bundle ξ. The Lie algebra of all endomorphisms of ξ over the identity of M, which is a Lie subalgebra of the above-mentioned algebra, corresponds to the subgroup of all automorphisms of ξ over the identity of M.

5. Covariance of Differential Operators

A scalar differential operator on a G-manifold M is said to be invariant if it commutes with the representation of G into $\mathsf{K}(M)$ defined by the action of G. The notion of invariance for a differential operator D from ξ to ξ', where ξ and ξ' are arbitrary vector bundles over M, is only defined when there is a natural action of G on ξ and ξ', for instance when ξ and ξ' are tensor bundles over M. In the general case we can define and study the notion of covariance [7].

5.1. *Definitions*

Let ξ and ξ' be vector bundles over a manifold M. Let D be a differential operator from ξ to ξ'.

Let G be a Lie group which acts on M. We shall say that D is covariant with respect to G or is G-covariant if G acts on the sections of ξ and ξ' in such a way that D commutes with these actions. More precisely, D is *G-covariant* if there exist C^∞ semi-linear representations $\mathcal{R}$ and $\mathcal{R}'$ of G into $\Gamma(\xi)$ and $\Gamma(\xi')$, semi-linear with respect to the representation of G into $\mathsf{K}(M)$ defined by the action of G on M, such that for all $\mu \in G$,

$$(\mathcal{R}'\mu)D - D(\mathcal{R}\mu) = 0.$$

Let $\mathcal{G}$ be a Lie algebra which acts on M. We shall say that D is *$\mathcal{G}$-covariant* if there exist derivative representations $\mathcal{M}$ and $\mathcal{M}'$ of $\mathcal{G}$ into $\Gamma(\xi)$ and $\Gamma(\xi')$, derivative with respect to the given representation of $\mathcal{G}$ by derivations of $\mathsf{K}(M)$, such that for all $X \in \mathcal{G}$,

$$(\mathcal{M}'X)D - D(\mathcal{M}X) = 0.$$

Let us remark that any scalar differential operator which is quasi-invariant with respect to $\mathcal{M}$ in the sense of Kostant [9] is $\mathcal{G}$-covariant in the above sense.

If $\mathcal{G}$ is the Lie algebra of the Lie group G, any G-covariant differential operator is clearly $\mathcal{G}$-covariant. If G is a connected Lie group, a differential operator D is G-covariant if and only if there exist C^∞ semi-linear representations $\mathcal{R}$ and $\mathcal{R}'$ of G into $\Gamma(\xi)$ and $\Gamma(\xi')$ such that D commutes with the differentials of $\mathcal{R}$ and $\mathcal{R}'$, that is to say if D commutes with 'integrable' derivative representations of $\mathcal{G}$ into $\Gamma(\xi)$ and $\Gamma(\xi')$.

5.2. *Covariance and invariance*

We shall see that in certain cases the covariance is not a much more general situation than the invariance.

From the results of 3.1, we have

PROPOSITION 14. Let ξ and ξ' be tensor bundles over M and let D be a differential operator from ξ to ξ'. If $H^1(G, A(\xi)) = \{1\}$ and $H^1(G, A(\xi')) = \{1\}$, D is G-covariant if and only if D commutes with representations which are equivalent to the usual representations of G into $\Gamma(\xi)$ and $\Gamma(\xi')$.

In the case of a scalar operator, we can use the theorem of Palais and Stewart and its consequence, Proposition 11, to prove

PROPOSITION 15. Let D be a scalar differential operator on a second countable manifold M. Let G be a compact semi-simple Lie group acting on M. D is $\mathcal{G}$-covariant if and only if D commutes with representations which are equivalent to the Lie derivation, namely if there exist functions u and u' such that for all $X \in \mathcal{G}$,

$$(X + Xu')D - D(X + Xu) = 0.$$

In this case, D is $\mathcal{G}$-covariant if and only if it is quasi-invariant with respect to a representation which is equivalent to the Lie derivation.

But if G is not compact or is not semi-simple, there may exist adders and even scalar adders which are not coboundaries. This may be the case for the conformal group of a Riemannian or pseudo-Riemannian manifold, where the divergence is a scalar adder which is not in general a coboundary. It is of particular interest to study the covariance of the standard differential operators under the group of conformal transformations. The above results can be used and lead to the introduction of the conformal bidegree of a differential operator [8], which is closely related to the physicists' conformal weight.

Université des Sciences et Techniques de Lille I

REFERENCES

[1] Bkouche, R., 'Idéaux mous d'un anneau commutatif. Applications aux anneaux de fonctions', *Comptes rendus Acad. Sc. Paris* **260**, série A (1965), 6496–6498.
[2] Bott, R., 'The Index Theorem for Homogeneous Differential Operators', in S. Cairns (ed.), *Differential and Combinatorial Topology*, Princeton University Press, 1965.
[3] Bourbaki, N., *Algèbre*, Chapitres 1 à 3, Hermann, Paris, 1970.
[4] Flato, M., Simon, J., and Sternheimer, D., 'Conformal Covariance of Field Equations', *Ann. of Phys.* **61** (1970), 78–97.
[5] Hermann, R., *Lie Groups for Physicists*, Benjamin, New York, 1966.
[6] Kosmann, Y., 'Dérivées de Lie des spineurs', *Ann. di Matematica Pura ed Appl. (IV)* **91** (1972), 317–395.
[7] Kosmann, Y., 'Groupes de transformations et covariance des opérateurs différentiels', *Comptes rendus Acad. Sc. Paris* **275**, série A (1972), 1235–1237.
[8] Kosmann, Y., 'Sur les degrés conformes des opérateurs différentiels', *Comptes rendus Acad. Sc. Paris* **280**, série A (1975), 229–232 et 'Degrés conformes des laplaciens et des opérateurs de Dirac', *ibid.*, 283–285.
[9] Kostant, B., 'Verma Modules and the Existence of Quasi-Invariant Differential Operators', Colloque de Marseille-Luminy, juillet 1974, Lecture Notes in Mathematics **466**, Springer-Verlag, Berlin, 1975.
[10] Lichnerowicz, A., 'Opérateurs différentiels invariants sur un espace homogène', *Ann. Scient. Ec. Norm. Sup., 3ème série*, **81** (1964), 341–385.
[11] Palais, R., and Stewart, T., 'The Cohomology of Differentiable Transformation Groups', *Amer. J. of Math.* **83** (1961), 623–644.
[12] Segal, G., 'Equivariant K-theory', *Publications mathématiques n° 34*, I.H.E.S., 1968, pp. 129–151.
[13] Serre, J. P., *Corps locaux*, Hermann, Paris, 1962.
[14] Varadarajan, V. S., *The Geometry of Quantum Theory*, Vol. 2, Van Nostrand, New York, 1970.

JACQUELINE LELONG-FERRAND

GEOMETRICAL INTERPRETATIONS OF SCALAR CURVATURE AND REGULARITY OF CONFORMAL HOMEOMORPHISMS

1. Introduction

To complete the solution I gave [5] of A. Lichnérowicz's conjecture concerning the conformal group of compact[1] riemannian manifolds, one had still to prove that every conformal homeomorphism of C^∞ manifolds is of C^∞ class. Up to now, it seems that this result has been proved only in the euclidean case by Y. G. Resetnyak [7] and F. W. Gehring [3]. We plan in this paper to solve the general case by a method analogous to the one of Resetnyak.

To understand the difficulty of this regularity problem, it is enough to give the history of the euclidean case: in 1850 Liouville proved that any conformal diffeomorphism of class C^3 of open sets of E^n is a product of inversions, thus of class C^∞. But one had apparently to wait for the works of P. Hartman [4] to extend this result to the class C^1 and it was only in 1960–61 that Y. G. Resetnyak and F. W. Gehring were able to handle the case of conformal homeomorphisms. Resetnyak's proof is based on the isoperimetrical inequality of E_n, Gehring's proof uses the technique of quasi-conformal maps; both reduce, like Hartman's proof, to the regularity of the solutions of an elliptic equation.

The case of riemannian manifolds raises new difficulties for several reasons: on one hand, no extremals for the conformal invariant $\int_M |\text{grad } u|^n \, d\tau$ are known *a priori* and the technique of 'conformal capacities' or of 'extremal lengths' gives less accurate results, which seems to exclude an extension of Gehring's method. On the other hand no simple isoperimetrical inequality is known in riemannian geometry, and one has no regularizing kernel which commutes with the Laplacian, so that Resetnyak's method does not extend straightforwardly. Anyhow the problem is cleared up – if not made easier – by the appearance of the scalar curvature (see §1 and 2) and our method will be based on 'approximate' isoperimetrical inequalities (§3 and 4) due to the fact that a manifold is 'approximately euclidean' in the neighbourhood of each point.

Cahen & Flato (eds.), Differential Geometry and Relativity, 91–105.

Notations. In what follows, we will denote by (M, g) and $(\bar{M}, \bar{g})$ two riemannian manifolds of dimension $n \geqslant 3$ and of class C^∞ (unless otherwise stated), by R and $\bar{R}$ their scalar curvatures, and by Δ, $\bar{\Delta}$ the Laplacians relative to these manifolds.

We will say that a differentiable map $\varphi : M \to \bar{M}$ is conformal if for any $x \in M$, the differential $\varphi'(x)$ is the zero map or a similarity from $T_x(M)$ onto $T_{\varphi(x)}(\bar{M})$; the ratio of this similarity will be denoted by $|\varphi'(x)|$.

On the other hand a homeomorphism φ from M onto $\bar{M}$ will be said to be conformal if

(i) φ is ACL^n (i.e. if φ is absolutely continuous on almost every coordinate curve relative to a given atlas)

(ii) for almost every $x \in M$, the differential $\varphi'(x)$ exists, and it is a similarity with non-zero ratio from $T_x(M)$ onto $T_{\varphi(x)}(\bar{M})$.

It is known that the inverse of such a homeomorphism is also conformal.

PROPOSITION 1.1. Let $\varphi : M \to \bar{M}$ be a conformal map of class C^3, and let $u(x) = |\varphi'(x)|^{(n-2)/2}$. If f is any numerical function of class C^2 on $\bar{M}$, then $f \circ \varphi$ satisfies

$$\Delta(uf \circ \varphi) - \frac{n-2}{4(n-1)} Ruf \circ \varphi = u|\varphi'|^2 \left(\bar{\Delta} f - \frac{n-2}{4(n-1)} \bar{R} f\right) \circ \varphi \tag{1.1}$$

in any point x of M such that $\varphi'(x) \neq 0$.

Proof. If $\varphi'(x) \neq 0$, there exists a neighbourhood V of x such that $\varphi|_V$ is a diffeomorphism of V onto $\varphi(V)$. One may then consider $\varphi|_V$ as an isometry from $(V, |\varphi'|^2 g)$ onto $(\varphi(V), \bar{g})$, which implies $(\bar{\Delta} f) \circ \varphi = \Delta_\varphi(f \circ \varphi)$, denoting by Δ_φ the laplacian of $(V, |\varphi'|^2 g)$.

An easy calculation gives, for any function F of class C^2 on V:

$$\Delta_\varphi F = \frac{1}{u|\varphi'|^2} (\Delta(uF) - F\Delta u) \tag{1.2}$$

and

$$\bar{R} \circ \varphi = |\varphi'|^2 \left(R - \frac{4(n-1)}{n-2} \frac{\Delta u}{u}\right) \tag{1.3}$$

(the two sides of (1.3) being equal to the scalar curvature of $(V, |\varphi'|^2 g)$).

From (1.2) and (1.3) one deduces that (1.1) is true at the point x.

COROLLARY 1. For every $x \in M$ such that $\varphi'(x) \neq 0$ one has:

$$\Delta u = \frac{n-2}{4(n-1)}(Ru - \bar{R} \circ \varphi\, u^{(n+2)/(n-2)}). \tag{1.4}$$

COROLLARY 2. If $\varphi : M \to \bar{M}$ is a conformal diffeomorphism of class C^3, then φ is of class C^∞.

Indeed, in this case, φ' is never zero; by a conformal change of metric[2] of class C^∞ on each of the manifolds M, $\bar{M}$, one may assume that $R = 0$, $\bar{R} = 0$; equation (4) then reduces to $\Delta u = 0$ and shows that u is of class C^∞; so is $|\varphi'|$; and a conformal change of metric brings us back to the case where φ is an isometry; the result then follows by applying Myers–Steenrod's theorem.

2. Geometrical interpretation of scalar curvature

Proposition 1.1 shows the interest, in conformal geometry, of the operator

$$\Gamma f = \Delta f - \frac{n-2}{4(n-1)} Rf.$$

Although it is unnecessary for what follows, we shall give an interpretation of this operator in terms of means, which makes possible to extend it to functions that are not of class C^2 and to explain geometrically the results stated above.

One has first of all:

PROPOSITION 2.1. Let f be a numerical function on M which admits a second differential at a. Let us denote by $B(a, r)$ and $S(a, r)$ the geodesic ball and sphere with center a and radius r; one has:

$$\frac{1}{v_n r^n}\int_{B(a,r)} f\, d\tau = f(a) + \frac{r^2}{2(n+2)}\left[\Delta f(a) - \frac{R}{3} f(a)\right] + o(r^2), \tag{2.1}$$

$$\frac{1}{n v_n r^{n-1}}\int_{S(a,r)} f\, d\sigma = f(a) + \frac{r^2}{2n}\left[\Delta f(a) - \frac{R}{3} f(a)\right] + o(r^2), \tag{2.2}$$

denoting by $v_n = (1/n)\omega_{n-1}$ the volume of the unit ball in E_n.

Proof. Let (x^i) be geodesic coordinates with center a such that $r^2 = \Sigma (x^i)^2$ and let $d\tau = \sqrt{g}\, dx^1 \dots dx^n$. Let us set

$$\sqrt{g} f(x) = f(a) + \sum a_i x_i + \sum b_{ij} x^i x^j + o(r^2).$$

One easily gets

$$\int_{B(a,r)} f \, d\tau = v_n r^n f(a) + \frac{1}{2} v_n \frac{r^{n+2}}{n+2} \sum b_{ii} + o(r^{n+2})$$

with

$$\sum b_{ii} = \sum \frac{\partial^2}{(\partial x^i)^2} (\sqrt{g} f)_a = \Delta f(a) - \frac{R}{3} f(a),$$

taking into account the fact that the scalar curvature of M at a satisfies

$$\sum \frac{\partial^2}{(\partial x^i)^2} (\sqrt{g})_a = -\frac{R}{3}.$$

Formula (2.1) then follows; (2.2) would be proved in a similar way.

Notations. For any measurable function f on M, any $a \in M$ and any $r > 0$, let

$$V_a(r, f) = \int_{B(a,r)} f^n \, d\tau \quad \text{and} \quad S_a(r, f) = \int_{S(a,r)} f^{n-1} \, d\sigma \qquad (2.3)$$

each time that the written integrals make sense. [If f is positive, $V_a(r, f)$ and $S_a(r, f)$ represent the volume of $B(a, r)$ and the area of $S(a, r)$ for the metric $f^2 g$.] From Proposition 2.1 one deduces:

PROPOSITION 2.2. If f is a positive numerical function on M which admits a second differential at a and if $f(a) \neq 0$, one has:

$$\lim_{r \to 0} \frac{[S_a(r, f)]^n - n^n v_n [V_a(r, f)]^{n-1}}{r^2 [S_a(r, f)]^n} = \frac{2(n-1)}{n^2 - 4} \left(\frac{\Delta (f^{(n-2)/2})}{f^{(n-2)/2}} - \frac{n-2}{4(n-1)} R \right)_a. \qquad (2.4)$$

This proposition immediately follows by applying (2.1) and (2.2) to the functions f^n and f^{n-1} and making use of the identity:

$$\frac{1}{n-1} \frac{\Delta(f^{n-1})}{f^{n-1}} - \frac{1}{n+2} \frac{\Delta(f^n)}{f^n} = \frac{4}{n^2 - 4} \frac{\Delta(f^{(n-2)/2})}{f^{(n-2)/2}}.$$

In the euclidean case, one recognizes a relation proved by Resetnyak [7].

Remark. Relation (2.4) may be written

$$\lim_{r\to 0} \frac{S_a^n(r,f) - n^n v_n V_a^{n-1}(r,f)}{r^2 S_a^n(r,f)} = -\frac{f^2(a)}{2n+4} R_f(a), \tag{2.5}$$

where $R_f(a)$ denotes the scalar curvature of $(M, f^2 g)$ at a, so also:

$$\lim_{r\to 0} \frac{S_a^n(r,f) - n^n v_n V_a^{n-1}(r,f)}{[S_a(r,f)]^{n+2/(n-1)}} = \frac{R_f(a)}{(2n+4)(nv_n)^{2/(n-1)}}. \tag{2.6}$$

From this one deduces the following extension of Proposition 2.2:

PROPOSITION 2.3. Relation (2.6) remains valid if, in definition (2.3) of S_a and V_a, one denotes by $B(a, r)$ and $S(a, r)$ the geodesic ball and sphere with center a and radius r relative to any metric $\bar{g} = \lambda^2 g$ conformal with g.

Proof. Let us denote by $V_{a,\lambda}$ and $S_{a,\lambda}$ the functions given by (2.3) after this modification, and set $\bar{V}_a$, $\bar{S}_a$ for the metric $\bar{g} = \lambda^2 g$. One has:

$$S_{a,\lambda}(r,f) = \bar{S}_a\left(r, \frac{f}{\lambda}\right), \qquad V_{a,\lambda}(r,f) = \bar{V}_a\left(r, \frac{f}{\lambda}\right);$$

thus, applying (2.5) to $(M, \bar{g})$ and f/λ and denoting by $\bar{R}$ the scalar curvature of $(M, \bar{g})$:

$$\begin{aligned}\lim_{r\to 0} \frac{S_{a,\lambda}^n(r,f) - n^n V_{a,\lambda}^{n-1}(r,f)}{[S_{a,\lambda}(r,f)]^{n+2/(n-1)}} &= -\frac{\bar{R}_{f/\lambda}(a)}{(2n+4)(nv_n)^{2/(n-1)}} \\ &= -\frac{R_f(a)}{(2n+4)(nv_n)^{2/(n-1)}}.\end{aligned}$$

Interpretation. Proposition 2.3 can be explained by the fact that the geodesic ball with center a and radius r in $(M, \lambda^2 g)$ is, to the order $o(r^3)$, also a geodesic ball in (M, g), with center $a - r^2/2 \operatorname{grad}(\log \lambda)_a$ and radius $r/\lambda(a)$: this fact can be established by expanding the geodesic distance in $(M, \lambda^2 g)$ or by considering the conformal transformation of the osculating euclidean space at a to (M, g) induced by this change of metric (this transformation maps spheres onto spheres).

Geometrical signification of results of §1

With notations of §1, let $\varphi : M \to \bar{M}$ be a conformal diffeomorphism. Then $V_a(r, |\varphi'|)$ is the volume of $\varphi(B(a, r))$ and $S_a(r, |\varphi'|)$ is the area of $\varphi(S(a, r))$, and by applying Proposition 2.2 one gets:

$$\lim_{r\to 0} \frac{S_a^n(r, |\varphi'|) - n^n v_n V_a^{n-1}(r, |\varphi'|)}{[S_a(r, |\varphi'|)]^{n+[2/(n-1)]}} = \frac{-1}{(2n+4)(nv_n)^{2/(n-1)}} \times \left(R(a) - \frac{4(n-1)}{n-2}\frac{\Delta u}{u}\right). \tag{2.7}$$

But, by Proposition 2.3, formula (2.7) remains valid if we replace $B(a, r)$ by $\varphi^{-1}[\bar{B}(\varphi(a), r)]$ and $S(a, r)$ by $\varphi^{-1}[\bar{S}(\varphi(a), r)]$ (geodesic ball and sphere of M relative to the metric $|\varphi'|^2 g$): in that case $S_a(r, |\varphi'|)$ is the area of $\bar{S}(\varphi(a), r)$ while $V_a(r, |\varphi'|)$ is the volume of $\bar{B}(\varphi(a), r)$, so that the left side of (2.7) is equal to

$$\frac{-\bar{R}(\varphi(a))}{(2n+4)(nv_n)^{2/(n-1)}}:$$

one gets back the result of Corollary 2 in Proposition 1.1. Anyhow this interpretation cannot be taken as a proof of the regularity of conformal homeomorphisms; indeed:

(a) to apply Proposition 2.3 with $\lambda = |\varphi'|$, one must know *a priori* that $|\varphi'|$ is of class C^2,

(b) stated propositions apply only if $\varphi'(a) \neq 0$.

We shall now study the general regularity problem by a direct method proceeding by successive steps; but we shall be led by the idea that the differential of a conformal homeomorphism satisfies relation (1.1), and that this relation follows from the behaviour of

$$S_a^n(r, |\varphi'|) - n^n v_n V_a^{n-1}(r, |\varphi'|)$$

when r tends to 0.

3. First order regularity

In this section, we prove that every conformal homeomorphism between manifolds of class C^2 is bilipschitzian. The proof will be in two steps (Propositions 3.3 and 3.4) and will depend on Lemmas 3.1 and 3.2. Here M and $\bar{M}$ will denote riemannian manifolds of dimension $n \geqslant 2$, δ_M and $\delta_{\bar{M}}$ their geodesic distances, $B(a, r)$ [and $\bar{B}(a, r)$]

the geodesic open ball with center a and radius r of M [and of $\bar{M}$]. To make short, we will say that a bijection h is k-bilipschitzian if h and h^{-1} are lipschitzian with ratio k.

LEMMA 3.1. Let K be a compact subset of M.

(a) If M is of class C^1, to each $\varepsilon > 0$ one can associate $\rho > 0$ such that, for any $a \in K$, the ball $B(a, \rho)$ admits a $(1+\varepsilon)$-bilipschitzian bijection onto an open set of E_n.

(b) If M is of class C^2, to each $\varepsilon > 0$ one can associate $\rho > 0$ such that, for each $a \in K$, there exists a bijection h_a from $B(a, \rho)$ onto an open set of E_n, the restriction of which to each ball $B(a, r)$, with $r \leqslant \rho$, is $(1+\varepsilon r)$-bilipschitzian.

(c) If M is of class C^3, there exist $\rho > 0$ and $A > 0$ such that for each $a \in K$, there exists a bijection h_a from $B(a, \rho)$ onto the euclidean ball $|x| < \rho$ the restriction of which to each ball $B(a, r)$, with $r \leqslant \rho$, is $(1+Ar^2)$-bilipschitzian, and maps $B(a, r)$ onto the euclidean ball $|x| < r$.

Proof. It is enough to prove this lemma for a compact subset K of a local chart (Ω, h) of M. Statement (a) follows from the uniform continuity of the metric tensor on K. If M is of class C^2, let us denote by $g^a_{ij}\, dx^i\, dx^j$ the osculating euclidean metric to (M, g) at a. From the uniform continuity of the $\partial g_{ij}/\partial x_k$ on K, one deduces, for all $\alpha > 0$, the existence of $\rho > 0$ such that ($a \in K$ and $\delta_M(a, x) \leqslant r \leqslant \rho$) implies for all i, j:

$$|g^a_{ij}(x) - g_{ij}(x)| \leqslant \alpha r.$$

Taking α small enough, these inequalities imply:

$$(\forall u \in \mathbb{R}^n) \qquad (1+\varepsilon r)^{-2} \leqslant \frac{g^a_{ij}(x)u^i u^j}{g_{ij}(x)u^i u^j} \leqslant (1+\varepsilon r)^2. \tag{3.1}$$

From this follows that the restriction h_a of h to $B(a, r)$ is $(1+\varepsilon r)$-bilipschitzian. If M is of class C^3, the mapping h_a given by normal coordinates of center a satisfies the claimed condition.

On the other hand, the isoperimetrical inequality of E_n immediately implies:

LEMMA 3.2. Let U be an open set of M which admits a k-bilipschitzian bijection onto an open set of E_n and let K be a compact subset of U. Then the volume V of K and the area S of ∂K satisfy (with notations of §2)

$$n^n v_n V^{n-1} \leq k^{2n(n-1)} S^n. \tag{3.2}$$

Before proceeding further, we first prove the following general lemma, which extends to riemannian manifolds a known property of quasi-conformal homeomorphisms of E_n (cf. [3] corollary of th. 11).

PROPOSITION 3.3. Let $M, \bar{M}$ be of class C^1, and $\varphi : M \to \bar{M}$ a k-quasiconformal homeomorphism, i.e. an ACL^n and almost everywhere differentiable homeomorphism, whose differential φ' and metric jacobian J_φ satisfy, for nearly all $x \in M : |\varphi'(x)|^n \leq k^n J_\varphi(x)$. Then, for all $\nu < k^{-n}$, φ satisfies locally a Hölder condition with exponent ν.

Proof. Let K be a compact subset of M: the number $\varepsilon \in]0, 1]$ being fixed, let ρ and $\bar{\rho}$ be the numbers associated with K and $\varphi(K)$ by Lemma 3.1a, and let $r_0 \in]0, \frac{1}{2}\rho]$ be such that, for all $a \in K$, one has:

$$\varphi[B(a, 2r_0)] \subset \bar{B}(\varphi(a), \bar{\rho}). \tag{3.3}$$

The point $a \in K$ being fixed, let h be a $(1+\varepsilon)$-bilipschitzian bijection from $B(a, 2r_0)$ onto an open set of E_n, such that $h(a) = 0$. Then $\psi = \varphi \circ h^{-1}$ is defined (at least) on the ball $|x| < r_0$ of E_n, and ψ is $k(1+\varepsilon)^2$-quasiconformal (because h is $(1+\varepsilon)^2$-quasiconformal). For all $r < r_0$, let us denote by β_r the ball $|x| < r$ of E_n, by $\delta(r)$ the geodesic diameter of $\psi(\partial\beta_r)$ in $\bar{M}$, and by $\tau(r)$ the volume of $\psi(\beta_r)$. By Lemma 4.1 of [5], one has, for all $R < r_0$:

$$\int_{R/2}^{R} \delta^n(r) \frac{\mathrm{d}r}{r} \leq k^n (1+\varepsilon)^{2n} C_n \tau(R), \tag{3.4}$$

where C_n denotes an absolute constant. Besides, the area $\sigma(r)$ of $\psi(\partial\beta_r)$ satisfies, for almost every $r \in]0, r_0]$:

$$\sigma^n(r) \leq \left[\iint_{\partial\beta_r} |\psi'|^{n-1}\, \mathrm{d}\sigma\right]^n \leq \left[\iint_{\partial\beta_r} |\psi'|^n\, \mathrm{d}\sigma\right]^{n-1} \int_{\partial\beta_r} \mathrm{d}\sigma.$$

Therefore, as ψ is $k(1+\varepsilon)^2$-quasiconformal:

$$\sigma^n(r) \leq \left[\iint_{\partial\beta_r} k^n (1+\varepsilon)^{2n} J_\psi\, \mathrm{d}\sigma\right]^{n-1} n v_n r^{n-1};$$

that is:

$$\sigma^n(r) \leqslant k^{n(n-1)}(1+\varepsilon)^{2n(n-1)}\left(\frac{\mathrm{d}\tau}{\mathrm{d}r}\right)^{n-1} n v_n r^{n-1}. \tag{3.5}$$

Finally, (3.3) implies the existence of a $(1+\varepsilon)$-bilipschitzian bijection from $\psi(\beta_{r_0})$ onto an open set of E_n; therefore, applying Lemma 3.2 to the compact set $\psi(\bar{\beta}_r)(r < r_0)$:

$$n^n v_n \tau^{n-1}(r) \leqslant (1+\varepsilon)^{2n(n-1)}\sigma^n(r). \tag{3.6}$$

From (3.5) and (3.6) one deduces

$$\frac{\tau'(r)}{\tau(r)} \geqslant \frac{n}{k^n r(1+\varepsilon)^{4n}};$$

thus, since τ is an increasing function:

$$(\forall\, r \in]0, r_0[) \qquad \frac{\tau(r)}{\tau(r_0)} \leqslant \left(\frac{r}{r_0}\right)^{\nu n},$$

with $\nu = k^{-n}(1+\varepsilon)^{-4n}$. Looking back at (3.4) with $R = 2r$, one sees that for $r < r_0/2$, there exists $t \in [r, 2r]$ such that

$$\delta^n(t) \leqslant \frac{k^n}{\log 2}(1+\varepsilon)^{2n} C_n \left(\frac{2r}{r_0}\right)^{\nu n}. \tag{3.7}$$

Since $\psi(\beta_{r_0})$ admits a $(1+\varepsilon)$-bilipschitzian bijection onto an open set of E_n, one sees easily that, for $t \in [0, r_0]$, the geodesic diameter $\Delta(t)$ of $\psi(\beta_t)$ in $\bar{M}$ satisfies $\Delta(t) \leqslant (1+\varepsilon)^2\delta(t)$ (because the geodesic diameter of a subset of E_n equals that of its boundary); and since $\Delta(t)$ is an increasing function of t, (3.7) implies:

$$\Delta^n(r) \leqslant (1+\varepsilon)^{2n}\delta^n(t) \leqslant \frac{k^n}{\log 2}(1+\varepsilon)^{4n} C_n \left(\frac{2r}{r_0}\right)^{\nu n}.$$

Consequently, there exists a constant γ, depending only on M, $\bar{M}$, φ and K, such that the relations $a \in K$ and $\delta_M(a, x) \leqslant r \leqslant \frac{1}{2}r_0$ imply

$$\delta_{\bar{M}}[\varphi(a), \varphi(x)] \leqslant \gamma r^\nu, \tag{3.8}$$

and $\nu = k^{-n}(1+\varepsilon)^{-4n}$ is arbitrarily close to k^{-n}, therefore the result. Making use of Proposition 3.3, and by a similar method, one has then:

PROPOSITION 3.4. *Let $M, \bar{M}$ be of class C^2 and $\varphi: M \to \bar{M}$ a conformal homeomorphism. Then φ is locally bilipschitzian.*

Proof. Let K be a compact subset of M. The number $\varepsilon \in]0, 1]$ being fixed, let ρ and $\bar{\rho}$ be the numbers associated with K and $\varphi(K)$ by Lemma 3.1b. Let then $r_0 > 0$, γ and $\nu > 0$ be such that ($a \in K$ and $\delta_M(a, x) \leqslant r < \frac{1}{2}r_0$) imply (3.8) and finally let $r_1 > 0$ satisfy $r_1 < \inf(r_0, \rho, 1)$ and $\gamma(2r_1) \leqslant \rho$. One can then repeat the above calculation with $k = 1$, replacing r_0 by r_1 and assuming this time that h is a bijection from $B(a, 2r_0)$ onto an open set of E_n, whose restriction to each ball $B(a, r)$, with $r < 2r_0$, is $(1 + \varepsilon r)$-bilipschitzian. Moreover, for $r < 2r_0$, $\varphi(B(a, r))$ admits a $(1 + \varepsilon\gamma r^\nu)$-bilipschitzian bijection onto an open set of E_n. Inequalities (3.4), (3.5) and (3.6) are replaced by the sharper inequalities:

$$(\forall R \in]0, r_1[) \qquad \int_{R/2}^{R} \frac{\delta^n(r)}{r} \mathrm{d}r \leqslant (1 + \varepsilon R)^{2n} C_n \tau(R), \tag{3.9}$$

$$\sigma^n(r) \leqslant (1 + \varepsilon r)^{2n(n-1)} [\tau'(r)]^{n-1} n v_n r^{n-1}, \tag{3.10}$$

$$n^n v_n \tau^{n-1}(r) \leqslant (1 + \varepsilon\gamma r^\nu)^{2n(n-1)} \sigma^n(r). \tag{3.11}$$

Therefore:

$$\frac{\tau'(r)}{\tau(r)} \geqslant \frac{n}{r(1 + \varepsilon r)^{2n}(1 + \varepsilon\gamma r^\nu)^{2n}}.$$

From (3.10) and (3.11) one deduces that τ satisfies, for nearly all $r < r_1$, a differential inequality of the form

$$\frac{\tau'(r)}{\tau(r)} \geqslant \frac{n}{r}(1 - \alpha r^\nu)$$

with α a constant, and thus for $r < r_1$ (since τ is increasing):

$$\frac{\tau(r)}{\tau(r_1)} \leqslant \left(\frac{r}{r_1}\right)^n \exp\left[\frac{n\alpha}{\nu}(r_1^\nu - r^\nu)\right];$$

from this one deduces the existence of a constant A such that $\tau(r) \leqslant Ar^n$. Coming back to (3.9) with $R = 2r$, one sees that for each $r < \frac{1}{2}r_1$, there exists $t \in [r, 2r]$ satisfying

$$\delta^n(t) \leqslant (1 + \varepsilon R)^{2n} \frac{C_n}{\log 2} Ar^n.$$

From there one deduces easily the existence of a constant λ, depending only on M, $\bar{M}$, φ and K, such that the relations $a \in K$ and $\delta_M(a, x) \leqslant r < \frac{1}{2}r_1$ imply $\delta_{\bar{M}}[\varphi(a), \varphi(x)] \leqslant \lambda r$, which shows that φ is locally lipschitzian. The same argument holds for φ^{-1}, therefore the result.

Remark. By an easy extension, one can show that the conclusion of Proposition 3.4 remains valid if one only assumes that M and $\bar{M}$ are of class C^1 and with riemannian metrics which satisfy locally a Hölder condition.

4. Higher order regularity

Keeping notations of §3 and assuming that $M, \bar{M}$ are of class C^∞, we will first prove that the distribution-laplacian $\Delta|\varphi'|^n$ is a measure, and next we will show that $u = |\varphi'|^{(n-2)/2}$ satisfies the elliptic equation (1.3). Regularity of φ will follow.

PROPOSITION 4.1. Let $M, \bar{M}$ be of class C^3 and $\varphi: M \to \bar{M}$ a conformal homeomorphism. For any compact subset K of M, there exist $\rho > 0$ and $C > 0$ such that for all $a \in K$, the increasing function

$$r \mapsto V_a(r) = \int_{B(a,r)} |\varphi'|^n \, d\tau$$

admits nearly everywhere on $]0, \rho[$ a derivative satisfying

$$\frac{V_a'(r)}{V_a(r)} \geq \frac{n}{r}(1 - Cr^2). \tag{4.1}$$

Proof. Let $\rho > 0$ satisfy Lemma 3.1c and small enough to have $S(a, r) = \partial B(a, r)$ for all $r \in]0, \rho[$. Let us denote by

$$S_a(r) - \int_{S(a,r)} |\varphi'|^{n-1} \, d\sigma$$

the area of $\varphi(S(a, r))$ for $r \in]0, \rho[$ one has

$$\int_{S(a,r)} d\sigma \leq (1 + Ar^2)^{n-1} n v_n r^{n-1}$$

therefore, for almost every $r \in]0, \rho[$:

$$\begin{aligned} S_a^n(r) &< \left[\int_{S(a,r)} |\varphi'|^n \, d\sigma\right]^{n-1} \int_{S(a,r)} d\sigma \\ &\leq [V_a'(r)]^{n-1} (1 + Ar^2)^{n-1} n v_n r^{n-1}. \end{aligned} \tag{4.2}$$

On the other hand let $\bar{\rho}>0$ and $\bar{A}>0$ be such that for all $b\in\varphi(K)$, there exists a bijection from $\bar{B}(b,\bar{\rho})$ onto an open set of E_n, whose restriction to each ball $\bar{B}(b,r)$, with $r<\bar{\rho}$, is $(1+\bar{A}r^2)$-bilipschitzian. In view of Proposition 3.4 there exists $k>0$ such that, for all $a\in K$, the restriction of φ to $B(a,\rho)$ is k-bilipschitzian; and (taking ρ smaller) one may assume $k\rho<\bar{\rho}$. One has then $\varphi(B(a,r))\subset\bar{B}(\varphi(a),\bar{\rho})$, and applying Lemma 3.2 to $\bar{M}$:

$$n^n v_n V_a^{n-1}(r)\leqslant(1+\bar{A}k^2r^2)^{2n(n-1)}S_a^n(r).$$

Comparing with (4.2) one gets:

$$\frac{V_a'(r)}{V_a(r)}>\frac{n}{r(1+Ar^2)(1+\bar{A}k^2r^2)^{2n}}.$$

Therefore (4.1) with $C=A+2n\bar{A}k^2$:

COROLLARY. For all $a\in K$, $J(a)=\lim_{r\to0}V_a(r)/v_nr^n$ exists; and for $r\in\,]0,\rho[$ one has:

$$V_a(r)\geqslant v_nr^nJ(a)\exp\left(\frac{-Cnr^2}{2}\right). \tag{4.3}$$

Proof. From (4.1) one deduces that the function

$$r\mapsto r^{-n}\exp\left(\frac{Cnr^2}{2}\right)V_a(r)$$

is increasing, which implies the existence of $J(a)$ and inequality (4.3).

The function J is obviously defined on all of K and satisfies $J(a)=|\varphi'(a)|^n$ at every point where φ' exists. Finally, if φ is k-bilipschitzian on K, one has $k^{-n}\leqslant J(a)\leqslant k^n$ on K.

PROPOSITION 4.2. The distribution-laplacian T of J is a measure.

Proof. If f is of class C^∞ with compact support on M, by Proposition 2.1 one has, with the above notations:

$$T(f)=\lim_{r\to0}\frac{2(n+2)}{v_nr^{n+2}}\int_M d\tau(x)\int_{B(x,r)}[J(y)-J(x)]f(x)\,d\tau(y)$$

$$=\lim_{r\to0}\frac{2(n+2)}{v_nr^{n+2}}\int_M\left[V_x(r)-v_nr^nJ(x)\left(1+r^2\frac{R(x)}{6(n+2)}\right)\right]f(x)\,d\tau(x).$$

Let then ρ and C be the numbers associated with the support K of f by Proposition 4.1. Applying inequality (4.3) one has, if f is positive:

$$T(f) \geqslant \lim_{r\to 0} \frac{2(n+2)}{r^2} \int_K \exp\left[\frac{-Cnr^2}{2} - 1 - \frac{r^2R(x)}{6(n+2)}\right] J(x)f(x)\,\mathrm{d}\tau(x).$$

Therefore, since J and R are bounded on K, there exists a constant α depending only on the support of f, such that:

$$T(f) \geqslant -\alpha \int_M |f(x)|\,\mathrm{d}\tau(x),$$

which implies that $T + \alpha\,\mathrm{d}\tau$ is a positive functional, thus a measure.

In what follows, we will denote by $H_p(M)$ [or $H_p(\bar{M})$] the space of those numerical functions on M [or $\bar{M}$] whose distribution-derivatives of order p belong to $H_0(M) = L^2_{\mathrm{loc}}(M)$ [or $H_0(\bar{M})$]. First of all one has:

PROPOSITION 4.3. If $\varphi: M \to \bar{M}$ is a conformal homeomorphism, then $|\varphi'| \in H_1(M)$; and if $f \in H_2(\bar{M})$, then $f \circ \varphi \in H_2(M)$.

Proof. The first assertion follows from the fact that $\Delta|\varphi'|^n$ is a measure and that $|\varphi'|$ and $|\varphi'|^{-1}$ are locally bounded. The second one follows from the identity:

$$\int_M \operatorname{grad}(f_1 \circ \varphi) \cdot \operatorname{grad}(f_2 \circ \varphi)|\varphi'|^{n-2}\,\mathrm{d}\tau = \int_{\bar{M}} \operatorname{grad} f_1 \cdot \operatorname{grad} f_2\,\mathrm{d}\tau,$$

which holds for $f_1, f_2 \in H_1(\bar{M})$. Indeed, let $f_1 \in H_2(\bar{M})$ and $f_2 \in H_1(\bar{M})$, where f_2 has compact support K on $\bar{M}$. If, to be precise, one assumes that φ^{-1} is k-bilipschitzian on K, one has:

$$\left|\int_M \operatorname{grad}(f_1 \circ \varphi) \cdot \operatorname{grad}(f_2 \circ \varphi)|\varphi'|^{n-2}\,\mathrm{d}\tau\right| = \left|\int_{\bar{M}} f_2 \Delta f_1\,\mathrm{d}\tau\right|$$

$$\leqslant \|f_2\|_{L^2(K)} \|\Delta f_1\|_{L^2(K)} \leqslant k^{n/2} \|f_2 \circ \varphi\|_{L^2(M)} \|\Delta f_1\|_{L^2(K)}.$$

By letting f_2 vary, this shows that $\operatorname{div}(\operatorname{grad}(f_1 \circ \varphi)|\varphi'|^{n-2}) \in H_0(M)$, which implies $f_2 \circ \varphi \in H_2(M)$.

Remark. More generally, if $f \in H_{p+1}(\bar{M})$ and if one assumes that $|\varphi'| \in H_p(M)$, then one deduces that $f \circ \varphi \in H_{p+1}(M)$ (cf. [1]).

THEOREM A. If M and $\bar{M}$ are two C^∞ riemannian manifolds of dimension $n \geqslant 2$, and if $\varphi: M \to \bar{M}$ is a conformal homeomorphism, then φ is of class C^∞.

Proof. Let us first assume $n \geqslant 3$. If λ is a numerical function >0 of class C^∞ on M, one knows that the scalar curvature R_λ of $(M, \lambda^2 g)$ is given by:

$$R_\lambda = \lambda^{-2}\left(R - 4\frac{n-1}{n-2}\frac{\Delta u}{u}\right)$$

with $u = \lambda^{(n-2)/2}$. Therefore

$$\Delta(\lambda^n) = \frac{n+2}{2n}\lambda^{-n}\operatorname{grad}^2(\lambda^n) + \frac{n}{2n-2}(R\lambda^n - R_\lambda\lambda^{n+2}).$$

More generally, let λ be a function >0 on M such that λ and λ^{-1} are locally bounded and the laplacian of $J = \lambda^n$ is a measure. The classical computation of the scalar curvature of $(M, \lambda^2 g)$ can then be interpreted in terms of distributions:

$\lambda \in H_1(M)$, the Christoffel symbols relative to $\lambda^2 g$ and to a system of local coordinates of M belong to $H_0(M)$, and the curvature tensor of $(M, \lambda^2 g)$ is a distribution of order 1. Applying these results with $J = |\varphi'|^n$, one sees that the measure $\Delta(|\varphi'|^n \, d\tau)$ satisfies

$$\Delta(|\varphi'|^n \, d\tau) = \frac{n+2}{2n}|\varphi'|^{-n}\operatorname{grad}^2(|\varphi'|^n)\, d\tau$$
$$+ \frac{n}{2n-2}(R|\varphi'|^n - \bar{R} \circ \varphi|\varphi'|^{n+2})\, d\tau.$$

This shows that the function $J = |\varphi'|^n$ can be modified on a set of measure zero in order to satisfy in the usual sense:

$$\Delta J = \frac{n+2}{2n}J^{-1}\operatorname{grad}^2 J + \frac{n}{2n-2}(RJ - \bar{R} \circ \varphi J^{1+(2/n)}). \tag{4.4}$$

From formula (4.4) one deduces, by induction, that $J \in H_p(M)$ for all $p \in N$: we know indeed already that $J \in H_1(M)$; and if $J \in H_p(M)$, the remark above shows that $\bar{R} \circ \varphi \in H_{p+1}(M)$, thus that $\Delta J \in H_p(M)$, which implies $J \in H_{p+1}(M)$.

If $n = 2$, one proves in the same way that $J = |\varphi'|^2$ satisfies $\Delta J = J^{-1}\operatorname{grad}^2 J$, and the same conclusion remains.

In any case, the remark above shows that the local coordinates of φ belong to $H_p(M)$ for all $p \in N$; and, since they are continuous, one deduces that they are of class C^∞.

Moreover, if $M, \bar{M}$ are C^ω, φ is also C^ω. (Since $|\varphi'|$ satisfies Equation 1.4.)

From Theorem A it follows that in the case of C^∞ manifolds, the notion of conformal homeomorphism is identical with that of conformal diffeomorphism. Returning to [5], we may state:

THEOREM B. If M is a compact C^∞ riemannian manifold, and if the group $C(M)$ of conformal diffeomorphisms of M is not compact, there exists a conformal diffeomorphism from M onto the standard sphere of the same dimension.

It is worthy of note that this theorem does not follow from the proof given by M. Obata concerning the connected component $C_0(M)$ of $C(M)$.

Université de Paris VI

NOTES

[1] Since then, the case of compact manifolds was solved by D. V. Alekseevski in [2]; but it seems that this lengthy and difficult proof has to be completed on certain points (cf. [8]).

[2] This change of metric problem amounts to the construction, on each manifold, of a solution of the equation $\Delta u - (n-2)/[4(n-1)]Ru = 0$, defined in the neighbourhood of an arbitrary point a and such that $u(a) > 0$. A proof of the existence of such a solution was communicated to me by H. Brezis. F. W. Warner told me of another proof by J. L. Kazdan.

REFERENCES

[1] Agmon, *Lectures on Elliptic Equations*, Van Nostrand.

[2] Alekseevski, D.-V., 'Groups of Conformal Transformations of Riemannian Spaces', *Mat. Sb.* **89**(131) (1972), 280–296; *Math. U.S.S.R. Sb.* **18** (1972), 285–301.

[3] Gehring, F.-W., 'Rings and Quasi-conformal Mappings in Space', *Trans. Amer. Math. Soc.* **103** (1962), 353–393.

[4] Hartman, P., 'On Isometries and on a Theorem of Liouville', *Math. Z.* **69** (1958), 202–210.

[5] Lelong-Ferrand, J., 'Transformations conformes et quasi-conformes des variétés riemanniennes compactes (démonstration de la conjecture de A. Lichnerowicz)', *Acad. Roy. Belg. Cl. Sci. Mem. Coll.* in 8° (2), 39, n° 5, 1971.

[6] Obata, M., 'The Conjectures on Conformal Transformations of Riemannian Manifolds', *Bull. Amer. Math. Soc.* **77** (1971), 269–270.

[7] Resetnyak, J.-G., 'On Conformal Mappings of a Space', *Dokl. Akad. Nauk, S.S.S.R.* **130** (1960), 1196–1198; *Soviet. Math. Dokl.* **1** (1960), 153–155.

[8] Yoshimatsu, Y., 'On a Theorem of Alekseevskii Concerning Conformal Transformations', *J. Math. Soc. Japan* **28** (1976), 278–289.

PAULETTE LIBERMANN

PFAFFIAN SYSTEMS AND TRANSVERSE DIFFERENTIAL GEOMETRY

In this paper, we first recall some results of a previous study [13] concerning Pfaffian systems which are not completely integrable; we introduce the derived system (in the sense of E. Cartan) as the kernel of the structure tensor of a Pfaffian system (obstruction to the integrability); using J. Martinet's methods [15], we prove the nullity of certain Pontjargin classes for totally regular systems (their successive derived systems are of constant rank). The same methods could be applied to the study of local equivalence of G-structures.

Then we develop the study of some geometrical structures which are used in theory of foliations. The transverse geometry was introduced by C. Ehresmann [5] and A. Haefliger [7] in a more general setting than ours. We introduce the notion of transverse k-frame and prove for transverse prolongations properties similar to those relative to ordinary prolongations; we recover some results of P. Molino [17] and I. Vaisman [21] concerning foliated bundles.

In the last part of the paper, we develop the theory of partial principal connections (which are a generalization of Cartan's connections [4]); connections attached to the G-structured Pfaffian systems of P. Molino [16] and on the bundle of transverse k-frames, lead to partial principal connections. We prove the transverse version of a theorem of A. Lichnerowicz [14] concerning locally reductive spaces.

Relations with Molino's transverse geometry and partial Γ_q-structures will be studied later [18].

Note. All manifolds and mappings are assumed to be C^∞. In the spaces of jets of order one, when no confusion is possible, we will drop the index 1.

1. On some properties of Pfaffian systems [13]

On a n-dimensional manifold M, a vector subbundle E (of rank $n-p$) of the tangent bundle TM defines a $G_{n-p,p}$-structure (i.e. a $G_{n-p,p}$-

Cahen & Flato (eds.), Differential Geometry and Relativity, 107–126.

principal subbundle H' of the principal bundle H of frames) where $G_{n-p,p}$ is the group of matrices $\begin{pmatrix} A & 0 \\ B & C \end{pmatrix}$ with $A \in L_{n-p}(= GL(n-p, \mathbf{R}))$, $C \in L_p$, $B \in L_{n-p,p}$ (space of linear mappings from $\mathbf{R}^{n-p}$ to $\mathbf{R}^p$). The group $G_{n-p,p}$ is the semi-direct product $(L_{n-p} \times L_p) \times N_{n-p,p}$, where $N_{n-p,p}$ is the abelian group of matrices $\begin{pmatrix} 1 & 0 \\ B & 1 \end{pmatrix}$.

Let P_Q be the principal bundle which is associated to the normal bundle $Q = TM/E$; there is a surjective morphism of principal bundles $H'(V) \to P_Q$ corresponding to the group morphism: $G_{n-p,p} \to L_p$.

The subbundle E can be defined by a Pfaffian system E^*; E^* is a vector subbundle of rank p of the cotangent bundle T^*M; if we split the bundle T^*M into the direct sum of E^* and a complement S^* (which is isomorphic to Q), we get a $(L_p \times L_{n-p})$ principal subbundle of the bundle $H^*(M)$ of coframes; so S^* being chosen, there is a bigraduation of the exterior algebra of T^*M; let $(\omega^1, \ldots, \omega^p)$ sections of E^*, $(\omega^{p+1}, \ldots, \omega^n)$ sections of S^* over an open set U of M, such that $(\omega^1, \ldots, \omega^p, \omega^{p+1}, \ldots, \omega^n)$ is a field of coframes; for $i = 1, \ldots, p$, we have $\mathrm{d}\omega^i = \mathrm{d}_1\omega^i + \mathrm{d}_2\omega^i + \mathrm{d}_3\omega^i$ with $\mathrm{d}_1\omega^i$ of type $(2,0)$, $\mathrm{d}_2\omega^i$ of type $(1,1)$, $\mathrm{d}_3\omega^i$ of type $(0,2)$; for any function f, we have $\mathrm{d}_3(f\omega^i) = f\,\mathrm{d}_3\omega^i$, so the forms

$$\Omega^i = \mathrm{d}_3\omega^i = \sum_{\alpha,\beta=p}^{n} A^i_{\alpha\beta}\omega^\alpha \wedge \omega^\beta$$

with $A^i_{\alpha\beta} + A^i_{\beta\alpha} = 0$ $(i = 1, \ldots, p)$ are the components of tensorial differential form, called the *canonical torsion* of the $(L_p \times L_{n-p})$ structure. If we take another complement S'^* to E^* (with sections $\bar{\omega}^\alpha = \omega^\alpha + \Sigma_{i=1}^p \lambda^\alpha_i \omega^i$ for $\alpha = p+1, \ldots, n$) we get a splitting $\mathrm{d} = \mathrm{d}'_1 + \mathrm{d}'_2 + \mathrm{d}'_3$ and $\mathrm{d}'_3\omega^i = \Sigma^n_{\alpha,\beta=p} A^i_{\alpha\beta}\bar{\omega}^\alpha \wedge \bar{\omega}^\beta$ with the same coefficients $A^i_{\alpha\beta}$; so d_3 and d'_3 define a same vector bundle morphism $\delta : E^* \to \Lambda^2(T^*M/E^*)$; it is the *structure tensor* of E^* (introduced by J. Martinet [15]). The system E^* is completely integrable if and only if $\delta = 0$.

The $A^i_{\alpha\beta}$ can be considered as a function on $H'(M)$, with values in $\mathbf{R}^p \times \Lambda^2(\mathbf{R}^{n-p})^*$, of type $\mathcal{R}(G_{p,n-p})$, $\mathcal{R}$ being the representation of $G_{p,n-p}$ in $\mathbf{R}^p \otimes \Lambda^2(\mathbf{R}^{n-p*})$ such that $\forall g \in G_{p,n-p}$, $\mathcal{R}(g) = \rho_2^*(g) \otimes \Lambda^2(\rho_1^*(g))^{-1}$ where ρ_1^* and ρ_2^* are the morphisms $G_{p,n-p} \to L_p$, $G_{p,n-p} \to L_{n-p}$. We get D. Bernard's structure tensor [1]. It can be proved [13]:

PROPOSITION 1.1. Let on M two complementary subbundles E^* and S^* of T^*M defining a $(L_p \times L_{n-p})$-principal subbundle φ^* of $H^*(M)$; there exists on φ^* principal connections inducing on P_{E^*} and P_{S^*} the canonical torsion (P_{E^*} and P_{S^*} are the principal bundles which are associated to E^* and S^*).

These connections have been defined in [9] for $n = 2p$ (almost 'paracomplex' structures). For any n, if we restrict to connections on P_{E^*} we get the 'adapted' connections of Martinet and in the integrable case, the connections of Bott.

Example. [9]

The quadric Q_6, submanifold of $\mathbb{R}^7$ defined by

$$(x^1)^2 + (x^2)^2 + (x^3)^2 + (x^4)^2 - (x^5)^2 - (x^6)^2 - (x^7)^2 = 1$$

admits two fields E and S of 3-planes which are complementary; these fields are not completely integrable; Q_6 can be identified to the homogeneous space $G_2'/S\mathring{U}_3$, where G_2' is the non compact form of the complex simple group admitting G_2 as real compact form, $S\mathring{U}_3$ (unimodular 'para-unitary group' in the sense of [9]) is the group of matrices $\begin{pmatrix} a & 0 \\ 0 & a' \end{pmatrix}$ where $a \in SL_3$ and a' is the contragredient of a; $S\mathring{U}_3$ is isomorphic to SL_3, but as a subgroup of L_6, $S\mathring{U}_3$ is of finite type, so there is a unique $S\mathring{U}_3$-connection for a given torsion. This connection on Q_6 is given by the fundamental form $(\omega^1, \ldots, \omega^6)$ and the connection form $(\omega_j^i, \omega_{j'}^{i'})$ with $\omega_j^i + \omega_{i'}^{j'} = 0$, $\omega_1^1 + \omega_2^2 + \omega_3^3 = 0$, where i, j run from 1 to 3 and $i' = i + 3$; the Pfaffian system E^* is defined by $\omega^1 = 0$, $\omega^2 = 0$, $\omega^3 = 0$, S^* by $\omega^4 = 0$, $\omega^5 = 0$, $\omega^6 = 0$; the torsion Ω^i and $\Omega^{i'}$ forms are given by

$$\Omega^1 = d\omega^1 - \omega^1 \wedge \omega_1^1 - \omega^2 \wedge \omega_2^1 - \omega^3 \wedge \omega_3^1 = 2\omega^5 \wedge \omega^6,$$

$$\Omega^2 = d\omega^2 - \omega^1 \wedge \omega_1^2 - \omega^2 \wedge \omega_2^2 - \omega^3 \wedge \omega_3^2 = 2\omega^6 \wedge \omega^4,$$

$$\Omega^3 = d\omega^3 - \omega^1 \wedge \omega_1^3 - \omega^2 \wedge \omega_2^3 - \omega^3 \wedge \omega_3^3 = 2\omega^4 \wedge \omega^5,$$

$$\Omega^4 = d\omega^4 - \omega^4 \wedge \omega_4^4 - \omega^5 \wedge \omega_5^4 - \omega^6 \wedge \omega_6^4 = 2\omega^2 \wedge \omega^3,$$

$$\Omega^5 = d\omega^5 - \omega^4 \wedge \omega_4^5 - \omega^5 \wedge \omega_5^5 - \omega^6 \wedge \omega_6^5 = 2\omega^3 \wedge \omega^1,$$

$$\Omega^6 = d\omega^6 - \omega^4 \wedge \omega_4^6 - \omega^5 \wedge \omega_5^6 - \omega^6 \wedge \omega_6^6 = 2\omega^1 \wedge \omega^2.$$

Remark that the curvature forms are given by

$$\Omega_j^i = -\Omega_{i'}^{j'} = -3\omega^{j'} \wedge \omega^i + \delta_i^j \sum_{i=1}^{3} \omega^{i'} \wedge \omega^i.$$

We get the 14 structure equations of G_2'.

We recall that the *class* of the Pfaffian system can be defined in the following way [15]: at every $x \in M$, the support H_x of $\delta(E_x^*)$ is the smallest vector subspace of $Q_x^* = T_x^*M/E_x^*$ such that $\delta_x(E_x^*) \subset \Lambda^2 H_x$; the class C_x is the dimension of $\hat{E}_x$, inverse image of H_x by the projection $T_x^*M \to Q_x^*$; we get $C_x \geqslant p$. If $C_x = p$ for every x, then E^* is completely integrable.

If the structure function is of a constant rank q, then the kernel $E_{(1)}^*$ of δ is a vector bundle of rank $p - q$; it is the *derived* system of E^* in the sense of E. Cartan [3]; the system E^* is completely integrable if and only if $E^* = E_{(1)}^*$.

PROPOSITION 1.2. If the structure tensor is of constant rank q, the tangent bundle TM admits a vector subbundle of rank q.

Proof. The direct sum $E_{(1)}^* \oplus S^*$ (where S^* is a complement to E^*) is a vector bundle of rank $(p - q) + (n - p)$; it defines a vector subbundle of TM of rank q.

If $p = n - 2$, then $\Lambda^2 Q_x^*$ is 1-dimensional and δ is of rank 0 or 1.

COROLLARY. If a compact manifold, with non zero Euler-Poincaré characteristic admits a 2-plane field, then there is at least one point where the structure tensor vanishes.

For instance, the projective space $P_{2n+1}(\mathbf{C})$ admits a fibration on the quaternionic space $P_n(H)$, the fibers being diffeomorphic to S_2; if there exists a non integrable 2-plane field, then the structure function δ cannot be of constant rank.

Remark. If we take any one of the 3-plane field defined previously on Q_6, the kernel of δ is 0 but Q_6 being diffeomorphic to $S_3 \times \mathbf{R}^3$ admits two 3-plane fields which are completely integrable.

If $E_{(1)}^*$ is of rank $p_1 > 0$, we can define its structure function δ_1 and its kernel $E_{(2)}^*$ and so on (if δ_1 is of constant rank).

We say that E^* is *totally regular* if all the derived systems are of constant rank. So we get a sequence $E^* \supset E_{(1)}^* \supset \cdots \supset E_{(r)}^*$ of vector subbundles of T^*M (the sequence of ranks being strictly decreasing) and there exists r such that $E_{(r)}^*$ satisfies one of the two conditions

(a) $E_{(r+1)}^* = E_{(r)}^*$ and $E_{(r)}^*$ is completely integrable;

(b) $E_{(r+1)}^*$ is of rank 0.

Let $\mathscr{I}_{(0)}, \mathscr{I}_{(1)}, \ldots, \mathscr{I}_{(r)}$ be the ideals generated by the systems

$E^*, E^*_{(1)}, \ldots, E^*_{(r)}$ in the exterior algebra of T^*M; for $i = 1, \ldots, r$, the differential $d\omega_{(i)}$ of a local section $\omega_{(i)}$ of $E^*_{(i)}$ belongs to the ideal $\mathcal{I}_{(i-1)}$.

It is possible to find connections $\mathscr{C}$ which are adapted to E^* such that the torsion of the restriction $\mathscr{C}_{(i)}$ of $\mathscr{C}$ to $E^*_{(i)}$ belongs to $\mathcal{I}_{(i-1)}$; using the Bianchi identities, it can be shown that the curvature forms of $\mathscr{C}_{(i)}$ belong to $\mathcal{I}_{(i-2)}$ for $i \geq 2$.

As $\text{Pont}^k (F)$, the homogeneous part of degree k, of the Pontjargin ring of a vector bundle F, is a polynomial of degree k with respect to the curvature, we get:

PROPOSITION 1.3. Let E^* be a totally regular Pfaffian system of rank $p = (p_0)$, $E^*_{(1)}, \ldots, E^*_{(r)}$ the derived systems of ranks $p_1, \ldots, p_r$. Then $\text{Pont}^k (E^*_{(i)}) = 0$ for $k > 2p_{i-2}$ and $i > 2$.

Martinet [15] has proved that $\text{Pont}^k (E^*_{(i)}) = 0$ for $k > 2q_i$ where q_i is the maximum of the class of $E^*_{(i)}$; our result only uses the rank of $\delta_{(i)}$.

If $E^*_{(r)} = E^*_{(r+1)}$, then $E^*_{(r+2)} = E^*_r$ and we get Bott result (who proved that in the integrable case the curvature forms belong to the ideal $\mathcal{I}$ generated by E^*).

Remark. In the case of the quadric Q_6, though the 3-plane fields E and S are not integrable, for each field the curvature forms belong to the ideal generated by the differential system defining this field.

Using E. Cartan's method, it can be proved [13]:

PROPOSITION 1.4. If a manifold M, of dimension $p+2$, admits a totally regular Pfaffian system E^* of rank p, such that the derived system $E^*_{(i)}$ is of rank $p-i$ and the system E^*_{p-i} is not completely integrable, in the neighborhood of every point, there are adapted coframes fields $(\bar{\omega}^1, \ldots, \bar{\omega}^{p+2})$ and distinguished coordinates $(x^1, \ldots, x^{p+1}, t)$ such that $\bar{\omega}^1 = dx^1 + x^2\, dt$, $\bar{\omega}^2 = dx^2 + x^3\, dt, \ldots$

$$\bar{\omega}^{p-1} = dx^{p-1} + x^p\, dt, \qquad \bar{\omega}^p = dx^p + x^{p+1}\, dt,$$

E^* being generated by $\omega^1, \ldots, \omega^p$, $E^*_{(i)}$ by $\omega^1, \ldots, \omega^{p-i}$. So all manifolds satisfying these conditions are locally equivalent to the manifold $\mathcal{I}J^p(R, V)$ of invertible p-jets from $\mathbf{R}$ to a one-dimensional manifold V.

In the same way, using derived systems we could get topological conditions for a manifold to be locally equivalent to a manifold of jets from $\mathbf{R}^m$ to a m'-manifold V.

Instead of Pfaffian structures, for G-structures which admit a canonical torsion tensor, the research of submanifolds on which the structure is integrable leads to derived systems (subbundles of T^*M on which the torsion vanishes). For instance, we know that almost complex and almost symplectic structures have a canonical torsion which vanishes if and only if the structure is integrable. Consider the case of an almost symplectic structure on a manifold M (of dimension $2n$); it is defined by a non degenerated 2-form Ω; if $(\omega^1, \ldots, \omega^{2n})$ are local forms such that locally Ω can be written $\omega^1 \wedge \omega^{n+1} + \cdots + \omega^n \wedge \omega^{2n}$ the torsion is defined by the local forms $\Omega^i = \frac{1}{3} e_i \lrcorner \mathrm{d}\Omega$ where e_i is the local vector field such that $e_i \lrcorner \Omega = \omega^i$; so at each point, the derived system is the dual with respect of the non degenerate form Ω of the characteristic system of $\mathrm{d}\Omega$. If the rank is constant, the derived system is completely integrable but to get a symplectic structure on a submanifold of dimension q we must have the conditions: $q = 2q'$, and the rank of the restriction of Ω is $2q'$.

The equivalence problem of almost complex structures with the use of derived systems has been studied in [9] for 4-dimensional manifolds.

2. Transversal structures - definitions

The notion of transverse jet has been introduced by C. Ehresmann [5]; we will give a slightly different definition.

We keep the notations of §1 and suppose that the subbundle E of TM is completely integrable; let $\mathcal{F}$ be the corresponding foliation. This foliation can be defined by an open cover (U_α) of M and a maximal set of submersions $f_\alpha : U_\alpha \to R^p$ (called *distinguished* submersions) such that for every $x \in U_\alpha \cap U_\beta$, there exists an element of the pseudogroup of diffeomorphisms of $\mathbf{R}^p$, with the condition

$$f_\alpha = \gamma_{\alpha\beta} \circ f_\beta;$$

then, $E|U_\alpha = \ker(Tf_\alpha)$.

To each distinguished submersion are associated distinguished coordinates $(x^1, \ldots, x^p, x^{p+1}, \ldots, x^n)$ such that $\mathrm{d}x^1, \ldots, \mathrm{d}x^p$ are local sections of E^*.

Given two foliated manifolds $(M', \mathcal{F}')$ and $(M, \mathcal{F})$ ($\mathcal{F}'$ and $\mathcal{F}$ being of codimensions p' and p), a local map f from an open set in M' to M is compatible with the foliations $\mathcal{F}'$ and $\mathcal{F}$ (more briefly is a *local morphism*) if and only if for every pair (f'_μ, f_α) of distinguished submersions on M' and M such that the sources of f'_μ and f_α have non empty intersection respectively with the source and target of g, there exists a map $g_{\alpha\mu}$ from an open set in $\mathbf{R}^{p'}$ to $\mathbf{R}^p$ such that $f_\alpha \circ g = g_{\alpha\mu} \circ f'$; the map $g_{\alpha\mu}$ will be called the (f'_μ, f_α)-projection of g.

The λ-jet of a mapping will mean the local jet (or germ) of this mapping. The *transverse* λ-jet (resp. the *transverse k-jet*) at $y \in M'$ of the local morphism g will be the equivalence class of local morphisms h from M' to M satisfying the following conditions:

(1) The source of h contains y and $g(y) = h(y)$;

(2) there exists a pair of distinguished submersions (f'_μ, f_α) such that the (f'_μ, f_α)-projections of g and h define the same λ-jet (resp. k-jet) at $f'_\mu(y)$; this notion is independent of the choice of the pair (f'_μ, f_α). For the transverse jets at y, we will use the notations $\hat{j}^\lambda_y g$ and $\hat{j}^k_y g$.

If we take distinguished coordinates $(y^1, \ldots, y^{p'}, y^{p'+1}, \ldots, y^{n'})$ and $(x^1, \ldots, x^p, x^{p+1}, x^n)$ around y and $g(y)$, the mapping g can be represented by

$$x^i = g^i(y^i, \ldots, y^{p'}) \qquad i = (1, \ldots, p),$$

$$x^a = g^a(y^1, \ldots, y^{n'}) \qquad a = (p+1, \ldots, n);$$

if y has coordinates $(y_0^1, \ldots, y_0^{p'}, \ldots, y_0^{n'})$, the transverse jet $\hat{j}^k_y g$ is represented by the values at $(y_0^1, \ldots, y_0^{n'})$ of the g^a's and the values at $(y_0^1, \ldots, y_0^{p'})$ of the g^i's and the partial derivatives of order $\leqslant k$ of the g^i's with respect to the y^j's.

Transverse jets of local morphisms can be composed.

A mapping f from a manifold M' to a foliated manifold $(M, \mathcal{F})$ is said to be *transversal* to $\mathcal{F}$ if the $(f_\alpha \circ f)$ are submersions; then f defines a foliation $f^*\mathcal{F}$ on M' and f is a morphism, so we may define the transverse jets of f.

A *transverse* λ*-frame* (resp. *transverse k-frame*) at $x \in M$ is the transverse jet $\hat{j}^\lambda_0 h$ (resp. $\hat{j}^k_0 h$) of a transversal mapping h from an open set in $\mathbf{R}^p$ to M, which maps 0 on $x = h(0)$. For a distinguished submersion f_α the transverse jet $\hat{j}^\lambda_x f_\alpha$ and the jet $j^\lambda_x f_\alpha$ can be identified. Notice that a transverse frame $\hat{j}^\lambda_0 h$ (resp. $\hat{j}^k_0 h$) has a right inverse which is the jet of a distinguished submersion mapping x on 0: if f_α is a distinguished submersion such that $h_\alpha = f_\alpha \circ h$ has a non empty source, then $j^\lambda_0 h_\alpha$ is the germ of a diffeomorphism and the inverse of $\hat{j}^\lambda_0 h$ is $(j^\lambda_0 h_\alpha)^{-1} \circ j^\lambda_x f_\alpha$ (the inverse is independent of the choice of f_α).

The set $\hat{H}^k(M, \mathcal{F})$ of all transverse k-frames is a principal bundle with structure group L_p^k acting on the right (L_p^k is the group of invertible k-jets of $\mathbf{R}^p$ into $\mathbf{R}^p$ with source and target at 0). The set $\hat{H}^{*k}(M, \mathcal{F})$ of all k-jets of distinguished submersions with target 0 (which was considered by A. Haefliger [7]) is a left principal bundle; it can be considered as the bundle of transverse k-coframes. In the case of a simple foliation (defined by a submersion $\pi: M \to V$), $\hat{H}^k(M, \mathcal{F})$ and $\hat{H}^{*k}(M, \mathcal{F})$ are the pull-backs $H^k(V)$ and $H^{*k}(V)$. We could define in the same way k-velocities and k-covelocities [5].

When $k = 1$, the previous notions can be defined also for non integrable subbundles: the transverse 1-jet $\hat{j}^1_y g$ of a local morphism from $(M', \mathcal{F}')$ to $(M, \mathcal{F})$ can be identified with the linear map $\hat{T}_y g: T_y M'/E'_y \to T_{g(y)}M/E_{g(y)}$ induced by the linear map $T_y g: T_y M' \to T_{g(y)}M$ which maps E'_y on $E_{g(y)}$. The bundle $\hat{H}^1(M, \mathcal{F})$ or more briefly $\hat{H}$) is the principal bundle P_Q associated to $Q = TM/E$, and $\hat{H}^{*1}(M, \mathcal{F})$ is the principal bundle $P^*_{E^*}$ associated to E^*; the set of transverse tangent vectors is Q.

Suppose again E to be integrable and $k \geqslant 1$. Let Γ be the pseudogroup of all local automorphisms of the foliated structure $(M, \mathcal{F})$ and Θ be the infinitesimal pseudogroup (cf. [10]) of all local Γ-vector fields (vector fields which generate one-parameter local groups f_t, where f_t belongs to Γ). There is a surjective functor F^k from the transitive differentiable groupoïd $J^k(\Gamma)$ of k-jets of all $f \in \Gamma$, to the transitive differentiable groupoïd $\hat{J}^k(\Gamma)$ of transverse k-jets; to each $f \in \Gamma$, there corresponds sections $j^k f$ and $\hat{j}^k f = F^k \circ j^k f$; a local section s_k: $U \subset M \to \hat{J}^k(\Gamma)$ can be written $s_k = \hat{j}^k f$, if and only if $\beta \circ s_k \in \Gamma$ and $s_k(x) = \hat{j}^k_x(\beta \circ s_k) \forall x \in U$ (β is the target for any groupoïd).

The pseudogroup Γ can be prolongated to a pseudogroup Γ^k of local automorphisms of $J^k(\Gamma)$; the prolongation f^k of f is the map: $y^k \to j^k_{\beta(y^k)} f \cdot y^k$ for every $y^k \in J^k(\Gamma)$ such that $\beta(y^k)$ belongs to the source of f. We deduce a pseudogroup $\hat{\Gamma}^k$ of local automorphisms of $J^k(\Gamma)$ defined by $\hat{f}^k = F^k(f^k)$. The groupoïd $\hat{J}^k(\Gamma)$ acts on the bundle $\hat{H}^k(M, \mathcal{F})$ as follows: for any pair $(\hat{h}^k, \hat{h}'^k)$ of transverse k-frames, there exists one and only one $\hat{y}^k \in \hat{J}^k(\Gamma)$ such that $\hat{h}'^k = \hat{y}^k \cdot \hat{h}^k$; $\hat{J}^k(\Gamma)$ is associated to $\hat{H}^k(M, \mathcal{F})$ in the sense of [11]. It acts also on $\hat{H}^{*k}(M, \mathcal{F})$ and on the bundle of transverse velocities and co-velocities. Following C. Ehresmann [5], we call *transverse k-prolongation* of $(M, \mathscr{F})$ any bundle (with base M) on which $\hat{J}^k(\Gamma)$ acts differentiably.

Any local Γ-vector field X defines a one-parameter local group $\hat{f}_t^k$ on $\hat{J}^k(\Gamma)$ and hence a one-parameter local group on all transverse k-prolongations of $(M, \mathscr{F})$; so X can be lifted to a local vector field of any transverse k-prolongation. In particular any local section of E is a Γ-vector field: on every transverse k-prolongation there is a lifting of E which is completely integrable. So *every transverse k-prolongation admits a foliation $\mathscr{F}^k$ of dimension $n-p$* (the rank of E); in other words, a *transverse k-prolongation is a foliated bundle* in the sense of P. Molino [17], I. Vaisman [21], Kamber-Tondeur [8]. For $k = 1$, these results were proved by P. Molino [17].

Notice that P. Molino [17] defined transverse prolongations of $(M, \mathscr{F})$ in the following way: he considered the bundle of transverse 1-frames $\hat{H}^1(M, \mathscr{F})$, then the bundle $\hat{H}^1(\hat{H}^1(M, \mathscr{F}), \mathscr{F}^1)$ and so on. With our definition $\hat{H}^k(M, \mathscr{F})$ is a principal bundle over $\hat{H}^{k-1}(M, \mathscr{F})$, with abelian structure group (the kernel of $L_p^k \to L_p^{k-1}$).

Our definitions and the results in §3 are related with those of B. Cenkl [3′]; remarks that, for $k > 1$, it is not possible, in the non integrable case, to define L_p^k bundles.

3. Properties of transverse frames

We will prove propositions which are the 'transverse' version of propositions concerning ordinary prolongations.

First we recall that for any transitive differentiable groupoïd Φ (with source and target mappings α and β from Φ to the manifold V of units), we have called [11] depl Φ (or space of infinitesimal 'displacements' in the sense of C. Ehresmann) the set of all tangent vectors to Φ, with origin belonging to V and which are α-vertical; depl Φ is a vector bundle over V and the set of germs of sections of depl Φ is a Lie algebra sheaf. If Φ is associated to a principal bundle $P \to V$ (with structure group G), there is a canonical vector bundle isomorphism from depl Φ to TP/G (the set of tangent vectors to P mod. right translations of G).

PROPOSITION 3.1. Let M be a manifold with a foliation $\mathscr{F}$ of codimension p; then there exists a canonical vector bundle isomorphism Ψ^k from the set $\hat{J}_k\Theta$ of transverse k-jets of Γ-vector fields to $T\hat{H}^k(M, \mathcal{F})/L^k_p$ (or to depl $\hat{J}^k(\Gamma)$).

The proof is similar to the one concerning the isomorphism of J_kT and TH^k/L^k_n [10]. From previous remarks any local Γ-vector field X defines a one-parameter local group of principal automorphisms of $\hat{H}^k(M, \mathcal{F})$; so X defines a local section σ of $T\hat{H}^k(M, \mathcal{F})/L^k_p \to M$. Using distinguished coordinates $(x^1, \ldots, x^p, x^{p+1}, \ldots, x^n)$, X is defined by functions $X^1, \ldots, X^p, X^{p+1}, \ldots, X^n$; X is a Γ-vector field if and only if the X^i's $(i = 1, \ldots, p)$ are only functions of the x^j's $(j = 1, \ldots, p)$ and the transverse jet $\hat{J}^k_x X$ is defined by the values at u (image of x in $\mathbb{R}^n$) of $X^1, \ldots, X^n, X^{p+1}, \ldots, X^r$ and the derivatives of order $\leqslant k$ of the X^i's with respect of the x^j's; as in [10], we deduce from the Schwarz lemma, that $\sigma(x)$ is defined by the transverse jet $\hat{j}^k_x X$; then we prove that Ψ^k is a linear isomorphism on each fiber.

Notice that the fibers of $\hat{J}_k\Theta$ and $T\hat{H}^k(M, \mathcal{F})/L^k_p$ are isomorphic to $\mathbb{R}^n \oplus l^k_p$ (where l^k_p is the Lie algebra of L^k_p).

If we 'drop' the functions $X^{p+1}, \ldots, X^n$, the Γ-vector field X defines a section s of the normal bundle $Q = TM/E$ which is foliated in the sense of I. Vaisman [21], i.e. s is constant of the leaves of $\mathcal{F}$; so we have a natural map: $\hat{J}_k\Theta \to J^f_kQ$(space of k-jets of foliated sections of Q) and we deduce:

PROPOSITION 3.1′. There is a canonical vector bundle isomorphism from the set of k-jets of foliated sections of $Q = TM/E$ to the bundle $Q\hat{H}^k(M, \mathcal{F})/L^k_p$ (where $Q\hat{H}^k(M, \mathcal{F})$ is the normal bundle $T\hat{H}^k/(T\Pi^k)^{-1}E$, Π^k being the projection of $\hat{H}^k(M, \mathcal{F})$ on M).

The fibers of J^f_kQ and $Q\hat{H}^k(M, \mathcal{F})/L^k_p$ are isomorphic to $R^p \oplus l^k_p$.

PROPOSITION 3.2. (cf. [6]). On $H^k(M, \mathcal{F})$ there exists a fundamental form $\hat{\omega}^k$ with values in $R^p \oplus l^{k-1}_p$; this form $\hat{\omega}^k$ is equivariant, vanishes on $(T\Pi^k)^{-1}E$ and is invariant under local diffeomorphisms of $\hat{H}^k(M, \mathcal{F})$ which are transverse prolongations of diffeomorphisms belonging to Γ.

See also [3′].

For $k = 1$, the fundamental form was introduced by P. Molino [16].

Proof. The groupoïd $\hat{J}^k(\Gamma)$ and therefore $\hat{H}^k$ act on the vector bundle $J^f_{k-1}Q$; a tangent vector Y^k to $\hat{H}^k$ at $\hat{h}_k$ projects on $Y^{k-1} \in T\hat{H}^{k-1}$; Y^{k-1} defines an element Z^{k-1} of $Q\hat{H}^k$; $\hat{h}_k$ sends the orbit of

Z^{k-1} under the action of L_p^{k-1} (which can be identified to an element of $J_{k-1}^f Q$) on an element of $\mathbf{R}^p \oplus l_p^{k-1}$; so we get $\hat{\omega}^k$ which vanishes on $(T\Pi^k)^{-1}E$; it is easy to prove that for any $\gamma \in L_p^k$ then $\gamma^*\hat{\omega}^k = \gamma^{-1}\hat{\omega}^k$.

Let H'^k be the principal subbundle of the bundle $H^k(M)$ of k-frames which is the prolongation of the $G_{n-p,p}$-subbundle H' of $H(M)$ defined by E (cf. §1): H'^k is isomorphic to $J_{x_0}^k(\Gamma)$ (set of all k-jets of elements of Γ with source at x_0); there is a fundamental form ω^k on H'^k with values in $\mathbf{R}^n \oplus g_{n-p,p}^{k-1}$ (where $g_{n-p,p}^{k-1}$ is the Lie algebra of the prolongation $G_{n-p,p}^{k-1}$ of $G_{n-p,p}$). To the functor $F^k : J^k(\Gamma) \to \hat{J}^k(\Gamma)$, there corresponds morphisms $f^{k-1} : \text{depl}_{x_0} J^{k-1}(\Gamma) \to \text{depl}_{x_0} \hat{J}^{k-1}(\Gamma)$, $C^k : H'^k \to \hat{H}^k$, $c^{k-1} : R^n \oplus g_{n-p,p}^{k-1} \to R^p \oplus l_p^{k-1}$; we have: $(C^k)^*\hat{\omega}^k = c^{k-1} \circ \omega^k$; as the prolongation to H'^k of any $g \in \Gamma$ leaves ω^k invariant, so are $(C^k)^*\hat{\omega}^k$ and $\hat{\omega}^k$.

We can define the bundle $\bar{\hat{H}}^2(M, \mathcal{F})$ of *semi-holonomic transverse 2-frames* in the following way; let g be a local mapping from a neighborhood of 0 in $\mathbf{R}^p$ to $\hat{H}(M, \mathcal{F})$, such that $\Pi \circ g$ (with $\Pi : \hat{H} \to M$) is transversal to $\mathcal{F}$ and satisfies the condition $\hat{J}^1(\Pi \circ g) = g(0)$; g being transversal to the foliation $\Pi^*\mathcal{F}$, we can define the transverse jet $\hat{j}_0^1 g$; this transverse jet will be called semi-holonomic transverse frame of order 2. It can be proved that $\bar{\hat{H}}^2$ is a principal bundle with structure group $\bar{L}_p^2$ (group of semi-holonomic 2-jets from $\mathbf{R}^p$ to $\mathbf{R}^p$ with source and target at 0); $\hat{H}^2$ is a subbundle of $\bar{\hat{H}}^2$.

By iteration, we can define $\bar{\hat{H}}^k(M, \mathcal{F})$ for any k and the groupoïd $\bar{\hat{\Phi}}^k(M, \mathcal{F})$ which is associated to $\bar{\hat{H}}^k(M, \mathcal{F})$; it is a semi-holonomic prolongation of $J^1(\Gamma)$.

If E is non-integrable, it is possible to define $\bar{\hat{H}}^2(M, E)$ with the condition on g: $T_0(\Pi \circ g)$ maps $\mathbf{R}^p$ on a complement to E_x (therefore $T_0(\Pi \circ g)$ defines an isomorphism $\mathcal{I}_x$ of R^p on $Q_x = T_xM/E_x$), but in general, there is no local map f from M to $\mathbf{R}^p$ such that $T_xf = \mathcal{I}_x^{-1}$; so it is not possible to define holonomic 2-frames. We can define $\bar{\hat{H}}^k(M, E)$ and the groupoïd $\bar{\hat{\Phi}}^k(M, E)$ which is a prolongation of Φ (groupoïd associated to $\hat{H}$) but not of $\hat{J}^1(\Gamma)$ (set of transverse 1-jets of local diffeomorphisms leaving E invariant).

Suppose again E to be integrable and g as above. The transverse jet $\hat{j}_0^1(\pi \circ g)$ has an inverse $j_x^1 f$ (where $x = \pi \circ g(0)$), f being a distinguished submersion; so $\hat{j}_x^1(g \circ f) = \hat{j}^1 g \cdot j_x^1 f$ is the transverse jet of a local section $s = g \circ f$. Conversely given a local section s of $\hat{H}$, there exists a local transversal map γ from $\mathbf{R}^p$ to M such that $\hat{j}_0^1 \gamma = s(x)$ and $g = s \circ \gamma$ is transversal to $\pi^*\mathcal{F}$ with $g(0) = \hat{j}_0^1(\pi \circ g)$; so

$\hat{j}_0^1(s \circ \gamma) \in \hat{H}^2$; this transverse jet is defined by $\hat{j}_x^1 s$ (as s is compatible with the foliations $\mathcal{F}$ and $\pi^*\mathcal{F}$, it makes sense to speak of the transverse jet of s); we deduce:

PROPOSITION 3.3. (cf. [11]). There is a canonical diffeomorphism from the bundle $\hat{\bar{H}}^2(M, \mathcal{F})$ of transverse semi-holonomic frames to the set $\hat{J}_1\hat{H}(M, \mathcal{F})$ of transverse 1-jets of local sections of $\hat{H}(M, \mathcal{F})$ (we consider on $\hat{H}$ the foliation $\pi^*\mathcal{F}$).

Remarks. (1) Any element of $\hat{J}_1\hat{H}$ can be considered as a bijective linear mapping from $Q_x = T_xM/E_x$ to $Q_h(\hat{H}) = T_h\hat{H}/(T_h\pi)^{-1}E_x$ (where $x = \pi(h)$).

(2) With this interpretation of $\hat{J}_1\hat{H}$, Proposition 3.3 still holds for non integrable E, but as we already remarked the inverse of a transverse frame is not in general the jet of a map from M to $\mathbf{R}^p$.

Suppose E to be integrable. As $\hat{H}^2$ is a subbundle of $\hat{\bar{H}}^2$, the equivariant form $\hat{\omega}^2$ on $\hat{H}^2$ can be extended to a fundamental form $\hat{\bar{\omega}}^2$ on $\hat{\bar{H}}^2$; so a global section $\hat{H} \to \hat{\bar{H}}^2$ which is a principal morphism defines a form on $\hat{H}$ with values in $\mathbf{R}^p \oplus l_p$; this form splits into the fundamental form with value in $\mathbf{R}^p$ and a connection form with values in l_p; as $\hat{\bar{\omega}}^2$ vanishes on $(T\pi^2)^{-1}E$, we can deduce that the connection form vanishes on the tangent bundle E^1 to the foliation $\mathcal{F}^1$ on $\hat{H}$, i.e. the connection lifts E onto the canonical lift E^1 defined in §2; the connection is *adapted* in the sense of P. Molino (as the restriction of the connection to E is flat, we have a principal E^*-foliated bundle in the sense of Kamber-Tondeur). See also Bott [2].

Conversely, a principal connection C on $\hat{H}$ defines a lifting $\hat{H} \to J_1\hat{H}$; so for every $h \in \hat{H}$, C defines a linear injective mapping C_h from $T_{\pi(h)}M$ to $T_h(\hat{H})$; if C is adapted, C_h maps $E_{\pi(h)}$ onto $E_h^1 \subset (T\pi)^{-1}E$; it induces a bijective mapping $Q_{\pi(h)}(M) \to Q_h(\hat{H})$, i.e. an element of $\hat{J}_1\hat{H}$; from Proposition 3.3 we get a lifting $\hat{H} \to \hat{\bar{H}}^2$ which is principal because C is principal. We have:

PROPOSITION 3.4. (cf. [11]). There is a bijective correspondence between the set of global sections $\hat{H}(M, \mathcal{F}) \to \hat{\bar{H}}^2(M, \mathcal{F})$ which are principal morphisms and the set of adapted connections on $\hat{H}(M, \mathcal{F})$.

Remarks. (1) The basic connections of Bott correspond to sections $\hat{H}(M, \mathcal{F}) \to \hat{H}^2(M, \mathcal{F})$; so the torsion of an adapted connection can be interpreted as the obstruction to a lifting of $\hat{H}$ in the bundle of holonomic 2-frames.

(2) We could also define *higher order connections*; a principal lifting $s : \hat{H}^k(M, \mathcal{F}) \to \hat{H}^{k+1}(M, \mathcal{F})$ defines on $\hat{H}^k$ an equivariant form $s^*\omega^k$ with values in $\mathbf{R}^p \oplus l_p^k$; by projection on l_p^k, we get a connection form. We could also define semi-holonomic connections.

Suppose the manifold M admits a *transverse projectable connection* in the sense of P. Molino [17], i.e. an adapted connection on $\hat{H}^1(M)$ which is locally the inverse image of a connection on the bundle of 1-frames of a local transverse manifold (P. Molino has proved that the obstruction to the existence of such a connection is a cohomology class called the 'Atiyah class'); the covariant derivative on the local transverse manifolds lifts to a *transverse covariant derivative* ∇ for transverse tensor fields; (if we use distinguished coordinates, the transverse derivative depends only on $x^1, \ldots, x^p$); by iteration of ∇, we get a higher order connection on M (this process was used by Vaisman [21]).

(3) In the case of a non integrable E, a lifting $\hat{H} \to \bar{\hat{H}}^2$ does not define a connection on $\hat{H}$; from Proposition 4.3, we deduce, for any complement S to E, a lifting of S but we have no canonical lifting of E; we have what could be called a semi-connection (this is a 'partial' connection in the sense of Kamber-Tondeur but we will use this word in a different meaning).

(4) Though in the non integrable case, there is no holonomic prolongation of $\hat{H}(M, E)$, if we reduce the group L_p to a subgroup G, the new structure may be transitive; this is the case of the quadric $Q_6 = G'/S\mathring{U}_3$ (cf. §1); the groupoïd $J^1(\Gamma')$ (where Γ' is the pseudo-group of local automorphisms of the $S\mathring{U}_3$-structure) is transitive; the group $S\mathring{U}_3$ being of finite type, $J^2(\Gamma')$ is isomorphic to $J^1(\Gamma)$. Notice that the principal $S\mathring{U}_3$-frame bundle being isomorphic to the group G_2', $J^1(\Gamma')$ is isomorphic to the trivial groupoïd $\bar{G}_2' \times Q_6$ (cf. [11]).

4. Partial connections (cf. [12])

Let $(\mathcal{P}, M, \Pi)$ be a fibered manifold (i.e. $\mathcal{P}, M$ are manifolds and Π is a surjective submersion $\mathcal{P} \to M$); let (P, M, Π) be a fibered sub-manifold (P is a submanifold of $\mathcal{P}$ such that $\Pi' = \Pi | P$ is also a surjective submersion). The dimensions of $M, \mathcal{P}, P$ will be denoted by n, $n + r$, $n + r'$ (we suppose $r' < r$).

Even if the fibered manifold is not a fiber space, we define a connection on $(\mathcal{P}, M, \Pi)$ as a lifting $\mathcal{P} \to J_1\mathcal{P}$ (space of 1-jets of sections).

A *partial connection* (related to the pair $(\mathfrak{P}, P)$) is a lifting:

$$C: P \to i^* J_1 \mathcal{P}$$

(where i is the injective map $P \to \mathcal{P}$). So for every $y \in P$, $C(y)$ can be considered as an injective linear mapping from $T_{\pi'(y)}M$ to $T_y\mathcal{P}$; the image $\mathcal{H}_y$ of $T_{\pi'(y)}M$ by $C(y)$ will be the 'horizontal space'.

The partial connection C will be said to be *adapted, transversal, strictly transversal, non adapted* at y if we have respectively:

(1) $\mathcal{H}_y \subset T_yP$ (C is a connection on P),

(2) $\mathcal{H}_y + T_yP = T_y\mathcal{P}$ (then $n \geqslant r - r'$),

(3) $\mathcal{H}_y \oplus T_yP = T_y\mathcal{P}$ (then $n = r - r'$),

(4) $\mathcal{H}_y \cap T_yP = 0$ (then $n \leqslant r - r'$);

if these properties occur respectively at each point, C will be said adapted, transversal, strictly transversal, non adapted.

We introduce the *deviation* $\mathcal{D}$ of C, such that C is adapted if and only if $\mathcal{D}$ is 0; $\mathcal{D}$ is the mapping

$$P \to L_p(\pi'^* TM, N(P))$$

defined by $\mathcal{D}(y) = q(y) \circ C(y)$ where q is the projection of i^*TP on the normal bundle $N(P) = i^*T\mathcal{P}/TP$.

At y, C is transversal, strictly transversal, non adapted if $\mathcal{D}(y)$ is surjective, bijective, injective.

A partial connection may not be the restriction of a connection on $\mathcal{P}$, but if $\mathcal{P}$ is a principal bundle and P a principal subbundle, *any principal partial connection on P can be extended to a principal connection* on $\mathcal{P}$.

A principal partial connection can be defined by a form ω on P, with values in the Lie algebra $T_e\mathcal{G}$ of the structure group $\mathcal{G}$ of $\mathcal{P}$, this form being the restriction of a connection form Ω on $\mathcal{P}$.

Conversely, given a G-principal bundle P and a Lie group $\mathcal{G}$ such that G is a closed subgroup of $\mathcal{G}$, a form on P with values in $T_e\mathcal{G}$ such that $R_g^*\omega = Adg^{-1}\omega$ for any right translation R_g (with $g \in G$) is a partial connection form, P being the subbundle of the bundle $\mathcal{P} = P \times_M \mathcal{G}$ (extension of P).

Given a principal partial connection C on P, for every $y \in P$, the kernel $\operatorname{Ker} \omega_y = \mathcal{H}_y \cap T_yP$ is also the kernel of the projection of $\mathcal{H}_y$ on the normal bundle $N(P)$; so if $s_y =$ dimension of $\operatorname{Ker} \omega_y$, we have:

$$\operatorname{rank} \omega_y = n + r' - s_y$$

$$\operatorname{rank} \mathcal{D}(y) = n - s_y;$$

ω_y is surjective if and only if $\mathcal{D}(y)$ is surjective; the 'horizontal' field $\mathcal{H}$ being invariant under right translations, s_y is constant along each fiber of P. We will say that *P is in regular position with respect of the principal connection Ω* on $\mathcal{P}$ (or that the partial connection ω is *regular*) if s_y is constant on P; this is the case, for instance, of adapted and strictly transversal partial connections.

From the previous study we get:

PROPOSITION 4.1. A regular principal partial connection defines on the principal bundle P a vector subbundle E_p of the tangent bundle TP; E_p being invariant under right translations of the structure group G can be extended to a subbundle $E_{\mathcal{P}}$ of $T\mathcal{P}$ and projects on a vector subbundle E of TM; the three vector subbundles E_p, $E_{\mathcal{P}}$, E have the same rank s; the codimension of E is equal to the rank of the deviation.

When the partial connection is adapted, we get the horizontal field on P and $E = TM$; when $\mathcal{H}_y \cap T_yP = 0$, then the rank of E_p, $E_{\mathcal{P}}$, E is 0; in particular we have shown [12] that *strictly transversal principal partial connections are Cartan connections* in the sense of C. Ehresmann [4]; non adapted partial connections correspond to weak Cartan connections [12].

PROPOSITION 4.2. For a regular transversal partial principal connection, the normal bundle $Q(E_p) = TP/E_p$ admits a trivialization.

Proof. As for every $y \in P$, the restriction ω_y of the partial connection form is surjective, ω defines an isomorphism $Q(E_p) \to P \times T_e\mathcal{G}$. In particular for strictly transversal partial connections, we recover the parallelism which is associated to a Cartan connection.

PROPOSITION 4.3. If a principal partial connection defines a subbundle E_p of TP which is integrable, then $E_{\mathcal{P}}$ and E are also integrable and P is a foliated bundle.

Proof. The translations of the structure group $\mathcal{G}$ lift the integral manifolds of $E_{\mathcal{P}}$ to integral manifolds of E.

As the restriction of $T\Pi' : TP \to TM$ to a horizontal subspace $\mathcal{H}_y$ is bijective, the restriction of Π' to the leaves of the foliation $\mathcal{F}_p$ is an immersion; the image of these leaves are integral manifolds of E.

Example 1. Cartan connections.

Example 2. Consider a *G-structured Pfaffian system in the sense of P. Molino*, i.e. a principal G-bundle P (with $G \subset L_m$) endowed with a fundamental form θ of rank p with values in R^m. For any connection form $\bar{\omega}$ on P, $\bar{\omega} + \theta$ is a regular partial principal connection. If $m = p$ (*quotient G-structures* in the sense of P. Molino), then the partial connexion is transversal [16].

Example 3. In particular if E is a subbundle of TM, the bundle $\hat{H}(M, E)$ of transverse 1-frames or any principal subbundle of $\hat{H}$ is a quotient G-structure. In this case Proposition 4.2 has been proved by P. Molino [17].

Example 3'. If E is an integrable subbundle of TM which defines a foliation $\mathcal{F}$, then for any k, the bundle $\hat{H}^k(M, \mathcal{F})$ of transverse k-frames can be endowed with a transversal principal partial connection: we have seen (§3) that a principal lifting s: $\hat{H}^k \to \hat{H}^{k+1}$ defines on $\hat{H}^k$ an equivariant form $s^*\hat{\omega}^k$ with values in $\mathrm{R}^p \oplus l_p^k$; as L_p^k acts on R^p, we define the group $\mathcal{G}^k$, semi-direct product of R^p and L_p^k; $\mathcal{G}^k$ admits $\mathrm{R}^p \oplus l_p^k$ as Lie algebra; by a previous remark, we get a partial connection, and the kernel E_p of the partial connection form is exactly the lift of E as defined in §2.

We could have replaced $\hat{H}^k$ by a subbundle.

In all the previous examples, the partial connection is regular; in Examples 1, 3, 3', it is transversal (hence the normal bundle $Q(E_p)$ is trivial).

Example 4. Let on manifold a G-structure in the ordinary sense, i.e. a G-subbundle H_G of the bundle $H(M)$ of 1-frames (where G is a Lie subgroup of L_n). Any torsionless connection on $H(M)$ defines a partial connection on H_G which may not be regular; here we are going back to the problem (cf. §1) of finding submanifolds of M on which the G-structure is integrable.

Besides the deviation, we have defined the pseudo-torsion [12] of any partial connection (relative to the pair $P, \mathcal{P}$); it is a map $\tau: P \to L_p(\Pi'^*TM, L^*VT\mathcal{P})/L_p(\Pi'^*TM, VTP)$ (where $VT\mathcal{P}$ and VTP are the kernels of $T\Pi: T\mathcal{P} \to TM$ and $T\Pi': TP \to TM$), defined in the following way: for any $y \in H_G$ and any $z \in (J_1P)_y$, $C(y) - z$ is an element of $L(T_{\Pi'(y)}M, VT\mathcal{P})$; its class in $L(T_{\Pi'(y)}M, VT\mathcal{P})/L(T_{\Pi'(y)}M, VTP)$ is independent of the choice of z;

the partial connection is adapted if and only if $\tau = 0$. In the case $\mathcal{P} = H$ and $P = H_G$, as VTH and VTH_G are diffeomorphic to $H \times (\mathbf{R}^n \otimes \mathbf{R}^{n^*})$ and $H \times \mathfrak{g}$ (where $\mathfrak{g}$ is the Lie algebra of G), we get a reduced pseudotorsion $\tau' : H_G \to \mathbf{R}^n \otimes (\otimes^2 \mathbf{R}^{n^*})/\mathfrak{g} \otimes \mathbf{R}^{n^*}$. If we consider the set of all torsionless connections on H, we get the *structure tensor* of H_G in the sense of C. Ehresmann; it is a map $H_G \to \mathbf{R}^n \otimes (\bigcirc^2 \mathbf{R}^{n^*}) \backslash \mathbf{R}^n \otimes (\otimes^2 \mathbf{R}^{n^*})/\mathfrak{g} \otimes \mathbf{R}^{n^*}$ (where $\bigcirc^2 \mathbf{R}^{n^*}$ means the symmetrical power of $\mathbf{R}^{n^*}$); if this 'tensor' is 'null', then H_G admits torsionless connections.

We could define in the same way higher order structure tensors.

5. On 'transverse parallelism'

Let M be a n-manifold with a subbundle E of TM of rank $n-p$; let Γ be the pseudogroup of local automorphisms of the structure defined by E. If the normal bundle $Q = TM/E$ admits a trivialization, this trivialization will be called transverse parallelism.

We have seen in §4 that transversal partial principal connections give rise to transverse parallelism.

We may extend to transverse parallelism notions concerning parallelism ([11] and [12]).

As a transverse parallelism is a diffeomorphism $\varphi : M \times \mathbf{R}^p \to Q$ such that φ_x (defined by $\varphi_x(u) = \varphi(x, u)$) is an isomorphism onto Q_x, we deduce a functor $F : M \times M \to \hat{\Phi}$, where $\hat{\Phi}$ is the groupoïd associated to the bundle of transverse 1-frames; $F(x, x') = \varphi_{x'} \circ \varphi_x^{-1}$. A '*transverse translation*' is a map $g \in \Gamma$ such that for any x in the source U of g, we have $\hat{j}_x^1 g = F(x, g(x))$. If we introduce the *reduced transverse differential* $\hat{D}_{\text{red}} g : U \to \text{Isom}(\mathbf{R}^p)$ defined by $\hat{D}_{\text{red}} g(x) = \varphi_{g(x)}^{-1} \circ \hat{j}_x^1 g \circ \varphi_x$, then $g \in \Gamma$ is a transverse translation if and only if $\hat{D}_{\text{red}} g = i\,\mathrm{d}_{\mathbf{R}^p}$.

A *transverse parallelism* is said to be *integrable* if for any pair (x, x'), there exists a transverse translation g such that $g(x) = x'$.

Examples of integrable transverse parallelisms

(1) Let $\bar{\omega} : M \to V$ be a fibered manifold such that V admits an integral parallelism (i.e. for any pair $(u, u') \in V \times V$, there exists a translation t such that $t(u) = u'$); the normal bundle to the simple foliation defined by the fibers of $\bar{\omega}$ can be identified to $\Pi^* TV$ so we have a diffeomorphism $\varphi : M \times \mathbf{R}^p \to M \times_V TV (= \Pi^* TV)$; if the paral-

lelism on V is defined by $\psi : V \times \mathbb{R}^p \to TV$, we have $\varphi(x, y) = (x, \psi_u y)$ where $u = \bar{\omega}(x)$.

(2) Let V be endowed with an integrable parallelism; let γ be the pseudogroup of local automorphisms of the parallelism; a family of distinguished submersions from M to V compatible with γ (γ-foliation in the sense of Haefliger) defines a foliation on M which admits an integrable transverse foliation.

Let V be a manifold on which we choose an affine connection c; let γ be the pseudogroup of local automorphisms of this connection; if a manifold M admits a γ-foliation $\mathcal{F}$, the distinguished submersions induce local connections on their sources which agree on the intersection of the sources; so we get a global connection $\mathcal{C}$ on $\hat{H}(M)$ which is a transverse projectable connection in the sense of P. Molino [17]; for instance if V is endowed with a Riemann structure, we take for c the Levi-Civita connection; we get a *foliated structure with transverse riemannian metric* in the sense of C. Ehresmann [5] or *bundle-like metric* in the sense of Reinhart-Pasternak [20], [19]. The transverse parallelism on $\hat{H}(M)$ induced by $\mathcal{C}$ is integrable if and only if the parallelism on $H(V)$ induced by c is integrable; we have the same results replacing $H(V)$ and $\hat{H}(M)$ by subbundles. We have proved in [12] that a connection c on $H(V)$ induces a connection c_G on a subbundle H_G such that H_G admits an integrable parallelism if and only if the curvature and torsion of c have null covariant derivatives; then the manifold V is *locally reductive* in the sense of A. Lichnerowicz (i.e. locally equivalent to a structure of reductive homogenous space) [14]; the corresponding structure on M could be called *transverse locally reductive structure*; it is a particular case of 'foliated structure with a locally homogeneous structure' in the sense of C. Ehresmann. We deduce

PROPOSITION 5.1. A foliated manifold $(M, \mathcal{F})$ admits a transverse locally reductive structure if and only if it admits a transverse projectable connection such that the curvature and the torsion have a null transverse covariant derivative.

The model of this structure is a fibered manifold $\bar{\omega} : M \to V$ with V a homogeneous reductive space.

Remark. In Example 3′ of transversal principal partial connections, the transverse parallelism has the following property: the normal bundle $Q(E_p)$ to the 'horizontal' foliation on $\hat{H}^k$ splits into the direct sum of trivial bundles $VT\hat{H}^k \oplus Q((T\Pi^k)^{-1}E)$, where $VT\hat{H}^k$ is the vertical bundle tangent to $\hat{H}^k$ and $Q((T\Pi^k)^{-1}E)$ is the normal bundle to the subbundle $(T\Pi^k)^{-1}E$; the last bundle (which is isomorphic to $\hat{H}^k \times R^p$) is also isomorphic to the inverse image of E^* in the bundle $\hat{H}^{*k}$ of k-transverse coframes; for $k = 1$, we recover a result of J. Martinet [15]. For $k = 1$, vector fields on $\hat{H}(M)$ which are projectable on $Q(E_p)$ and correspond by the transverse parallelism to a fixed element of R^p are *basic* vector fields [17]; the projections on M of the trajectories of the basic vector fields could be called *transverse geodesics*. In the case of a transverse projectable connection, these transverse geodesics are inverse images of geodesics.

REFERENCES

[1] Bernard, D., *Thèse. Annales Inst. Fourier* **10** (1950), 151–170.

[2] Bott, R., 'Lectures on Characteristic Classes and Foliations', *Lectures Notes in Math* **279** (1972), 1–94, Springer.

[3] Cartan, E., (a) 'Systèmes de Pfaff à cinq variables', *Ann. Sc. Ecole Normale* (1910). Oeuvres Complètes, p. 927; (b) 'Equivalence absolue de certains systèmes différentiels', *Bull. Soc. Math. France* (1914). Oeuvres Complètes, p. 1133; (c) 'Systèmes indéterminés d'équations différentielles', *Jour. für Math.* (1915). Oeuvres Complètes, p. 1169.

[3′] Cenkl, B., Colloque Exotique, Lille, 1973.

[4] Ehresmann, C., 'Les connexions infinitésimales', *Colloq. Topologie Bruxelles 1950.* Liège (1951), 29–55.

[5] Ehresmann, C., (a) 'Veriétés feuilletées', *Rendiconti di Math.*, Serie V, vol. X, Fax. 1–2, Roma, 1951; (b) 'Structures feuilletées', *Proceedings of the Fifth Canadian Mathematical Congress*, pp. 109–172. Montréal, 1961.

[6] Guillemin, V., and Sternberg, S., 'Deformation Theory of Pseudogroup Structures', *Memoirs Amer. Math. Society* **64** (1966).

[7] Haefliger, A., (a) 'Structures feuilletées et cohomologie à valeurs dans un faisceau de groupoïdes', *Comm. Math. Helv.* **32** (1958) 4, 248–329; (b) 'Sur les classes caractéristiques des feuilletages', Séminaire Bourbaki, exposé 412 (1972).

[8] Kamber, F., and Tondeur, P., 'Characteristic invariants of foliated bundles', (Mimeographed) Colloq. Exotique, Lille, 1973.

[9] Libermann, P., These, Strasbourg, 1953. 'Sur le probleme d'equivalence de certaines structures infinitésimales régulières', *Anali Mat. Pura Applic.* **36** (1954), 27–120.

[10] Libermann, P., 'Pseudogroupes infinitésimaux', *Colloq. Intern. CNRS*, Lille, 1958; *Bull. Sc. Math. France* **87** (1959), 409–425.

[11] Libermann, P., (a) 'Sur les prolongements des fibrés principaux', Séminaire analyse globale Montréal, 1969); (b) 'Groupoïdes differentiables et presque parallélisme', *Symposia Math. Ist. Naz. Alta Mate. Roma* **10** (1972), 59–93.

[12] Libermann, P., 'Parallélismes', *Journ. of Diff. Geometry* **8**(4), (Dec. 1973).

[13] Libermann, P., 'Sur quelques propriétés des systèmes de Pfaff'. In honor Prof. Bompiani; to appear in *Bolletino U.M.I.*

[14] Lichnerowicz, A., *Géométrie des groupes de transformations*, Dunod, Paris, 1958.

[15] Martinet, J., 'Classes caractéristiques des systèmes de Pfaff', Colloque exotique, Lille, 1973. *Coll. Géom. Diff. Santiago de Compostela, Lecture Notes* **392** (1974), 30–37.

[16] Molino, P., (a) 'Connexions et G-structures sur les variétés feuilletées', *Bull. Sciences Math.* **92** (1968), 59–63; (b) 'Intégrabilité et presque intégrabilité des G-structures généralisées', *Bull. Sciences Math.* **92** (1968), 97–109.

[17] Molino, P., (a) 'Feuilletages et classes caractéristiques', *Symposia Math. t. Naz. Alta Mate. Roma* **X** (1972), 199–209; (b) 'Feuilletages à connexion transverse projetable', *Topology* **12**, (1973) 317–325; (c) 'La classe d'Atiyah d'un feuilletage', *C.R. Acad. Sc. Paris* **278**, Série A (4 Mars 1974), 719–721.

[18] Molino, P., (a) 'Sur la géométrie transverse des feuilletages', Mimeographed; (b) *C.R. Acad. Sc.* **281**, 203, Août 1975.

[19] Pasternak, J., 'Foliations and Compact Lie Group Actions', *Comment. Math. Helvet.* **46** (1971), 467–477.

[20] Reinhart, B., 'Foliated Manifolds with Bundle-like Metrics', *Ann. Math.* **69** (1959), 119–132.

[21] Vaismann, I., 'Sur l'existence des opérateurs différentiels feuilletés à symbole donné, *C.R. Acad Sc. Paris* **276**, Série A (25 Avril 1973), 1165–1168.

Added in proof. As B. Cenkl pointed out to me, there is another paper dealing with derived systems:

Tanaka, N., 'On Differential Systems, Graded Lie Algebras and Pseudogroups', *J. Math. Kyoto Univ.* **10** (1970), 1–82.

PIERRE MOLINO

THE CURVATURE CLASS OF AN ALMOST-COMPLEX MANIFOLD

The purpose of this paper is to investigate a natural generalization of the $\bar{\partial}$-cohomology of an analytic complex manifold in the case of an almost-complex manifold. The structures are assumed to be C^∞.

Let M be an almost-complex manifold. If α is a form of type (p, q), we denote by α_T its 'class modulo torsion'. Then the class modulo torsion $\bar{\partial}_T\alpha_T$ of the $(p, q+1)$-component of $\mathrm{d}\alpha$ depends only on α_T, and we obtain an operator:

$$\bar{\partial}_T : \bigwedge_T^{p,q}(M) \to \bigwedge_T^{p,q+1}(M).$$

In the case $p = 0$, we obtain a complex, whose cohomology is analogous to the cohomology 'along a distribution' of a real manifold, as defined in [4].

If $p \neq 0$, we do not generally have $\bar{\partial}_T \circ \bar{\partial}_T = 0$. But this property is true for a particular class of almost-complex manifolds: let us consider a connection ω in the complex tangent bundle of M. ω is *basic* if its torsion has no components of type $(1, 1)$. When ω is basic, the class modulo torsion Ω_T, of the $(0, 2)$-component of its curvature is independent of ω and will be referred to as the *curvature class of M*. We can now verify that if $\Omega_T = 0$, $\bar{\partial}_T \circ \bar{\partial}_T = 0$, and we obtain in this way a '*$\bar{\partial}$-cohomology modulo torsion*' *for almost-complex manifolds without curvature class.*

Moreover, if $\Omega_T = 0$, we can define as in [1] an *ATIYAH class modulo torsion*, $a_T(M)$. This class is the cohomological obstruction to the existence of a basic connection ω which is also *curvature basic*, that is to say that its curvature has no components of type $(1, 1)$.

Finally we observe that an analogous study can be made for real or complex almost-product structures. It would be interesting to obtain the relation between these notions and various well-known generalizations of hermitian or kählerian geometry [see for example [5–7]].

Cahen & Flato (eds.), Differential Geometry and Relativity, 127–134.

1. THE COHOMOLOGY $H_T^{0,*}(M)$ FOR AN ALMOST-COMPLEX MANIFOLD

1.1. *Let M be an almost-complex manifold of (real) dimension* 2*n*. If $\theta^1, \ldots, \theta^n$ are complex valued 1-forms which locally define the almost-complex structure on *M*, we have:

$$d\theta^i = \sum_j A_j^i \wedge \theta^j + T^i, \tag{1}$$

where $T^1, \ldots, T^n$ are (0, 2)-forms. These forms are the (local) components of the torsion form *T*.

We denote by I_T the ideal of complex valued forms on *M* generated locally by $T^1, \ldots, T^n$. If $I_T^{p,q} = I_T \cap \wedge^{p,q}$, the space $\wedge_T^{p,q} = \wedge^{p,q}/I_T^{p,q}$ is the space of (p, q)-*forms modulo torsion.* If $\alpha \in \wedge^{p,q}(M)$, the corresponding class modulo torsion will be denoted by α_T. We easily have:

LEMMA 1. The class modulo torsion of the $(p, q+1)$-component of $d\alpha$ only depends on α_T. This class will be denoted by $\bar{\partial}_T\alpha_T$.

By this lemma, we obtain a map:

$$\bar{\partial}_T : \bigwedge_T^{p,q}(M) \to \bigwedge_T^{p,q+1}(M). \tag{2}$$

In the particular case of a complex analytic manifold ($I_T = 0$) we obtain the classical $\bar{\partial}$-operator. The main purpose of this paper is to study the $\bar{\partial}_T$-operator.

1.2. *Let us now consider the ideal* $\mathcal{I}_T$ *of complex valued form on M generated locally by* $\theta^1, \ldots, \theta^n$, $T^1, \ldots, T^n$. If $\wedge_C^q(M)$ is the space of complex valued *q*-forms on *M*, and if $\mathcal{I}_T^q = \mathcal{I}_T \cap \wedge_C^q$, we observe first that:

$$\bigwedge_T^{0,q}(M) = \bigwedge_C^q(M)/\mathcal{I}_T^q. \tag{3}$$

Hence, in the case $p = 0$, the $\bar{\partial}_T$-operator is obtained from the usual differential d by passing to the quotient by the ideal $\mathcal{I}_T$. The situation is analogous to the case in [4] where a cohomology 'along a distribution' of a (real) manifold is introduced by passing to the quotient by the characteristic ideal of the distribution.

In particular, in this case, we have easily $\bar{\partial}_T \circ \bar{\partial}_T = 0$, and we obtain a cohomology:

$$H_T^{0,*}(M) = \bigoplus_{q=1}^{n} H_T^{0,q}(M). \tag{4}$$

$H_T^{0,q}$ is the space of *$\bar{\partial}$-cohomology modulo torsion of type* $(0, q)$.

This construction is functorial with respect to the morphisms of almost-complex manifolds.

1.3. *Let $E(M, G)$ be a principal bundle on M* with G as complex structural group.

If α is a tensorial q-form on E of type $(0, q)$ with values in a complex vector space W, we define as for scalar forms its class modulo torsion α_T. ω being a connection form on E, we denote by ω_T the partial connection defined by its class modulo tensorial 1-forms of type $(1, 0)$. Then the class $\bar{\nabla}_T\alpha_T$ modulo torsion of the $(0, q+1)$-component of the absolute differential $\nabla\alpha$ depends only on ω_T and α_T. The curvature Ω_T of ω_T is by definition the class modulo torsion of the $(0, 2)$-component of the curvature of ω. ω_T is *flat* if $\Omega_T = 0$.

DEFINITION 1. E is adapted to the almost-complex structure if it is equipped with a flat partial connection ω_T.

We can verify that ω_T is flat if and only if it can be extended to a connection ω on E whose curvature has no components of type $(0, 2)$.

When E is an adapted principal bundle, with ω_T as flat partial connection, the operator $\bar{\nabla}_T$ defines a $\bar{\nabla}_T$-cohomology on the spaces $\wedge_T^{0,q}(M, E_W)$ of tensorial forms on E of type $(0, q)$ with values in W (modulo torsion). The corresponding cohomology spaces are denoted by $H_T^{0,q}(M, E_W)$.

2. The curvature class $C(M)$

2.1. *We consider now the $GL(n, \mathbf{C})$-principal bundle $B_0(M)$* of complex frames of the amost complex manifold M. B_0 is endowed with a fundamental form θ_0, that is to say a tensorial 1-form with values in $\mathbf{C}^n$ defined by the usual method.

If ω is a connection on B_0, its torsion Σ is the absolute differential $\nabla\theta_0$. We say that ω *is basic* if its torsion has no components of type $(1, 1)$. In this case we can write:

$$\Sigma = \Sigma_{2,0} + T, \tag{5}$$

where T is the torsion of the almost-complex structure.

For basic connections, the associated partial connection ω_T (modulo 1-forms of type $(1,0)$) is obviously independent of the connection.

DEFINITION 2. The curvature class $C(M)$ of the almost-complex manifold M is the curvature Ω_T of the partial connection ω_T.

Hence, we have: $C(M) \in \wedge_T^{0,2}(M, L(B_0))$, where $L(B_0) = B_{0gl(n,\mathbf{C})}$ is the Lie algebra bundle associated with B_0 with $gl(n, \mathbf{C})$ as typical fibre.

If $C(M) = 0$, we say that M is an almost-complex manifold without curvature class. In this case, B_0 equipped with the basic partial connection ω_T is an adapted principal bundle in the sense of Definition 1.

As indicated in Section 1.3, $\Omega_T = 0$ if and only if the basic partial connection ω_T can be extended to a connection ω on B_0, whose curvature has no $(0,2)$-components. Such an extension is a *basic connection*, hence we can state:

PROPOSITION 1. $C(M) = 0$ if and only if M admits a basic connection ω, whose curvature is of the following type:

$$\Omega = \Omega_{20} + \Omega_{11}. \tag{6}$$

Such a connection will be referred to as a *basic connection of the first special type*:

2.2. *As examples of manifolds without curvature class* we shall give here only the following two trivial cases:

(a) If M is a *complex analytic manifold*, then $C(M) = 0$. Moreover in this case all the basic connections ($\Sigma = \Sigma_{2,0}$) are of the first type.

(b) If $n = 2$ and if *the torsion* T *is everywhere* $\neq 0$, then $C(M) = 0$: in this case the ideal $I_T^{0,2}$ coincides with $\wedge^{0,2}$.

3. THE $\bar{\partial}_T$-COHOMOLOGY FOR ALMOST COMPLEX MANIFOLDS WITHOUT CURVATURE CLASS

3.1. *We assume now that* $C(M)=0$.

PROPOSITION 2. We have for an arbitrary p:

$$\bar{\partial}_T \circ \bar{\partial}_T = 0 \tag{7}$$

where $\bar{\partial}_T$ is the map defined in Section 1.1.

Proof. As ω is a basic connection of the first type, we obtain (from (5)):

$$\Omega \wedge \theta_0 = \nabla \Sigma_{20} + \nabla \mathrm{T},$$

and by taking the (1, 2) components:

$$0 = \nabla_{1,2}\Sigma_{20} + \nabla_{1,2}T,$$

but the class modulo torsion of $\nabla_{1,2}\Sigma_{20}$ is obviously zero. Hence the class modulo torsion of $\nabla_{1,2}T$ is zero. If $\mathrm{d}_{1,2}T$ is the (1, 2)-component of $\mathrm{d}T$, its class modulo torsion is a well defined class $[\mathrm{d}_{1,2}T]_T \in \wedge_T^{1,2}(M, T(M))$, and we have:

$$[\mathrm{d}_{1,2}T]_T = 0. \tag{8}$$

If α is a (p, q)-form on M,

$$\mathrm{d}\alpha = \mathrm{d}_{p-1,q+2}\alpha + \mathrm{d}_{p,q+1}\alpha + \mathrm{d}_{p+1,q}\alpha + \mathrm{d}_{p+2,q-1}\alpha.$$

The $(p, q+2)$-component of $\mathrm{d}^2\alpha = 0$ is given by:

$$\mathrm{d}_{p,q+2}(\mathrm{d}_{p-1,q+2}\alpha) + \mathrm{d}_{p,q+2}(\mathrm{d}_{p,q+1}\alpha) + \mathrm{d}_{p,q+2}(\mathrm{d}_{p+1,q}\alpha) = 0$$

modulo torsion this component gives:

$$\bar{\partial}_T \circ \bar{\partial}_T(\alpha_T) + [\mathrm{d}_{p,q+2}(\mathrm{d}_{p-1,q+2}\alpha)]_T = 0$$

But $\mathrm{d}_{p-1,q+2}\alpha = \Sigma_i (\beta_{p-1,q})_i \wedge T^i$, hence

$$\mathrm{d}_{p,q+2}(\mathrm{d}_{p-1,q+2}\alpha) = \sum_i (\mathrm{d}_{p,q}\beta_i) \wedge T_i + \sum_i \beta_i \wedge \mathrm{d}_{1,2}T^i$$

and by (8) we obtain (7). Q.E.D.

Now, ∂_T will define a $\bar{\partial}$-*cohomology modulo torsion* ($\bar{\partial}_T$-cohomology). The corresponding cohomology spaces are denoted by:

$$H_T^{*,*}(M) = \bigoplus_{p,q} H_T^{p,q}(M). \tag{9}$$

If $p = 0$ we obtain the cohomology introduced in Section 1.1. If $T = 0$ this cohomology is the usual $\bar{\partial}$-cohomology.

3.2. *Let $E(M, G)$ be a complex principal bundle as in Section* 1.3.

We assume that E is adapted to the almost complex structure of M. The corresponding partial flat connection is denoted by ω_T.

If α is a tensorial form on E of type (p, q) with values in W, we define as in Section 1.3 the class modulo torsion α_T of α, and the $(p, q+1)$-form modulo-torsion $\bar{\nabla}_T\alpha_T$.

Then as for scalar forms, we have:

$$\bar{\nabla}_T \circ \bar{\nabla}_T = 0, \tag{10}$$

and we can define a $\bar{\nabla}_T$-cohomology for the tensorial forms (modulo torsion) of an arbitrary type on E. Let $H^{p,q}(M, E_W)$ be the corresponding cohomology spaces.

4. The ATIYAH class modulo torsion of an adapted principal bundle

We always assume that $C(M) = 0$.

4.1. *Let $E(M, G)$ be as in Section* 3.3.

ω_T is the partial flat connection given on E. A connection ω on E is adapted if it extends the partial connection ω_T. If ω and ω' are adapted, $\omega' - \omega = \pi$ is a tensorial form of type $(1, 0)$.

ω is *adapted of the first special type* if its curvature Ω has no components of type $(0, 2)$. By Section 1.3, such a connection always exists. Then:

$$\Omega = \Omega_{2,0} + \Omega_{1,1}.$$

Hence, by taking the (1,2)-components of the absolute differentials we obtain:

$$0 = \nabla_{1,2}\Omega_{2,0} + \nabla_{1,2}\Omega_{1,1}.$$

But the class modulo torsion of $\nabla_{1,2}\Omega_{2,0}$ is zero. Hence:

$$[\nabla_{1,2}\Omega_{1,1}]_T = 0,$$

and by definition, this is equivalent to

$$\bar{\nabla}_T[\Omega_{11}]_T = 0. \tag{11}$$

Hence, the class modulo torsion of Ω_{11} is a $\bar{\nabla}_T$-cocycle in $\wedge_T^{1,1}(M, L(E))$, where $L(E)$ is the Lie algebra bundle associated with E.

PROPOSITION 3. The cohomology class $a_T(E)$ defined by $[\Omega_{1,1}]_T$ in $H_T^{1,1}(M, L(E))$ is independent of the connection ω. $a_T(E)$ is by definition the *ATIYAH class modulo torsion* of E.

Proof. If $\omega' = \omega + \pi$ where π is of type $(1, 0)$ and ω' is adapted of the first special type, then:

$$\Omega' = \Omega + \nabla\pi + \tfrac{1}{2}[\pi, \pi],$$
$$\Omega'_{1,1} = \Omega_{1,1} + \nabla_{1,1}\pi,$$

and by passing to classes modulo torsion:

$$[\Omega'_{1,1}]_T = [\Omega_{1,1}]_T + \bar{\nabla}_T\pi. \qquad \text{Q.E.D.}$$

4.2. *The ATIYAH class modulo torsion as an obstruction.*

An adapted connection ω on E is *adapted of the second special type* if its curvature Ω has no components of type $(1, 1)$.

THEOREM: E admits an adapted connection of the second special type if and only if its ATIYAH class modulo torsion vanishes.

Proof. If ω' is adapted from the second special type and ω adapted from the first special type, $\omega' - \omega = \pi$. Then:

$$\Omega' = \Omega + \nabla\pi + \tfrac{1}{2}[\pi, \pi] \tag{12}$$
$$\Omega'_{1,1} = 0 = \Omega_{1,1} + \nabla_{1,1}\pi,$$

hence $[\Omega_{1,1}]_T = \bar{\nabla}_T\pi$ and $a_T(E) = 0$.

Conversely, if $a_T(E) = 0$ and Ω is the curvature of an adapted connection of the first special type,

$$[\Omega_{1,1}]_T = \bar{\nabla}_T\pi \qquad \text{where} \quad \pi \in \overset{1,0}{\bigwedge}(M, L(E)).$$

If $\omega' = \omega - \pi$, its curvature Ω' verifies

$$\Omega'_{1,1} = \Omega_{1,1} - \nabla_{1,1}\pi.$$

Hence $[\Omega'_{1,1}]_T = 0$. But this implies that $\Omega'_{1,1} = 0$ since $I_T^{1,1} = 0$ with the notation of Section 1.1.

Now ω' is obviously an adapted connection of the second special type. Q.E.D.

4.3. *The case of the (complex) bundle of frames.*

Let us consider the particular case where $E = B_0(M)$. By definition the ATIYAH class modulo torsion of B_0 will be referred to as the *ATIYAH class modulo torsion of M* and will be denoted by $a_T(M)$.

$a_T(M) \in H_T^{1,1}(M, L(B_0))$ is the obstruction to the existence of a *basic connection of the second special type*, that is to say a connection ω on $B_0(M)$ which is:

$$\textit{Torsion basic:} \quad \Sigma = \Sigma_{2,0} + T,$$
$$\textit{Curvature basic:} \quad \Omega = \Omega_{2,0} + \Omega_{0,2}.$$

Moreover, since $C(M)$ is assumed to be zero, the class modulo torsion of $\Omega_{0,2}$ is zero. Hence for such a connection, if it exists, we have a tensorial 1-form π of type $(1,0)$ on B_0 with values in the Lie algebra $gl(n, C)$ such that, if we denote also by π the corresponding map:

$$\pi : TM \to L(B_0)$$

$$\text{we have} \quad \begin{cases} \Sigma = \Sigma_{2,0} + T \\ \Omega = \Omega_{2,0} + \pi \circ T. \end{cases}$$

$\omega - \pi$ is then a basic connection of the first special type.

Basic connections of the second special type are the natural generalization of *analytic connections* in the case of a complex analytic manifold.

Université des Sciences et Techniques du Languedoc

REFERENCES

[1] Atiyah, M., 'Complex Analytic Connections on Fibre Bundles', *Trans. Am. Math. Soc.* **85** (1957), 181–207.
[2] Baum-Bott, 'Singularities of Holomorphic Foliations', *J. of Differential Geometry* **7** (1972), 279–342.
[3] Kamber-Tondeur, 'Invariant Differential Operators and the Cohomology of Lie Algebra Sheaves', *Mem. Am. Mathe. Soc.* **113** (1971).
[4] Molino, P., 'Γ_q-structures partielles et classes de Bott-Haefliger', (preprint).
[5] Vaisman, I., 'From the geometry of hermitian foliated manifolds', *Bull. Math. Soc. M. de Roumanie* **17** (1973), 71–100.
[6] Vidal-Vidal-Costa, E., 'Special Connections and Almost Foliated Metrics', *J. of Differential Geometry* **8** (1973), 297–304.
[7] Yano, K., *Differential Geometry on Complex and Almost Complex Spaces*, Pergamon Press, Oxford, 1965.

PHAM MAU QUAN

PSEUDOGROUPS AND LINEAR CONNECTIONS ON A BANACH FIBRE BUNDLE

THE LIE ALGEBRA OF LOCAL INFINITESIMAL TRANSFORMATIONS OF A VECTOR BUNDLE

1. *Local infinitesimal transformations*

Let $\pi : E \to M$ be a locally trivial Banach fibre bundle, of class C^k ($k \geqslant 2$), with connected base M modelled on a Banach space B and fibre a Banach space E. A local infinitesimal transformation, or l.i.t., of this vector bundle is a local vector field Z on E generating germs of 1-parameter groups of local isomorphisms of $\pi : E \to M$.

PROPOSITION 1.1. *A vector field Z of E is a l.i.t. of $\pi : E \to M$, if and only if in any local trivialization $\rho = E_U \to U \times \mathsf{E}$, Z admits the representation (X, A) where X is a vector field on U, $X = \pi_* Z$ and $A : U \to \operatorname{End} \mathsf{E}$ is a map of class C^{k-1}.*

Let Φ_t be a germ of one parameter groups of isomorphisms at $z = (x, \xi)$. It is defined in the local trivialization by $(s_t, S_t) : (x, \xi) \to (s_t x, S_t(x) \cdot \xi)$, where s_t is the projection of Φ_t and $S_t(x)$ is an isomorphism of E. One has

$$\left(\lim_{t \to 0} \frac{s_t x - x}{t}, \lim_{t \to 0} \frac{S_t(x) \cdot \xi}{t}\right) = (X(x), A(x) \cdot \xi), \tag{1.1}$$

where $A(x) \in \operatorname{End} \mathsf{E}$, which means that the field Z generating Φ_t admits the local representation (X, A).

Conversely, if a vector field Z of E admits the local representation (X, A), the local flow of the differential system

$$\frac{\mathrm{d}x}{\mathrm{d}t} = X(x), \qquad \frac{\mathrm{d}\xi}{\mathrm{d}t} = A(x) \cdot \xi$$

gives the germs of 1-parameter groups of local isomorphisms of $\pi : E \to M$.

Cahen & Flato (eds.), Differential Geometry and Relativity, 135–145.

PROPOSITION 1.2. *If Z, W are two l.i.t. with local representations* (X, A), (Y, B), *the bracket* $[Z, W]$ *has a local representation*

$$[(X, A), (Y, B)] = ([X, Y], [A, B] + D_X B - D_Y A),$$

where D is the differential in the local chart one considers.

One proves this property by using the lemma

LEMMA. *The bracket of two vector fields Z, W on a differentiable manifold is defined in a local chart by*

$$[Z, W] = D_Z W - D_W Z. \tag{1.2}$$

If Φ_t is the germ of 1-parameter group generated by Z one has $\Phi_t(z) = z + tZ(z) + t\alpha(t, z)$, where $\alpha(t, z)$ tends to 0 when t tends to 0. For any t, Φ_t is a local diffeomorphism, by derivation one has

$$D\Phi_t(z) = Id + tDZ(z) + tA(t, z),$$

where $A(t, z)$ tends to 0 when t tends to 0. One has then

$$[Z, W] = (\mathscr{L}_Z W) = \lim_{t\to\infty} \frac{W(\Phi_t(z)) - D\Phi_t(z) \cdot W(z)}{t}$$

$$= \lim_{t\to 0} \frac{W(\Phi_t(z)) - W(z)}{t} - \lim_{t\to 0} (DZ(z) + A(t, z)) \cdot W(z);$$

that is $[Z, W](z) = D_Z W(z) - D_W Z(z)$ and the lemma follows.

2. *Pseudogroups and exact sequences of associated Lie algebras*

The set of local infinitesimal transformations of the vector bundle $\pi: E \to M$ forms a sheaf of Lie algebras $\mathscr{L}$. It generates the pseudogroup of all local isomorphisms of π. This pseudogroup $P(\mathscr{L})$ leaves $\mathscr{L}$ invariant. The same occurs for any subsheaf $\mathfrak{L}$ of $\mathscr{L}$. The projection of $\mathfrak{L}$ defines a sheaf of Lie algebras $\mathscr{G}$ which generates a pseudogroup $G(\mathscr{G})$ on M, identical with the projection of $P(\mathfrak{L})$. Let $\pi_{\mathfrak{L}}: \mathfrak{L}E \to E$ be the vector bundle such that the fibre over any point $z \in E$ is formed by the Lie algebra of germs of l.i.t. of $\mathfrak{L}$. The local sections of this vector bundle give back the sheaf $\mathfrak{L}$. One has the property:

PROPOSITION 2.1. *With any sheaf* $\mathfrak{L}$ *of Lie algebras of l.i.t. of E is canonically associated the exact sequence of Lie algebra bundles*

$$0 \to \mathcal{V}E \xrightarrow{j} \mathfrak{L}E \xrightarrow{i} \pi^*\mathcal{G}M \to 0, \tag{2.1}$$

where $\mathcal{V}E$ is the subbundle of germs of vertical l.i.t. of $\mathfrak{L}$ and $\pi^*\mathcal{G}M$ is the inverse image on $\pi : E \to M$ of the fibre bundle $\tau_{\mathcal{G}} : \mathcal{G}M \to M$ associated with the sheaf $\mathcal{G}$ projection of $\mathfrak{L}$ and where the morphisms i, j are defined locally by

$$i(x, \xi, X, A) = (x, \xi, X) \quad \textit{and} \quad j(x, \xi, A) = (x, \xi, 0, A). \tag{2.2}$$

The proof is obvious.

If the Lie algebra $\mathfrak{L}$ is transitive, so is the pseudogroup $P(\mathfrak{L})$, and one has val $\mathfrak{L}_z = T_zE$, val $\mathcal{G}_x = T_xM$, val $\mathcal{V}_z = V_zE$, factorizing one gets the classical exact sequence of fibre bundles:

$$0 \to VE \xrightarrow{j} TE \xrightarrow{i} \pi^*TM \to 0, \tag{2.3}$$

where VE is the subbundle of vertical vectors of TE and π^*TM is the inverse image of $\tau : TM \to M$ over $\pi : E \to M$. The morphisms i, j are defined locally by:

$$i(x, \xi, y, \eta) = (x, \xi, y) \quad \text{and} \quad j(x, \xi, \eta) = (x, \xi, 0, \eta). \tag{2.4}$$

PROPOSITION 2.2. *For the sheaf $\mathcal{L}$ of all l.i.t. of E, the exact sequence* (2.1) *is identical with the exact sequence of sheafs of germs of sections of the sequence* (2.3), *i.e. with*

$$0 \to \underline{VE} \xrightarrow{j} \underline{TE} \xrightarrow{i} \underline{\pi^*TM} \to 0$$

For any subsheaf $\mathfrak{L} \subset \mathcal{L}$, $\mathcal{V}E$, $\mathfrak{L}E$, $\pi^*\mathcal{G}M$ are subbundles of $\underline{VE}$, $\underline{TE}$, $\pi^*\underline{TM}$.

3. *Action of a Lie pseudogroup*

Let G be a Lie pseudogroup generated by a Lie subalgebra $\mathcal{G}$ of the l.i.t. of M. We denote by G_x the germ of Lie group defined by those elements s the source of which contains x, and by $\mathcal{G}_x$ its Lie algebra.

DEFINITION. The pseudogroup G is said to act on the left on the vector bundle $\pi : E \to M$ if a function $R : G \times E \to E$ is given, satisfying the conditions:

(1) for any s with U as source, the map $z \to R(s, z)$ is a local isomorphism $R(s, \cdot) : E_U \to E$,

(2) for any s, t which can be composed, one has,

$$R(tos, z) = R(t, R(s, z)),$$

(3) $R(Id, z) = z$.

For any $x \in M$, G_x acts on $\pi: E \to M$ in the neighbourhood of the fibre E_x as a Lie group. G is said to act differentiably if G_x depends differentiably on x and if G_x acts differentiably. If the base M is paracompact, the action R of G may be defined by a representation of G in the pseudogroup of local isomorphisms of $\pi: E \to M$. This is in particular the case if G is a Lie group acting transitively on M which is then a homogeneous space.

DEFINITION. With the action R defined locally by $(s, x, \xi) \to (s(x), S(s, x) \cdot \xi)$, where $S(s, x)$ is an isomorphism of E, is canonically associated an action R^0 of G which acts on the right on the fibres (but on the left on the base M). R^0 is defined locally by

$$(s, x, \xi) \to (s(x), S(s^{-1}, x), \xi).$$

THEOREM 3.1. *Any action R of a transitive pseudogroup $G(\mathcal{G})$ of M on the vector bundle $\pi: E \to M$ induces a homomorphism $v: \mathcal{G} \to \mathcal{L}$ of the Lie algebra $\mathcal{G}$ into the Lie algebra $\mathcal{L}$ of the l.i.t. of $\pi: E \to M$, defined locally by $v(X) = (X, \Gamma(X))$, where $\Gamma([X, Y]) = [\Gamma(X), \Gamma(Y)]$.*

For any fixed x_0 of M, the action R defines a homomorphism $R_{x_0}: G_{x_0} \to \mathrm{Isom}_{x_0} E$. The differential of R_{x_0} is a homomorphism between the Lie algebras of the germs of l.i.t. $v_{x_0}: \mathcal{G}_{x_0} \to \mathrm{End}_{x_0} E$ defined locally by $\dot{X} \to (\dot{X}, \Gamma_{x_0}(\dot{X}))$ and one has:

$$[\dot{X}, \dot{Y}] \to ([\dot{X}, \dot{Y}]), [\Gamma_{x_0}(\dot{X}), \Gamma_{x_0}(\dot{Y}]). \tag{3.1}$$

The pseudogroup G being transitive, val $\mathcal{G}_{x_0} = T_{x_0}M$. Factorizing one defines in every point $x \in M$ a linear map $\Gamma(x): T_xM \to \mathrm{End}\, E_x$. One has so defined a 1-form Γ on M with values in End $E \to M$. This being true for any point $x \in M$, one has $\Gamma([X, Y])(x) = (\Gamma_x([\dot{X}, \dot{Y}])(x)$, where $\dot{X}$, $\dot{Y}$ are the germs in x of the l.i.t. X, Y. From this and (3.1) it follows:

$$v: [X, Y] \to ([X, Y], [\Gamma(X), \Gamma(Y)]).$$

Remark. If one considers the action on the right R^0 associated with

R on gets the homomorphism v^0 such that

$$v^0([X, Y]) = ([X, Y], -[\Gamma(X), \Gamma(Y)]).$$

4. *Infinitesimal transforms defined by an action R.*

The results are symmetrical for R and R^0. In what follows we consider the action on the right R^0 associated with R. If $\xi : M \to E$ is a differentiable local section of $\pi : E \to M$, the section $\mathscr{L}_X\xi$ defined by:

$$(\mathscr{L}_X\xi)(x) = \lim_{t=0} \frac{\xi(s_t(x)) - R(s_t^{-1}, x) \cdot \xi(x)}{t}, \tag{4.1}$$

where s_t is the germ of 1-parameter group associated with X, is called the infinitesimal transform of ξ with respect to the l.i.t. $X \in \mathscr{G}$.

PROPOSITION 4.1. *In a local chart one has:*

$$\mathscr{L}_X\xi = D_X\xi + \Gamma(X) \cdot \xi, \tag{4.2}$$

where D_X is the derivative with respect to X in this local chart.

Proof. Proof 4.1 may be written

$$(\mathscr{L}_X\xi)(x) = \lim_{t\to 0} \frac{\xi(s_t x) - \xi(x)}{t} - \lim \frac{R(s^{-1}, x)_t \cdot \xi(x) - \xi(x)}{t}.$$

Taking limits and using Theorem 3.1 one gets (4.1).

By functoriality the action $L(R^0)$ induced by R^0 on the vector bundle $L(\pi) : L(E) \to M$ is defined as

$$(s, x, A) \to (s(x), R(s^{-1}, x) \circ A \circ R(s, x)).$$

This enables us to define the infinitesimal transform $\mathscr{L}_X A$ of a section $A : M \to L(E)$, i.e. of a differentiable endomorphism on the base of $\pi : E \to M$ by the formula:

$$(\mathscr{L}_X A) \cdot \xi = \mathscr{L}_X(A \cdot \xi) - A \cdot \mathscr{L}_X\xi,$$

where ξ is a section of $\pi : E \to M$. From this follows:

PROPOSITION 4.2. *The infinitesimal transform of an endomorphism A with fixed base of $\pi : E \to M$ is given in a local chart by*

$$\mathscr{L}_X A = D_X A = [\Gamma(X), A].$$

THEOREM 4.3. *For any l.i.t. X defined by G, $\Gamma(X)$ is invariant by G, i.e. for any $X, Y \in \mathcal{G}: \mathcal{L}_X \Gamma(Y) = 0$.*

We must show that for any local transformation s_t of the germ of 1-parameter group generated by X, we have

$$R(s_t, x) \circ \Gamma(Y)(x) \circ R(s_t^{-1}, x) = \Gamma(Y)(x).$$

But:

$$\Gamma(Y)(x) = \lim_{v \to 0} \frac{R(u_v, x) - Id_x}{v},$$

where u_v is the germ of 1-parameter group generated by Y. It is thus sufficient to show that:

$$\lim_{v \to 0} \frac{R(s_t, x) \circ R(u_v, x) \circ R(s_t^{-1}, x) - Id_x}{v} = \Gamma(Y)(x).$$

Using the definition of the pseudogroup G, we see that the left side is equal to:

$$\lim_{v \to 0} \frac{R(s_t \circ u_v \circ s_t^{-1}, x)_t - Id_x}{v} = \Gamma(s_t \cdot Y)(s_t^{-1}(x)).$$

The Lie algebra $\mathcal{G}$ being invariant by G, we deduce from this that $\Gamma(s_t \cdot Y(s_t^{-1}(x))) = \Gamma(Y)(x)$ and the theorem follows.

5. *Splitting of an exact sequence*

Let H be the smallest Lie pseudogroup of local isomorphisms of $\pi: E \to M$ invariant by the action R of G. We shall study its sheaf $\mathfrak{L}$ of Lie algebras.

PROPOSITION 5.1. *The Lie algebra $\mathfrak{L}$ of l.i.t. of $\pi: E \to M$ which generate H is defined locally by*

$$\mathfrak{L} = \{(X, A): X \in \mathcal{G}, \quad \mathcal{L}_X A = 0\} \tag{5.1}$$

H being the minimal pseudogroup invariant by the action R of G, its projection must be G. Therefore X is an element of $\mathcal{G}$. To show that $\mathcal{L}_X A = 0$, it is sufficient to show that:

$$R(s_t, x) \circ S_u(x) \circ R(s_t^{-1}, x) = S_u(x),$$

where S_u is the 1-parameter group generated by $A(x)$ and s_t is the germ of 1-parameter group generated by X on M. The proposition follows from a proof similar to that of Theorem 4.3.

THEOREM 5.2. *There exists a splitting of the exact sequence of Lie algebras bundles associated with:*

$$0 \to \mathscr{V}E \xrightarrow{i} \mathfrak{L}E \xrightarrow{j} \pi^* \mathscr{G}M \to 0. \tag{5.2}$$

The morphisms $v : \pi^*\mathscr{G}M \to \mathfrak{L}E$ *and* $w : \mathfrak{L}E \to \mathscr{V}E$ *of the splitting are defined locally by* $v(x, \xi, X) = (x, \xi, X, -\Gamma(X))$ *and* $w(x, \xi, X, A) = (x, \xi, A + \Gamma(X))$.

Proof. v and w are obviously morphisms of underlying vector bundles such that $v \circ i = Id_{\pi^*\mathscr{G}M}$ and $j \circ w = Id_{vE}$. v is injective and $\operatorname{Im} v = \{(x, \xi, X, -\Gamma(X)) \mid X \in \mathscr{G}_x\}$, w is surjective and $\operatorname{Ker} w = \{(x, \xi, X, -\Gamma(X)) \mid X \in \mathscr{G}_x\}$. So we have $\operatorname{Im} v = \operatorname{Ker} w$ and $\mathfrak{L}E = \operatorname{Im} v \oplus \operatorname{Ker} w$. The inverse exact sequence is:

$$0 \leftarrow \mathscr{V}E \xleftarrow{w} \mathfrak{L}E \xleftarrow{v} \pi^* \mathscr{G}M \leftarrow 0. \tag{5.3}$$

We still have to show that v and w are Lie algebra homomorphisms of the fibre bundles. We will check this by direct computation. First we have:

$$v([X, Y]) = ([X, Y], -\Gamma([X, Y])),$$
$$[v(X), v(Y)] = ([X, Y], [\Gamma(X), \Gamma(Y)] - D_X\Gamma(Y) + D_Y\Gamma(X)).$$

By Theorem 4.3, $D_X\Gamma(Y) - [\Gamma(X), \Gamma(Y)] = 0$ and $D_Y\Gamma(X) - [\Gamma(Y), \Gamma(X)] = 0$. The second equality becomes:

$$[v(X), v(Y)] = ([X, Y], -[\Gamma(X), \Gamma(Y)]).$$

By Theorem 3.1, $[\Gamma(X), \Gamma(Y)] = \Gamma([X, Y])$. From this we deduce $v([X, Y]) = [v(X), v(Y)]$. Similarly we have:

$$w([(X, A), (Y, B)]) = [A, B] + [\Gamma(X), \Gamma(Y)] + D_XB - D_YA,$$
$$[w(X, A), w(Y, B)]$$
$$= [A, B] + [\Gamma(X), \Gamma(Y)] + [\Gamma(X), B] - [\Gamma(Y), A].$$

By hypothesis and by Proposition 5.1, $D_XB - [\Gamma(X), B] = 0$ and $D_YA - [\Gamma(Y), A] = 0$; the second equality becomes:

$$[w(X, A), w(Y, B)] = [A, B] + [\Gamma(X), \Gamma(Y)] + D_XB - D_YA.$$

From this we deduce $w([X, A), (Y, B)]) = [w(X, A), w(Y, B)]$.

Taking the local sections of the Lie algebra bundles of the exact sequence (5.2) one gets the sheafs of Lie algebras of the corresponding l.i.t. of $\pi: E \to M$ and the preceding theorem can be restated as follows:

THEOREM 5.3. *Let R be an action of a transitive Lie pseudogroup $G(\mathcal{G})$ of M on the vector bundle $\pi: E \to M$. The Lie algebra $\mathfrak{L}$ of l.i.t. associated with the smallest pseudogroup of local isomorphisms of $\pi: E \to M$ invariant by R, admits a local decomposition in direct sum of two Lie subalgebras*

$$\mathfrak{L} = \mathfrak{H} + \mathcal{V},$$

where $\mathfrak{H} = \{(X, \Gamma(X)) | X \in \mathcal{G}\}$ is the Lie subalgebra of horizontal l.i.t. and $\mathcal{V} = \{(0, A) | \forall X \in \mathcal{G}, \mathcal{L}_X A = 0\}$ is the Lie subalgebra of vertical l.i.t.

Application to linear connections

6. *Linear connection defined by the action of a pseudogroup*

We shall show that the action R of a transitive Lie pseudogroup $G(\mathcal{G})$ of M defines a linear connection on the vector bundle $\pi: E \to M$ more precisely.

THEOREM 6.1. *Any differentiable action R of a transitive pseudogroup $G(\mathcal{G})$ of the base M induces a linear connection on the vector bundle $\pi: E \to M$; the associated covariant derivative ∇ is determined by the infinitesimal transformation defined by the action R, i.e. for any local section ξ of $\pi: E \to M$, for any $y \in T_x M$, one has*

$$\nabla_y \xi = (\mathcal{L}_X \xi)(x), \tag{6.1}$$

where X is a l.i.t. of $\mathcal{G}$ such that $X(x) = y$.

Indeed, the action R of G defines the exact sequence of Lie algebra bundles (5.2) and a splitting of this exact sequence. The pseudogroup G being transitive, one has $\operatorname{val}_z(\pi^* \mathcal{G} M) = (\pi^* TM)_z$, $\operatorname{val}_z \mathfrak{L}E = T_z E$ and $\operatorname{val}_z \mathcal{V}E = V_z E$. Factorizing, we get the exact sequence of vector bundles

$$0 \to VE \xrightarrow{i} TE \xrightarrow{j} \pi^* TM \to 0,$$

and a corresponding splitting of this sequence defined by the morphisms $i: \pi^* TM \to TE$ and $j: TE \to VE$ defined locally by

$$v(x, \xi, y) = (x, \xi, y, -\Gamma(x)(y, \xi)),$$
$$w(x, \xi, y, \eta) = (x, \xi, \eta + \Gamma(x)(y, \xi)),$$

where $\Gamma(x): \mathsf{B} x \mathsf{E} \to \mathsf{E}$ is a bilinear continuous map.

It is known that the preceding splitting defines a linear connection on $\pi: E \to M$. If ξ is a differentiable local section of $\pi: E \to M$, the covariant derivative of ξ with respect to the vector $y \in T_X M$ is given by:

$$\nabla_y \xi = D_y \xi + \Gamma(x)(y, \xi(x)).$$

Let X be a l.i.t. of $\mathcal{G}$ such that $X(x) = y$. The infinitesimal transform of ξ with respect to X is given by Proposition 4.1, namely

$$\mathcal{L}_X \xi = D_X \xi + \Gamma(X) \cdot \xi.$$

At the point x, one has

$$(\mathcal{L}_X(\xi)(x) = D_{X(x)} \xi + \Gamma(x)(X(x), \xi(x)) = D_y \xi + \Gamma(x)(y, \xi(x)).$$

From this one deduces (6.1).

COROLLARY 6.2. *The covariant derivative with respect to any vector field X of M of a section ξ of π : E → M is the infinitesimal transform of ξ induced by the germs of 1-parameter groups of local isomorphisms* (s_t, S_t) *of π : E → M generated by* $(X, -\Gamma(X))$. *In the case where X is a generator of the pseudogroup G, it is the infinitesimal transform induced by the action R of G.*

Indeed, in a local trivialization, one has:

$$\lim_{t \to 0} \frac{\xi(s_t(x)) - S_t(x) \cdot \xi(x)}{t}$$
$$= \lim_{t \to 0} \frac{\xi(s_t(x)) - \xi(x)}{t} - \lim_{t \to 0} \frac{S_t(x) \cdot \xi(x) - \xi(x)}{t}$$
$$= (D_X \xi)(x) + (\Gamma(X) \cdot \xi)(x).$$

From this we deduce the first property. If X is a generator of G, we have $S_t(x) = R(s_t^{-1}, x)$ and the second property follows.

7. *The Lie algebra of all affine infinitesimal transformations*

Let $\pi: E \to M$ be a Banach vector bundle with a linear connection ω defined by the splitting ω: $TE \to VE$ of the exact sequence $0 \to VE \to$

$TE \to \pi^*TM \to 0$. An affine transformation of (π, ω) is an automorphism of E which leaves invariant the splitting. Hence it leaves invariant the field of horizontal subspaces defined by ω. An affine l.i.t. of (π, ω) is an assignment of a local vector field Z of E, which generates germs of 1-parameter groups of affine transformations of (π, ω).

Let us also denote by ω the 1-form on E with values in VE defined by the splitting morphism $\omega : TE \to VE$. One shows that a vector field Z of E is an affine l.i.t. if and only if

$$\mathcal{L}_Z\omega = 0. \tag{7.1}$$

THEOREM 7.1. *The affine l.i.t. of the vector bundle $\pi : E \to M$ with a linear connection ω, form a lie algebra $\mathfrak{L}$ such that for any pair Z, W with local representations $(X, A)(Y, B)$ one has:*

$$\nabla_Y A = 0, \qquad \nabla_X \Gamma(Y) = 0, \qquad \Gamma([X, Y]) = [\Gamma(X), \Gamma(Y)]. \tag{7.2}$$

Condition (7.1) is indeed equivalent to $\mathcal{L}_Z\omega(W) - \omega([Z, W]) = 0$. Using the local representations (X, A) and (Y, B) of Z and W, one gets the equivalent condition

$$-(D_Y A - [\Gamma(Y), A]) - (D_X\Gamma(Y) \\ - [\Gamma(X), \Gamma(Y)]) + \Gamma([X, Y]) - [\Gamma(X), \Gamma(Y)] = 0.$$

This being verified for any pair $(X, A), (Y, B)$, relation (7.2) follows.

THEOREM 7.2. *The Lie algebra $\mathfrak{G}$ of all affine l.i.t. of a vector bundle $\pi : E \to M$ with a linear connection ω admits a decomposition $\mathfrak{G} = \mathfrak{H} \oplus \mathcal{V}$, where $\mathfrak{H}$ is the set of elements of the form $(X, -\Gamma(X))$ and $\mathcal{V}$ the set of elements of the form $(0, A + \Gamma(X))$.*

Indeed any element (X, A) of $\mathfrak{G}$ decomposes in a unique way as a horizontal component $(X, -\Gamma(X))$ and a vertical component $(0, A + \Gamma(X))$ and one checks that the elements of the form $(X, -\Gamma(X))$ and $(0, A + \Gamma(X))$ form two closed Lie subalgebras of $\mathfrak{G}$.

We can now prove the converse of theorem 6.1.

THEOREM 7.3. *Given a linear connection ω on a Banach vector bundle $\pi : E \to M$, if the Lie algebra of all affine l.i.t. of π is transitive,* *there exists a transitive Lie pseudogroup of transformations of M and an action R of this pseudogroup on* $\pi : E \to M$ *which induces the given linear connection* ω.

Indeed, if the Lie algebra $\mathfrak{G}$ of affine l.i.t. is transitive, it generates a Lie pseudogroup $H(\mathfrak{G})$ of transformations of $\pi : E \to M$ the projection of which defines a transitive pseudogroup $G(\mathfrak{G})$ of M. The elements $(X, -\Gamma(X))$ of $\mathfrak{G}$ are the horizontal lifts of the elements X of $\mathscr{G}$. Integrating the differential system

$$\frac{\mathrm{d}x}{\mathrm{d}t} = X(x), \qquad \frac{\mathrm{d}\xi}{\mathrm{d}t} = -\Gamma(X)(x, \xi),$$

one gets the flow (s_t, S_t) which defines the action R of G on the vector bundle. One checks that R induces the linear connection ω.

Université de Paris-Nord

REFERENCES

[1] Eliasson, H., 'Geometry of Manifolds of Maps', *J. Diff. Geom.* (1967).
[2] Kobayashi, S. *Transformation Groups in Differential Geométry*, Springer Verlag, 1972.
[3] Libermann, P., 'Pseudogroupes infinitésimaux attachés aux pseudogroupes de Lie', *Bull. S.M.F.* **87** (1959).
[4] Lichnerowicz, A., *Géométrie des Groupes de Transformation*, Dunod, 1958.
[5] Lichnerowicz, A., *Théorie Globale des Connexions et Groupes d'Holonomie*, Ed. Cremonese, Rome, 1955.
[6] Pham Mau Quan, 'Pseudogroupes de transformations et connexions linéaires sur un fibré vectoriel', *C.R.A.S.* **276** (1973).
[7] Pham Mau Quan, 'Sur l'algèbre de Lie des transformations infinitésimales locales affines d'un fibré vectoriel banachique', *C.R.A.S.* **279** (1974).
[8] Singer, I. M. and Sternberg, S., The Infinite Groupes of Lie and Cartan', *J. Ana. Math* **15** (1965).
[9] Kamber, F. and Tondeur, P., 'Invariant Differential Operators and the Cohomology of Lie Algebra Sheaves', *Mem. A.M.S.* **113** (1971).
[10] Vilms, J., Connections on Tangent Bundles. *J. diff. Geom.* **1** (1967).

S. A. ROBERTSON

MOBILITY IN CATEGORIES AND METRIC SPACES

This paper is concerned with the way in which a morphism relates the automorphisms of its domain to those of its codomain. Although the ideas in what follows originated in attempts to understand the notions of *symmetry group* and *free mobility* for bodies in space, it turns out that the 'mobile' and 'comobile' morphisms that we define occur throughout mathematics. Indeed, they are to be found in any categories between which there is a pair of adjoint functors.

My plan here is to explain the definitions of mobile and comobile morphisms, and the associated notions of mobility and comobility objects, to illustrate these by describing a number of examples in familiar categories, to prove some general existence theorems, and to characterise the complete simply-connected spaces of constant curvature as the only Riemannian manifolds that are mobility objects in the metric category.

Thanks are due to David Chillingworth and Ali Alemzadeh for much helpful discussion.

1. Homogeneity and mobility

The axiom of free mobility states that a body in space can be moved from any position in space to any other [1, 2, 3, 8]. From a categorical point of view, we can formulate this idea as follows.

Let $\mathscr{C}$ be a category, and A, X objects in $\mathscr{C}$. We say that two monomorphisms μ, μ' from A to X are *equivalent* iff there are automorphisms α of A and ξ of X such that $\xi\mu = \mu'\alpha$.

$$\begin{array}{ccc} A & \xrightarrow{\mu} & X \\ {\scriptstyle\alpha}\downarrow & & \downarrow{\scriptstyle\xi} \\ A & \xrightarrow[\mu']{} & X \end{array}$$

Then a $\mathscr{C}$-object X is said to be *homogeneous* or to *satisfy the axiom*

Cahen & Flato (eds.), Differential Geometry and Relativity, 147–158.

of free mobility iff, for every $\mathscr{C}$-object A, any two $\mathscr{C}$-monomorphisms from A to X are equivalent. The problem in each category $\mathscr{C}$ is to find the homogeneous objects. For example, in the metric category (the category of distance-preserving maps between metric spaces), the simply-connected complete spaces of constant curvature are homogeneous.

Dually, we say that a $\mathscr{C}$-object Y is *cohomogeneous* iff for any $\mathscr{C}$-object B, any two epimorphisms η, η' from Y to B are equivalent in the sense that there exist automorphisms β, η of B and Y such that

$$\begin{array}{ccc} B & \xleftarrow{\eta} & Y \\ {\scriptstyle\beta}\downarrow & & \downarrow{\scriptstyle\eta} \\ B & \xleftarrow[\eta']{} & Y \end{array}$$

Obviously, if X is homogeneous in a category $\mathscr{C}$, then its dual X^* is cohomogeneous in the dual category $\mathscr{C}^*$.

We now consider the question of what is meant by the symmetry of an object or a morphism. If X is a $\mathscr{C}$-object, then the symmetry properties of X are embodied in the group Aut X of all $\mathscr{C}$-automorphisms of X. To make a corresponding statement about morphisms, we recall that there is a category $\mathscr{C}^\dagger$ whose objects are the morphisms of $\mathscr{C}$, and in which a morphism from $X \xrightarrow{f} Y$ and $Z \xrightarrow{g} W$ is a commutative diagram

$$\begin{array}{ccc} X & \xrightarrow{f} & Y \\ {\scriptstyle\varphi}\downarrow & & \downarrow{\scriptstyle\psi} \\ Z & \xrightarrow[g]{} & W \end{array} \tag{1}$$

where φ and ψ are $\mathscr{C}$-morphisms. If

$$\begin{array}{ccc} Z & \xrightarrow{g} & W \\ {\scriptstyle\xi}\downarrow & & \downarrow{\scriptstyle\eta} \\ U & \xrightarrow[h]{} & V \end{array} \tag{2}$$

is another $\mathscr{C}^{\dagger}$-morphism, then the composite of (1) and (2) is the diagram

$$\begin{array}{ccc} X & \xrightarrow{f} & Y \\ {\scriptstyle \xi\varphi}\downarrow & & \downarrow{\scriptstyle \eta\psi} \\ U & \xrightarrow[h]{} & V \end{array}$$

It follows that the symmetry properties of any $\mathscr{C}$-morphism $X \xrightarrow{f} Y$ are embodied in the group $\mathrm{Aut}^{\dagger} f$ of all $\mathscr{C}^{\dagger}$-automorphisms of f. Thus the elements of $\mathrm{Aut}^{\dagger} f$ are the commutative diagrams

$$\begin{array}{ccc} X & \xrightarrow{f} & Y \\ {\scriptstyle \alpha}\downarrow & & \downarrow{\scriptstyle \beta} \\ X & \xrightarrow[f]{} & Y \end{array} \qquad (3)$$

where $\alpha \in \mathrm{Aut}\, X$ and $\beta \in \mathrm{Aut}\, Y$. There is, therefore, a group monomorphism

$$\theta : \mathrm{Aut}^{\dagger} f \to \mathrm{Aut}\, X \times \mathrm{Aut}\, Y$$

which sends the diagram (3) to the pair (α, β). It is convenient to identify $\mathrm{Aut}^{\dagger} f$ with its image under θ and so to regard $\mathrm{Aut}^{\dagger} f$ as a subgroup of $\mathrm{Aut}\, X \times \mathrm{Aut}\, Y$.

The projections $\pi_X : \mathrm{Aut}\, X \times \mathrm{Aut}\, Y \to \mathrm{Aut}\, X$ and $\pi_Y : \mathrm{Aut}\, X \times \mathrm{Aut}\, Y \to \mathrm{Aut}\, Y$ restrict to $\mathrm{Aut}^{\dagger} f$ to give homomorphisms $f_X : \mathrm{Aut}^{\dagger} f \to \mathrm{Aut}\, X$ and $f_Y : \mathrm{Aut}^{\dagger} f \to \mathrm{Aut}\, Y$. Thus $f_X(\alpha, \beta) = \alpha$ and $f_Y(\alpha, \beta) = \beta$.

We say that f is *mobile* iff there is a homomorphism $\varphi : \mathrm{Aut}\, X \to \mathrm{Aut}^{\dagger} f$ such that $\pi_X \sigma = 1_{\mathrm{Aut}\, X}$. That is, f is mobile iff π_X is a retract homomorphism, or, equivalently, iff π_X has a right inverse φ. If such a homomorphism φ exists, then it must be of the form $\varphi(\alpha) = (\alpha, \sigma(\alpha))$, for some homomorphism $\sigma : \mathrm{Aut}\, X \to \mathrm{Aut}\, Y$. We then call σ a *mobilisation* of f.

Dually, we say that f is *comobile* iff there is a homomorphism $\psi : \mathrm{Aut}\, Y \to \mathrm{Aut}^{\dagger} f$ such that $\pi_Y \psi = 1_{\mathrm{Aut}\, Y}$, that is iff π_Y is a retract homomorphism (equivalently, iff π_Y has a right inverse ψ). Any such ψ must be of the form $\psi(\beta) = (\tau(\beta), \beta)$, for some homomorphism $\tau: \text{Aut } Y \to \text{Aut } X$, called a *comobilisation* of f.

It is an immediate consequence of these definitions that a $\mathscr{C}$-morphism $X \xrightarrow{f} Y$ is mobile iff the dual $\mathscr{C}^*$-morphism $X^* \xleftarrow{f^*} Y^*$ is comobile. Likewise, if $X \xrightarrow{f} Y$ is a mobile monomorphism, then every mobilisation of f is a group monomorphism. If f is a comobile epimorphism, then every comobilisation of f is a monomorphism. In what follows, we shall be concerned mostly with mobile monomorphisms and comobile epimorphisms.

We remark that if X is a homogeneous $\mathscr{C}$-object, and $A \xrightarrow{f} X$ is a subobject of X, then any automorphism $A \xrightarrow{\alpha} A$ of A extends to some automorphism $X \xrightarrow{\xi} X$ of X. That is, $\xi f = f\alpha$. However, even if we choose a unique such $\xi = \mu(\alpha)$ for each $\alpha \in \text{Aut } A$, we do not know that $\mu: \text{Aut } A \to \text{Aut } X$ is a homomorphism. The logical connexion between homogeneity and mobility is therefore far from simple. To lend force to this remark, we discuss the question of whether the inclusion of the circle S^1 in the 2-disc D^2 is mobile in the topological, metric and smooth categories. The arguments outlined here are due to D. R. J. Chillingworth and P. M. D. Furness.

Suppose then that $D^2 = \{x \in R^2 : \|x\| \leq 1\}$ is the unit disc in R^2, and $S^1 = \partial D^2$ is the unit circle. Let $f: S^1 \to D^2$ denote the inclusion. In the category Met of distance-preserving maps between metric spaces, f is a mobile monomorphism with respect to the metrics induced by the standard Euclidean metric of R^2. A particular mobilisation $\sigma: \text{Iso } S^1 \to \text{Iso } D^2$ is given by putting $\sigma(g)(tx) = tg(x)$, for all $x \in S^1$, $g \in \text{Iso } S^1$ and $0 \leq t \leq 1$. A similar construction shows that f is mobile in the category Top of continuous maps between topological spaces. (In fact, the inclusion of any space A in the cone $X = CA$ on A is mobile in the topological category.)

However, if we consider the category Diff of C^1-maps between C^1-manifolds, then we shall find that there is no (continuous) mobilisation of f, although it is well-known [5] that every C^1-diffeomorphism of S' to itself extends to a C^1-diffeomorphism of D^2 to itself.

For suppose that $\sigma: \text{Diff } S^1 \to \text{Diff } D^2$ is a continuous mobilisation of $f: S^1 \to D^2$. Then the inclusion $\iota: SO_2 \to \text{Diff } S^1$ determines a continuous SO_2 action φ on D^2, and this action has a unique fixed point. To see this, observe that each orbit of φ is either a point or a (topological) circle. Now there is at least one fixed point x, and the winding number of any other orbit of φ about z is well-defined. Since φ is continuous, the winding number gives a continuous integer-valued function on D^2. But the winding number of the orbit $\partial D^2 = S^1$ is nonzero, and so there is no other fixed point of φ.

Denote $\sigma(\lambda)$ by λ' for each $\lambda \in \text{Diff}\, S^1$, and let G denote the subgroup of Diff S^1 consisting of all C^1-diffeomorphisms of S^1 that commute with the reflexion α in $0 \in R^2$ (rotation through π). Then for each $g \in G$.

$$g'(x) = g'(\alpha'(x)) = (g\alpha)'(x) = (\alpha g)'(x) = \alpha'(g'(x)).$$

Hence $g'(x) = x$. Thus G acts on D^2 under the mobilisation σ with fixed point x.

Next define a continuous homomorphism

$$\theta : G \to GL(2, R)$$

by $\theta(g) = Dg'(x)$. It follows that $\theta(\alpha) = -I$, and so we can define a continuous homomorphism

$$\theta_* : G/\langle\alpha\rangle \to GL(2, R)/\langle -I\rangle \cong GL(2, R)$$

for which $\theta_*(g\langle\alpha\rangle) = \theta(g)\langle -I\rangle$. Now θ_* is nonzero because $\theta(SO_2) < \theta(G)$ is nonzero. Also ker θ_* is nonzero because $GL(2, R)$ is finite-dimensional. Hence $G/\langle\alpha\rangle$ has a closed nontrivial normal subgroup. But the covering map $S^1 \to S^1/\langle\alpha\rangle$ gives a continuous isomorphism from $G/\langle\alpha\rangle$ to Diff S^1, and the latter is simple [4]. Thus no continuous mobilisation of $f: S^1 \to D^2$ exists.

2. Examples

In this section we list a number of examples of mobile and comobile morphisms. Most of these are familiar in other contexts. They serve to illustrate the definitions given above, and to motivate the theorems that follow in §3.

(1) The discussion of the inclusion $S^1 \hookrightarrow D^2$ suggests the more general problem of whether the inclusion $\partial M \hookrightarrow M$ is mobile in the smooth category Diff, where M is a compact connected C^1-manifold and ∂M is connected. Here we observe that the inclusion $f: \partial M \to M$ is *comobile* in Diff. For we can define a comobilisation $\tau: \text{Diff}\, M \to \text{Diff}\, \partial M$ by $\tau(g) = g|\partial M$.

(2) Let X be a connected C^∞ manifold. Then the projection $\pi: TX \to X$ of the tangent bundle TX of X is a comobile epimorphism in the category of C^∞ maps between C^∞ manifolds. A comobilisation $\tau: \mathrm{Diff}^\infty X \to \mathrm{Diff}^\infty TX$ of π is given by $\tau(\alpha) = T\alpha$. The zero section $\zeta: X \to TX$ is mobile, and τ is also a mobilisation of ζ. We note that τ defines an action of Diff X on TX, with just two orbits (namely $\zeta(X)$ and its complement). The group ker π_X is the group of all C^∞ diffeomorphisms of TX keeping every fibre setwise fixed, while ker ζ_X is the group of all diffeomorphisms of TX keeping $\zeta(X)$ pointwise fixed.

(3) Let X be a metric space, and Y its completion. Then the natural isometric embedding $f: X \to Y$ is a mobile monomorphism in the category Met of distance-preserving maps between metric spaces. A mobilisation $\sigma: \mathrm{Iso}\, X \to \mathrm{Iso}\, Y$ is given by $\sigma(\varphi) = \psi$, where ψ is the unique extension of φ to a self-isometry $\psi: Y \to Y$.

(4) Let X be a normed linear space with norm ν. Then there is an induced norm ν^{**} on the second dual X^{**} of X. The natural embedding $f: X \to X^{**}$ given by $f(x)(\lambda) = \lambda(x)$, $x \in X$, $\lambda \in X^*$, is a mobile monomorphism in the category of norm-preserving linear maps between normed linear spaces.

(5) In the category Top_* of base-point preserving continuous maps between connected pointed topological spaces, the standard embedding $f: X \to \Sigma X$ of a space X into its (reduced) suspension ΣX is a mobile monomorphism. Furthermore, the standard map $g: X \to \Omega \Sigma X$ assigning to $x \in X$ the loop given by the generator of ΣX through x is a mobile monomorphism. And the map $h: \Sigma \Omega X \to X$, assigning to the point (γ, t) with parameter t above the loop $\gamma: S^1 \to X$ the point $x = \gamma(t)$ of X, is a comobile epimorphism. (Here $t \in [0, 1]$ and S^1 denotes the reals mod 1.).

(6) In the category of maps between sets, a morphism is mobile iff it is either a constant map or a monomorphism.

(7) Every monomorphism is mobile and every epimorphism is comobile in the category of linear maps between real linear spaces.

3. Functorial constructions

The results of this section show that some of the above examples arise from quite general functorial constructions.

LEMMA. Let $\mathscr{C}$ and $\mathscr{D}$ be categories, and $F:\mathscr{C}\to\mathscr{D}$ a functor. Suppose that φ is a natural transformation from F to the identity functor $1_{\mathscr{C}}$. Then for each $\mathscr{C}$-object X, $F(X)\xrightarrow{\varphi(X)}X$ is comobile. Also if ψ is a natural transformation from $1_{\mathscr{C}}$ to F, then $X\xrightarrow{\psi(X)}F(X)$ is mobile.

Proof. Suppose that F is covariant. Then for any $\mathscr{C}$-morphism $X\xrightarrow{f}Y$, $\varphi(Y)F(f)=f\varphi(X)$, and we can define a comobilisation $\tau:\mathrm{Aut}\,X\to\mathrm{Aut}\,F(X)$ of the morphism $\varphi(X)$ by $\tau(\alpha)=F(\alpha)$. The other cases are dealt with similarly.

Now recall that examples of such natural transformations are provided by the units and counits of pairs of adjoint functors. Thus if $F:\mathscr{C}\to\mathscr{D}$ and $G:\mathscr{D}\to\mathscr{C}$ are adjoint functors, then there is a natural transformation η (called the *unit* of (F, G)) from $1_{\mathscr{C}}$ to GF, and a natural transformation ε (the *counit* of (F, G)) from FG to $1_{\mathscr{D}}$. (See, for example, [7] page 81.) Thus the lemma has the following consequence.

COROLLARY. Let $F:\mathscr{C}\to\mathscr{D}$ and $G:\mathscr{D}\to\mathscr{C}$ be adjoint functors, with unit η and counit ε. Then for any $\mathscr{C}$-object X and any $\mathscr{D}$-object Y, the morphism $X\xrightarrow{\eta(X)}GF(X)$ is mobile and the morphism $FG(Y)\xrightarrow{\varepsilon(Y)}Y$ is comobile.

(In fact, $\eta(X)$ is a monomorphism, and $\varepsilon(Y)$ is an epimorphism [7].)

4. Mobility and homogeneity in metric spaces

We have observed in Example (6) above that every monomorphism in the category of sets is mobile. That is to say, any set X is so highly symmetrical that there is no obstruction to extending the group of automorphisms of any subset of X to a subgroup of the group of automorphisms of X itself. The same statement holds for the category of (real) linear spaces and linear maps, as indicated in Example (7). In other categories, this high degree of symmetry may be possessed by some, but not all, objects. Thus in the category Met, every monomorphism into a space with the discrete metric is mobile, while the inclusion of, say, the open interval $]-1, 1[$ into the half-open interval $[-1, 1[$ is not mobile with respect to the standard metric $d(x, y)=|x-y|$.

We say that an object X in a category $\mathscr{C}$ is a *mobility object* iff every monomorphism with codomain X is mobile. Dually, we say that X is a *comobility object* iff every epimorphism with domain X is comobile. For each category $\mathscr{C}$, there is the problem of determining the mobility objects and the comobility objects. Thus every set is a mobility object in the category of maps between sets, while every linear space in the real linear category is both a mobility object and a comobility object.

We now explore the relation between mobility objects and homogeneous objects in the metric category Met. For this purpose, we introduce the notions of *n-homogeneity* and *finite homogeneity*, as follows.

Let n be a positive integer, and let $Z_n = \{1, 2, \ldots, n\}$. A metric space (X, d) is said to be *n-homogeneous* iff, for any metric δ on Z_n, any two isometric embeddings of (Z_n, δ) in (X, d) are equivalent in the sense of §1. We say that (X, d) is *finitely homogeneous* iff it is n-homogeneous for every positive integer n.

It is convenient to use the term *mobility space* for a metric space that is a mobility object in the metric category. Thus (X, d) is a mobility space iff every isometric embedding $f:(A, \delta) \to (X, d)$ is mobile.

THEOREM 1. Every mobility space is finitely homogeneous.

Proof. We proceed by induction on n. Let (X, d) be a mobility space. Trivially, any map of $Z_1 = \{1\}$ in X is an isometric embedding. Let $f(1) = a_1$, $g(1) = b_1$ be two such maps. Consider the subset $S = \{a_1, b_1\}$ of X. Then the isometry group $\operatorname{Iso} S$ is isomorphic to the cyclic group of order two, and its generator is the involution $\varphi : S \to S$, $\varphi(a_1) = b_1$, $\varphi(b_1) = a_1$. Since (X, d) is a mobility space, φ extends to an involution $\tilde{\varphi}$ of (X, d), under some mobilisation $\sigma : \operatorname{Iso} S \to \operatorname{Iso}(X, d)$. Thus $\tilde{\varphi} f = g 1_{Z_1}$, and 1_{Z_1}, is the only isometry of Z_1. So (X, d) is 1-homogeneous.

Suppose next that (X, d) is r-homogeneous, for some $r \geq 1$. Let $f : Z_{r+1} \to X$, $g : Z_{r+1} \to X$ be isometric embeddings for some metric δ on Z_{r+1}, with $f(j) = a_j$, and $g(\varphi(j)) = b_i$, for some isometry $\varphi \in \operatorname{Iso}(Z_{r+1}, \delta)$, $j = 1, \ldots, r+1$. Let $\iota : Z_r \to Z_{r+1}$ be the inclusion, and $\delta' = \delta | Z_r \times Z_r$. Then $f' = f\iota$ and $g' = g\varphi\iota$ are isometric embeddings of (Z_r, δ') in (X, d), with $f'(i) = a_i$, $g'(i) = b_i$, $i = 1, \ldots, r$. Thus there are isometries $\psi : Z_r \to Z_r$ and $\theta : X \to X$ such that $\theta f' = g' \psi$.

Suppose that $\psi(a_{r+1}) = b_*$, and define $g'': Z_{r+2} \to X$ by $g''(j) = b_j$, $j = 1, \ldots, r+1$, $g''(r+2) = b_*$. Then g'' induces a metric δ'' on Z_{r+2}, with respect to which the involution $h: Z_{r+2} \to Z_{r+2}$, given by $h(i) = i$, $i = 1, \ldots, r$, $h(i+1) = i+2$, $h(i+2) = i+1$ is an isometry. Since (X, d) is a mobility space, h maps under some mobilisation of g'' to an isometry ρ of X. Then $\omega = \rho\theta$ is an isometry of X, and $g\varphi = \omega f$. Hence, by induction on r, (X, d) is finitely homogeneous.

The next result shows that certain subsets of a mobility space, namely the spheres round a point, are themselves mobility spaces. Let (X, d) be a metric space. Then for each $x \in X$ and each $\varepsilon > 0$, the subset $S(x, \epsilon) = \{y \in X : d(x, y) = \varepsilon\}$ is called the *sphere* in X with *radius* ε and *centre* x. We give $S = S(x, \varepsilon)$ the induced metric $d_S = d|S \times S$.

THEOREM 2. Any sphere (S, d_S) in a mobility space (X, d) is a mobility space.

Proof. Let $S = S(x, \varepsilon)$ be nonempty. Let A be a nonempty subset of S, and put $B = A \cup \{x\}$. We denote the isometry group of each subspace (Y, d_Y) of (X, d) by $I(Y)$. Define a monomorphism $\alpha: I(A) \to I(B)$ by $\alpha(g)(a) = g(a)$ for $a \in A$ and $\alpha(g)(x) = x$, for all $g \in I(A)$. Since (X, d) is a mobility space, there is a mobilisation $\sigma: I(B) \to I(X)$ of the inclusion $i: B \to X$. We want to use σ to construct a mobilisation σ_A of the inclusion $j: A \to S$. Let $g \in I(A)$. Since $g' = \sigma(\alpha(g))$ keeps x fixed, it follows that $g'|S \in I(S)$ and $g'|A \in I(A)$. In fact, $g'|A = g$. Hence we can define the required mobilisation σ_A by $\sigma_A(g) = \sigma(\alpha(g))|S$.

Before going on to apply these results to Riemannian manifolds, we give an example of how the conclusion of Theorem 1 can be strengthened when we make some additional assumptions about the metric topology. We give the isometry group of (X, d) the compact-open topology.

THEOREM 3. Let (X, d) be a separable mobility space, and suppose that $G = \mathrm{Iso}\,(X, d)$ is compact. Then (X, d) is homogeneous.

Proof. Let $F: A \to X$, $g: A \to X$ be isometric embeddings of (A, δ) in (X, d), and let $\varphi: A \to A$ be an isometry of A onto itself. Then φ is a homeomorphism of A with respect to the metric topology of δ. Since (X, d) is separable, there is a countable dense subset Δ of X.

Thus $\Delta_f = \Delta \cap f(A)$ and $\Delta_g = g\varphi f^{-1}(\Delta_f)$ are countable dense subsets of $f(A)$ and $g(A)$ respectively. Further, since X is Hausdorff and f, g and φ are continuous, the map $\psi = g\varphi f^{-1}: f(A) \to f(B)$ is uniquely determined by its restriction ψ_* to Δ_f.

If Δ_f is finite, then ψ_* and hence ψ extends to an isometry θ of (X, d), by Theorem 1. Suppose then that Δ_f is countably infinite, say $\Delta_f = \{x_1, \ldots, x_n, \ldots\}$. For each positive integer n, put $\Delta_n = \{x_1, \ldots, x_n\}$. Then $\psi_n = \psi|\Delta_n$ extends to $\theta_n \in G$, for each n, again by Theorem 1. Since G is compact, there is a convergent subsequence of the sequence (θ_n), with limit $\theta \in G$. Thus θ is an extension of ψ. Hence $\theta f = g\varphi$ as required.

5. Riemannian manifolds

In this section we consider a smooth ($= C^\infty$) m-dimensional Riemannian manifold M, and give M the associated metric d. Thus (M, d) is a metric space. We call M a *mobility manifold* iff (M, d) is a mobility space.

THEOREM 4. Let M be a smooth connected Riemannian m-manifold. Then M is a mobility manifold iff it is a complete simply-connected space of constant curvature.

Proof. It is a straightforward exercise to verify that flat Euclidean m-space, the standard m-sphere S^m of radius $R > 0$, and any real hyperbolic m-space are mobility manifolds. It remains only to eliminate all other Riemannian manifolds.

By Theorem 1, any mobility m-manifold M is 2-homogeneous. Now these were completely determined by Wang [10] and Tits [9] (see also Wolf [11]). By Theorem 2, the geodesic sphere $S = S(x, \varepsilon)$ in M with centre $x \in M$ and sufficiently small radius $\varepsilon > 0$ is a mobility subspace with respect to the induced metric d_S obtained from d by restriction. Now for $\varepsilon > 0$ sufficiently small, S is a smoothly embedded $(m-1)$-sphere, and the inclusion of S in M induces a smooth Riemannian tensorfield, and hence a metric δ, on S. Of course d_S and δ are distinct metrics (see Figure 1). However, the fact that M is a mobility manifold implies that (with the obvious notation) $I(S, d_S)$ is a subgroup of $I(S, \delta)$. For if $g \in I(S, d_S)$, then g extends as in the proof of Theorem 2 to an element g_* of $I(M)$ keeping x fixed, and hence g_* restricts to an isometry of (S, δ).

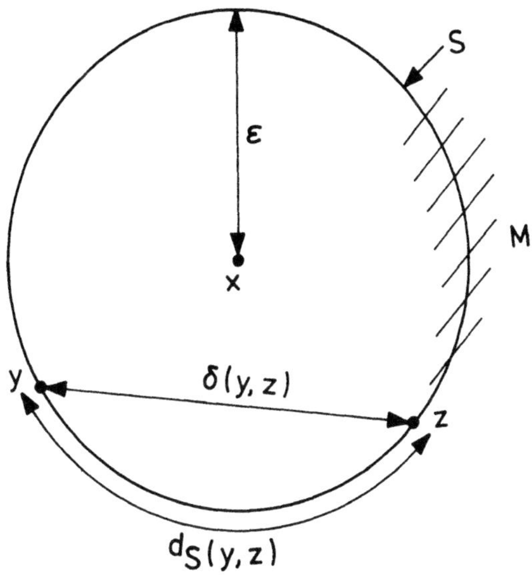

Fig. 1.

By Theorems 1 and 2, (S, d_S) is 2-homogeneous. Hence (S, δ) is also 2-homogeneous and so S, as a Riemannian $(m-1)$-manifold, is isometric to a standard $(m-1)$-sphere with elliptic Riemannian structure. Thus M has constant sectional curvature. Now the only 2-homogeneous spaces of constant curvature are those listed in the theorem, together with the standard real projective spaces. We have only to show that the latter are not mobility manifolds.

To see that this is the case, consider the real projective m-space P^m of diameter $\frac{1}{2}\pi$. Then we can embed the set of vertices of an equilateral Euclidean plane triangle of side $\frac{1}{3}\pi$ in P^m either as a set of collinear points or as a set of noncollinear points. This proves that P^m is not 3-homogeneous, and hence not a mobility manifold for $m \geqslant 2$. Of course $P^1 = S^1$.

The history of the ideas that I have touched upon in this article is long and interesting, and the list of references is intended to supply sources of further historical information. For the latter purpose, the books of Busemann [2], Blumenthal [1] and Laugwitz [6], and Russell's essay [8] are particularly useful.

The University of Southampton

REFERENCES

[1] Blumenthal, L. M., *Distance Geometry*, Oxford, 1953.
[2] Busemann, H., 'Metric Methods in Finsler Spaces and in the Foundations of Geometry', *Ann. of Math. Study* **8**, Princeton, 1942.
[3] Cartan, E., *Leçons sur la géométrie des espaces de Riemann*, 2nd edition, Gauthier-Villars, Paris, 1951.
[4] Epstein, D. B. A., 'The Simplicity of Certain Groups of Homeomorphisms', *Comp. Math.* **22** (1970), 165–173.
[5] Kervaire M. and Milnor, J. W., 'Groups of Homotopy Spheres I', *Ann. of Math.* **77** (1963), 504–537.
[6] Laugwitz, D., *Differential and Riemannian Geometry*, Academic Press, New York, 1965.
[7] MacLane, S., *Categories for the Working Mathematician*, Springer, New York, 1971.
[8] Russell, B. A. W., *An Essay on the Foundations of Geometry*, Cambridge, 1897, (Dover reissue, New York, 1956).
[9] Tits, J., 'Sur certaines classes d'espaces homogènes de groupes de Lie', Memoir 29, Belgian Academy of Sciences, 1955.
[10] Wang, H. W., 'Two Point Homogeneous Spaces', *Ann. of Math.* **55** (1952), 177–191.
[11] Wolf, J. A., *Spaces of Constant Curvature*, McGraw-Hill, New York, 1967.

T. J. WILLMORE

THE TENSION FIELD OF MAPS OF RIEMANNIAN MANIFOLDS

1. Introduction

The concept of the tension field of a mapping of riemannian manifolds was introduced by J. Eells and J. H. Sampson in [3]. In the particular case when the mapping is an isometry, the tension field becomes none other than the mean curvature vector field. A harmonic map is one for which the tension field is zero, so that an isometric harmonic map has zero mean curvature, i.e. the image is a minimal surface. Considerable interest has been shown during the last decade about the value of the square of the length of the mean curvature vector for isometric immersions. However this is clearly a special case of more general results concerning the square of the length of the tension field of maps (in general, not isometric) of riemannian manifolds.

In §2 we obtain fundamental local formulas involving the tension field. In §3 we give a brief summary of known results about isometric maps and indicate some natural problems involving tension fields in the more general case. In §4 we consider totally geodesic maps, and obtain a global theorem about such maps.

2. Fundamental local formulas

Since the first draft of this paper was written, a paper by S. S. Chern and S. Goldberg 'On the Volume-Decreasing Property of a Class of Real Harmonic Mappings' [2] has appeared, so we have rewritten this section making use of their notation and several of their formulas.

Let M and N be smooth oriented riemannian manifolds of dimension m and n respectively. Let ω_i, $i = 1, 2, \ldots, m$, and ω^*_α, $\alpha = 1, 2, \ldots, n$, be linear differential forms in M and N respectively, such that the metrics can be expressed locally as

$$ds^2_M = \omega_1^2 + \cdots + \omega_m^2,$$
$$ds^2_N = \omega_1^{*2} + \cdots + \omega_n^{*2}.$$

Cahen & Flato (eds.), Differential Geometry and Relativity, 159–167.

The structure equations in M are

$$d\omega_i = \sum_j \omega_j \wedge \omega_{ji},$$

$$d\omega_{ij} = \sum_k \omega_{ik} \wedge \omega_{kj} + \Omega_{ij},$$

$$\omega_{ij} + \omega_{ji} = 0.$$

If we denote by (e_i) the orthonormal frame dual to the coframe (ω_j), the connexion D in the tangent bundle is given by

$$De_i = \sum_j \omega_{ij} e_j.$$

The Ω_{ij} are the curvature forms, related to the Riemann Christoffel curvature tensor by

$$\Omega_{ij} = -\frac{1}{2} \sum_{k,l} R_{ijkl} \omega_k \wedge \omega_l,$$

where

$$R_{ijkl} = -R_{jikl} = -R_{ijlk},$$
$$R_{ijkl} = R_{klij},$$
$$R_{ijkl} + R_{iklj} + R_{iljk} = 0.$$

Similar equations are valid in N.

Let $f: M \to N$ be a smooth mapping. Then the pull-back of ω^*_α is given by

$$f^* \omega^*_\alpha = \sum_i a_{\alpha i} \omega_i. \tag{2.1}$$

In what follows it is convenient to simplify (2.1) by deleting f^*, i.e. by rewriting this as

$$\omega^*_\alpha = \sum a_{\alpha i} \omega_i. \tag{2.2}$$

If we take the exterior derivative of (2.2) and use the structure equations we find

$$Da_{\alpha i} \equiv da_{\alpha i} + \sum_j a_{\alpha j} \omega_{ji} + \sum_\beta a_{\beta i} \omega^*_{\beta\alpha} = \sum a_{\alpha ij} \omega_j, \tag{2.3}$$

where

$$a_{\alpha ij} = a_{\alpha ji}. \tag{2.4}$$

The tension field τ of the map f is given by

$$\tau = \sum_{i,\alpha} a_{\alpha ii} e^*_\alpha, \tag{2.5}$$

where (e^*_α) is the orthonormal frame dual to (ω^*_α).

By differentiating (2.3) and using the structure equations we find

$$\sum Da_{\alpha ij} \wedge \omega_j = \sum a_{\alpha j} \Omega_{ji} + \sum a_{\beta i} \Omega^*_{\beta\alpha} \tag{2.6}$$

where

$$Da_{\alpha ij} = \mathrm{d}a_{\alpha ij} + \sum a_{\beta ij} \omega^*_{\beta\alpha} + \sum a_{\alpha kj} \omega_{ki} + \sum a_{\alpha ik} \omega_{kj}. \tag{2.7}$$

Writing

$$Da_{\alpha ij} = \sum a_{\alpha ijk} \omega_k \tag{2.8}$$

we find from (2.6) the 'Ricci identity'

$$a_{\alpha ijk} - a_{\alpha ikj} = -\sum a_{\alpha l} R_{likj} - \sum a_{\beta i} a_{\gamma k} a_{\delta j} R^*_{\beta\alpha\gamma\delta}. \tag{2.9}$$

Finally, we obtain from (2.4) and (2.9) the Laplacian

$$\Delta a_{\alpha i} \equiv \sum_k a_{\alpha ikk} = \sum a_{\alpha kki} + \sum a_{\alpha l} R_{li} - \sum a_{\beta k} a_{\gamma k} a_{\delta i} R^*_{\beta\alpha\gamma\delta}. \tag{2.10}$$

The equations obtained in this section are well known, and are summarized in [2]. We make use of these in the subsequent sections.

3. Known results of isometric maps and analogous problems involving tension fields

From (2.5) we have

$$\tau^2 = \sum_\alpha \left(\sum_i a_{\alpha ii} \right)^2. \tag{3.1}$$

We define the total tension of the map f by

$$T(f) = \int_M \tau^m * 1 \tag{3.2}$$

where $*\ 1$ denote the volume element of M. Here we assume that the integral in (3.2) converges, which is certainly the case when M is compact.

In the particular case when f is an isometry, (3.2) reduces to

$$T(f) = \int_M H^m * 1 \tag{3.3}$$

where H is the mean curvature. In particular when $m = 2$ and $N = E^3$, (3.3) reduces to

$$T(f) = \int_M H^2 * 1. \tag{3.4}$$

A problem which has aroused interest is to compute the infimum of $T(f)$ in (3.4), for variable f ([4], [5], [6] and [1]). The main results known in this special case are

THEOREM 3.1. For any closed surface M, $T(f) \geqslant 4\pi$, equality holding only when M is S^2 and $f(S^2)$ is the (round) euclidean sphere.

THEOREM 3.2. Let M be a torus $S^1 \times S^1$ and $f(M)$ be a surface of revolution in E^3. Then $T(f) \geqslant 2\pi^2$, equality holding only when $f(S^2)$ is an anchor ring whose generating circles have radii in the ratio $\sqrt{2}:1$.

THEOREM 3.3. Let M be a torus $S^1 \times S^1$ and $f(M)$ be a tubular surface obtained by rotating a circle around a closed curve γ so that its centre traverses γ and the plane of the circle always lies in the normal plane of γ. Then $T(f) \geqslant 2\pi^2$, equality holding only in the special case of the anchor ring whose generating radii are in the ratio $\sqrt{2}:1$.

In 1968 I made the following conjecture:

Let M be a torus $S^1 \times S^1$ and let f be any smooth immersion of M in E^3. Then $T(f) \geqslant 2\pi^2$.

This conjecture has been made more plausible by results obtained since then, but the conjecture remains unresolved.

If we consider variations of (3.4) for normal deformations, the Euler equation is known to be

$$\Delta H + 2H(H^2 - K) = 0. \tag{3.5}$$

Both the integral (3.4) and the corresponding equation (3.5) are known to be invariant under conformal transformations of E^3. Moreover, it is known that there are closed surfaces of arbitrary genus g which satisfy (3.5).

Naturally analogous problems are associated with the infimum of $T(f)$ obtained from Equation (3.2), and also in the particular case

$$T(f) = \int_M \tau^2 * 1, \tag{3.6}$$

where $m = 2$.

An obvious infimum occurs when $\tau = 0$, i.e. the map is harmonic. However it is known from [2] that if f is isometric, then f is harmonic if and only if it is a minimal immersion. It is known from recent work of B. Lawson [4], that minimal immersions exist for surfaces of arbitrary genus when $N = S^3$. However it is well known that harmonic isometric maps of a compact manifold M into E^n do not exist. Little seems known in the general case, and the problem reduces to that of the existence of harmonic maps.

4. Totally geodesic maps

We return to the equations of §2. We recall that the condition for the map f to be harmonic is

$$\sum_i a_{\alpha ii} = 0. \tag{4.1}$$

In particular, if

$$a_{\alpha ij} = 0, \tag{4.2}$$

the map f is said to be *totally geodesic*. Evidently every totally geodesic map is harmonic but not conversely. We now obtain a global theorem and give conditions when a harmonic map is totally geodesic. In particular we prove

THEOREM 4.1. Let M be a compact connected riemannian manifold and let $f: M \to N$ be a smooth immersion into a riemannian manifold N. Let $Q=\Sigma_{\alpha,i} a_{\alpha i}\Delta a_{\alpha i}$ be non-negative everywhere on M. Then f is a totally geodesic map.

Let

$$\begin{aligned} h &= \sum_{\alpha,i} a_{\alpha i}^2 \\ &= \operatorname{trace}(AA^T), \end{aligned} \tag{4.3}$$

where A denotes the matrix $(a_{\alpha i})$.

It follows that h is a scalar-valued invariant defined over M, independent of the choice of bases (e_i), (e^*_α).

If we take the exterior derivative of (4.3) and use (2.3) we get

$$\begin{aligned} \mathrm{d}h &= 2\sum_{\alpha,i} a_{\alpha i}\,\mathrm{d}a_{\alpha i} \\ &= 2\sum_{\alpha,i,j} a_{\alpha i}a_{\alpha ij}\omega_j - 2\sum_{\alpha,i,j} a_{\alpha i}a_{\alpha j}\omega_{ji} - 2\sum_{\alpha,\beta,i} a_{\alpha i}a_{\beta i}\omega^*_{\beta\alpha}. \end{aligned}$$

The last two terms of the right hand member vanish due to the skew-symmetry of ω_{ji} and $\omega^*_{\beta\alpha}$. Thus

$$\mathrm{d}h = 2\sum_{\alpha,i,j} a_{\alpha i}a_{\alpha ij}\omega_j. \tag{4.4}$$

Writing

$$\mathrm{d}h = \sum_k h_k\omega_k \tag{4.5}$$

it follows from (4.4) that

$$h_k = 2\sum_{\alpha,i} a_{\alpha i}a_{\alpha ik}. \tag{4.6}$$

We take the exterior derivative of (4.6) to get

$$\mathrm{d}h_k = 2\sum_{\alpha,i} (\mathrm{d}a_{\alpha i}a_{\alpha ik} + a_{\alpha i}\,\mathrm{d}a_{\alpha ik}).$$

Using (2.3), (2.7), (2.8) this simplifies to

$$\mathrm{d}h_k = 2\sum_{\alpha,i,j} [(a_{\alpha ij}a_{\alpha ik} + a_{\alpha i}a_{\alpha ikj})\omega_j - a_{\alpha i}a_{\alpha ij}\omega_{jk}]. \tag{4.7}$$

Also by taking the exterior derivative of (4.5) we have

$$0=\sum_k (\mathrm{d}h_k \wedge \omega_k + h_k\, \mathrm{d}\omega_k)$$
$$=\sum_k \left(\mathrm{d}h_k \wedge \omega_k + h_k \sum_j \omega_j \wedge \omega_{jk}\right)$$
$$=\sum_k \left(\mathrm{d}h_k - \sum_j h_j \omega_{kj}\right) \wedge \omega_k.$$

Writing

$$\mathrm{d}h_k - \sum_j h_j \omega_{kj} = \sum_j h_{kj}\omega_j$$

we have

$$h_{kj} = h_{jk}.$$

Using (4.6) we have

$$\mathrm{d}h_k - 2\sum_{\alpha,i,j} a_{\alpha i} a_{\alpha ij} \omega_{kj} = \sum h_{kj}\omega_j. \tag{4.8}$$

From (4.7) and (4.8) we have, eliminating $\mathrm{d}h_k$,

$$2\sum_{i,j,\alpha} [(a_{\alpha ij} a_{\alpha ij} + a_{\alpha i} a_{\alpha ikj})\omega_j - a_{\alpha i} a_{\alpha ij} \omega_{jk}]$$
$$=2\sum_{i,j,\alpha} a_{\alpha i} a_{\alpha ij} \omega_{kj} + \sum_j h_{kj}\omega_j.$$

Thus

$$h_{kj} = 2\sum_{\alpha,i} (a_{\alpha ij} a_{\alpha ik} + a_{\alpha i} a_{\alpha ikj}),$$

and thus

$$\Delta h = \sum_k h_{kk} = 2\sum_{\alpha,i,k} (a_{\alpha ik} a_{\alpha ik} + a_{\alpha i} a_{\alpha ikk})$$
$$= 2\sum_{\alpha,i,k} (a_{\alpha ik})^2 + 2\sum_{\alpha,i} a_{\alpha i} \Delta a_{\alpha i}.$$

Writing

$$Q = \sum a_{\alpha i} \Delta a_{\alpha i},$$

we have thus proved that

$$\Delta h = 2\sum_{\alpha,i,k} (a_{\alpha ik})^2 + 2Q. \tag{4.9}$$

Since the hypothesis of the theorem states that $Q \geqslant 0$, it follows from the lemma of E. Hopf that $\Delta h = 0$ and h is a constant function. Thus

$Q=0$ everywhere and $\Sigma_{\alpha,i,k}\, a_{\alpha ik}^2 = 0$, from which $a_{\alpha ik}=0$ and the map is thus totally geodesic. This completes the proof of the theorem.

We note† from (4.4) that if f is given totally geodesic, then h is a constant function. Moreover from (2.8) it follows that $a_{\alpha ij}=0$ implies $\Delta a_{\alpha i}=0$ and hence $Q=0$.

From (2.10) we have

$$Q=\sum_{\alpha,k,i} a_{\alpha kki}a_{\alpha i}+\sum_{\alpha,k,i} a_{\alpha k}a_{\alpha i}R_{ki}-\sum_{\alpha,\beta,\gamma,\delta,k,i} a_{\alpha i}a_{\beta k}a_{\gamma k}a_{\delta i}R^*_{\beta\alpha\gamma\delta}. \tag{4.10}$$

Let us now assume that the map f is harmonic. Then the first term in the expression of Q given by (4.10) vanishes. Thus $Q\geqslant 0$ implies

$$\sum_{\alpha,k,i} a_{\alpha k}a_{\alpha i}R_{ki}\geqslant\sum a_{\alpha i}a_{\beta k}a_{\gamma k}a_{\delta i}R^*_{\beta\alpha\gamma\delta}. \tag{4.11}$$

THEOREM 4.2. Let M, N be smooth connected orientable riemannian manifolds with M compact. Let $f: M\to N$ be a harmonic map. If

$$\sum_{\alpha,k,i} a_{\alpha k}a_{\alpha i}R_{ki}\geqslant\sum a_{\alpha i}a_{\beta k}a_{\gamma k}a_{\delta i}R^*_{\beta\alpha\gamma\delta}$$

everywhere on M, then f is totally geodesic.

We note that $\Sigma_\alpha\, a_{\alpha k}a_{\alpha i}$ and $\Sigma_i\, a_{\alpha i}a_{\delta i}$ are both symmetric matrices which are positive semi-definite. If R_{ki} is positive semi-definite, then so is the left hand member of (4.11). Moreover, if N has a metric of constant curvature k, the right hand member of (4.11) has the same sign as k. But Theorem 4.2 implies that if (4.11) is satisfied everywhere, then $Q=0$ and hence (4.11) is an equality. As a corollary we conclude that there are no harmonic maps of a compact Ricci-positive semi-definite M into a manifold N of constant negative curvature. Again we are forced to consider the problem of the existence of harmonic maps, as at the end of §3.

University of Durham

† This was pointed out to me by Julian Havil in the course of preparing his Ph.D. Thesis at Durham.

REFERENCES

[1] Chen, B. Y., *Journal Lond. Math. Soc.* **6** (1973), 321–325.
[2] Chern, S. S. and Goldberg, S. I., 'On the Volume-Decreasing Property of a Class of Real Harmonic Mappings', *Amer. J. Math.* **97** (1975), 133–147.
[3] Eells, J. and Sampson, J. H., 'Harmonic Mappings of Riemannian Manifolds', *Amer. J. Math.* **86** (1964), 109–160. M.R. 29 # 1603.
[4] Lawson, B., 'Global Analysis', *Proc. Symp. Pure Math. XV*, Berkeley, Calif. (1968), 275–282.
[5] Willmore, T. J., An. Sti. Univ., Al. I. Cuza: Iassi, la. Mat. 11B (1965), 493–496. M.R. 34 # 1940.
[6] Willmore, T. J., *Journal Lond. Math. Soc.* **3** (1971), 307–310.
[7] Willmore, T. J., *Lecture Notes in Mathematics*, No. 392, Springer-Verlag, 1974.

PART II

MATHEMATICAL PHYSICS

F. BAYEN

CONFORMAL INVARIANCE IN PHYSICS

1. Introduction

In the last decade mathematical physicists have been especially interested by the group of conformal transformations of space-time – the so-called conformal group. What are the main physical reasons for the increasing interest in such an old subject [1]?

(a) The experimental discovery that, for high-energy and large momentum transfer inelastic electron-proton scattering, form factors may be scale invariant. This fact, which confirmed Bjorken scaling laws [2], has suggested to study the constraints imposed by dilatation and conformal invariance on high-energy limit of scattering amplitudes. For a two-body scattering it was found [3] that, once scale invariance is satisfied, the scattering amplitude for four-scalar particles is automatically conformal invariant. However, in the same paper it was shown that conformal invariance yields nontrivial constraints on the scattering of spinning particles.

In quantum field theory the assumption of conformal invariance is very promising. It allows to determine apart from a constant Wightman's 2- and 3-point functions [4]:

$$w_2(x_1, x_2) = \langle 0|\varphi_1(x_1)\varphi_2(x_2)|0\rangle \quad \text{and}$$
$$w_3(x_1, x_2, x_3) = \langle 0|\varphi_1(x_1)\varphi_2(x_2)\varphi_3(x_3)|0\rangle$$

once the dimensions of the fields are known. When $n \geqslant 4$, the form of the n-point function is severely restricted by conformal invariance but is not unique. The reason is the existence for four points x_1, x_2, x_3, x_4 of space-time of two conformal invariant 'anharmonic ratios':

$$\xi = \frac{(x_1 - x_2)^2(x_3 - x_4)^2}{(x_1 - x_3)^2(x_2 - x_4)^2}, \qquad \eta = \frac{(x_1 - x_4)^2(x_2 - x_3)^2}{(x_1 - x_3)^2(x_2 - x_4)^2}.$$

Thus any function of ξ and η will be conformal invariant.

Finally for a wide family of Lagrangian scale, invariance implies conformal invariance (see Ref. [5] and Section 3).

Cahen & Flato (eds.), Differential Geometry and Relativity, 171–195.

(b) Segal [6] has proposed a chronogeometric space-time model $\tilde{M}$ which is locally identical with Minkowski space, but globally distinct. The manifold $\tilde{M}$, homeomorphic to $\mathbb{R} \times S^3$, is the universal covering space of an homogeneous space of the group $SO_0(4, 2)$ which is locally isomorphic to the conformal group. The group of causality-preserving symmetries of $\tilde{M}$ is an infinite-sheeted covering of the conformal group. This model implies among other cosmological features a non-Doppler redshift. It gives also an explanation of the anomalous apparent velocity of certain quasi-stellar objects.

(c) The conformal group is locally isomorphic to SU (2, 2) the Lie algebra of which is a unification [7] of the Poincaré Lie algebra with the Lie algebra su (2, 1). The group SU (2, 1) was an alternative to the internal symmetry group SU (3) and contains as a maximal compact subgroup the invariance group U (2) of strong interactions. These facts allowed the following interpretation: the Poincaré subgroup (with its representations) describes the kinematics of the particles and SU (2, 1) enables one to classify the particle within multiplets. The role of the unification was to predict the discrete mass spectrum of the hadrons. This was achieved [8] with the help of the technique of group contraction.

(d) It is well known that the Lie algebra so (4, 2) appears in the treatment of the classical and quantum mechanical nonrelativistic Kepler problem. In both cases the symmetry group of the problem is generated by the angular-momentum and the Runge-Lenz vector (SO (4) for bound states). In the quantum case there exists a unitary irreducible representation of the group SO (4, 1) which describes all the bound states of the hydrogen atom [9]. This representation may be extended in the same Hilbert space to a unitary representation of $SO_0(4,2)$. This enlargement introduces additional operators which may be identified with interaction operators [10]. This fact enables one to treat the corresponding interactions (e.g. the Stark effect) in a purely algebraic manner, i.e. without any computation of definite integrals. Furthermore, this 'dynamical group' role of $SO_0(4,2)$ is connected to the solution of various problems [11].

Other topics connected with the conformal group are:

(1) infinite-component wave equations in hadron physics [12], in order to correlate the huge number of masses and spins of resonances;
(2) twistor theory in flat and curved space-time [13];
(3) the link between invariance with respect to the (infinite-

dimensional) group of general relativity and the joined invariance with respect to affine and conformal group [14].

All this points show that the subject is now widespread, in fact more than three hundred papers in ten years. Therefore, no attempt to completeness will be done here except for the problem of conformal invariance of field equations where, moreover, some new facts will be given.

2. Mathematical results

2.1. *Some standard facts about the conformal group*

A conformal transformation in Minkowski space M is an analytic transformation $x \mapsto x'$ defined in a maximal open set of M and which is such that $ds^2 = 0$ implies $ds'^2 = 0$ (i.e. light cones are mapped into light cones). The metric which will be used in this article is $(+,-,-,-)$:

$$ds^2 = g_{\mu\nu}\, dx^\mu\, dx^\nu = (dx^0)^2 - (dx^1)^2 - (dx^2)^2 - (dx^3)^2.$$

The set of all these transformations will be denoted by C. They are generated by

(a) The transformation of the Poincaré group $\mathcal{P} : x \mapsto x' = \Lambda x + b = Px$, where $\Lambda \in SO_0(3,1)$, $b \in M$, and $P = (b, \Lambda)$.
(b) Positive dilatations $x \mapsto x' = \lambda x$ with $\lambda > 0$.
(c) Special conformal transformations (accelerations) $x \mapsto x' = \omega(a,x)^{-1}(x + ax^2)$ where $a \in M$, $\omega(a,x) = 1 + 2ax + a^2x^2$ and $a^2 = g_{\mu\nu}a^\mu a^\nu$, $ax = g_{\mu\nu}a^\mu x^\nu$.
(d) Space reflection $x = (x^0, x^1, x^2, x^3) \mapsto x' = (x^0, -x^1, -x^2, -x^3)$ and time reversal $x \mapsto x' = (-x^0, x^1, x^2, x^3)$.

The transformations (a), (b), and (c) generate the connected component C_0 of the identity in C. In the sequel C_0 will be called the conformal group. C_0 is a fifteen-dimensional Lie group and is the lowest dimensional semi-simple Lie group containing $\mathcal{P}$. A representation of the Lie algebra of C_0 is obtained by Cartan's method:

$$\begin{cases} M_{\mu\nu} = x_\mu \partial_\nu - x_\nu \partial_\mu & \text{Lorentz transformations} \\ P_\mu = \partial_\mu & \text{Translations} \\ D = x^\rho \partial_\rho & \text{Dilatations} \\ K_\mu = x^2 \partial_\mu - 2x_\mu D & \text{Special conformal transformations} \end{cases}$$

with the following commutation relations

$$\begin{cases} [M_{\mu\nu}, M_{\rho\sigma}] = g_{\mu\sigma}M_{\nu\rho} + g_{\nu\rho}M_{\mu\sigma} - g_{\mu\rho}M_{\nu\sigma} - g_{\nu\sigma}M_{\mu\rho} \\ [M_{\mu\nu}, P_\rho] = g_{\nu\rho}P_\mu - g_{\mu\rho}P_\nu \\ [P_\mu, P_\nu] = 0 \\ [M_{\mu\nu}, D] = 0 \\ [P_\mu, D] = P_\mu \\ [K_\mu, K_\nu] = 0 \\ [M_{\mu\nu}, K_\rho] = g_{\nu\rho}K_\mu - g_{\mu\rho}K_\nu \\ [K_\mu, D] = -K_\mu \\ [K_\mu, P_\nu] = 2(g_{\mu\nu}D + M_{\mu\nu}). \end{cases}$$

The Lie algebra of C_0 is of type $\mathfrak{D}_3$ and is isomorphic to $\mathrm{su}(2,2) \approx \mathrm{so}(4,2)$. The generators $M_{\mu\nu}$, P_μ, D and $M_{\mu\nu}$, K_μ, D respectively generate two eleven-parameter subgroups W and W^* which are isomorphic and called Weyl groups. W and W^* are conjugate by the involutive inner automorphism of $C_0: g \mapsto \exp(\pi X) g \exp(-\pi X)$, where $X = \frac{1}{2}(P_0 - K_0)$. The universal covering groups of $\mathscr{P}$ and W or W^* will be denoted by $\bar{\mathscr{P}}$ and $\bar{W}$. ($\bar{W}$ is a semi-direct product of $\mathrm{SL}(2,\mathbb{C}) \times \mathbb{R}$ by a four-dimensional invariant translation subgroup: $\bar{W} = [\mathrm{SL}(2,\mathbb{C}) \times \mathbb{R}] \cdot \mathbb{R}^4$). Let us denote by Λ the usual homomorphism $\mathrm{SL}(2,\mathbb{C}) \mapsto \mathrm{SO}_0(3,1)$. The action of $\mathrm{SL}(2,\mathbb{C}) \times \mathbb{R}$ on the translations is then given by $(t, A) \cdot a = e^t \Lambda(A) a$, where $t \in \mathbb{R}$, $A \in \mathrm{SL}(2,\mathbb{C})$, and $a \in M$. Besides in $\mathrm{su}(2,2)$ one has the following two important relations:

$$\mathrm{Ad}(\exp(tD))P_\mu = e^{-t}P_\mu, \tag{2.1.1}$$

$$\mathrm{Ad}(\exp(a^\nu K_\nu))P_\mu = P_\mu + 2a^\nu(g_{\nu\mu}D + M_{\nu\mu}) + (a^2\delta_\mu^\nu - 2a_\mu a^\nu)K_\nu. \tag{2.1.2}$$

The three groups C_0, $\mathrm{SO}_0(4,2)$, and $\mathrm{SU}(2,2)$ are locally isomorphic. More precisely the centers of $\mathrm{SO}_0(4,2)$ and of $\mathrm{SU}(2,2)$ have respectively two and four elements. One has $C_0 \approx \mathrm{SO}_0(4,2)/Z_2 \approx \mathrm{SU}(2,2)/Z_4$. We briefly study the homomorphism $\varphi: \mathrm{SO}_0(4,2) \to C_0$, $\theta: \mathrm{SU}(2,2) \to \mathrm{SO}_0(4,2)$, $\psi: \mathrm{SU}(2,2) \to C_0$.

To begin with, we construct the conformal transformation corresponding to $\varphi(g)$ for any $g \in \mathrm{SO}_0(4,2)$. Let $\mathbb{P}^5$ be the five-dimensional projective space and $\pi: R^6 - \{0\} \to \mathbb{P}^5$ the canonical map. Consider the pointed cone $Q_0^5 = \{\eta \in \mathbb{R}^6 - \{0\} | \eta^a \eta_a = (n^0)^2 - (n^1)^2 - (\eta^2)^2 - (\eta^3)^2 - (\eta^4)^2 + (\eta^5)^2 = 0\}$ and its image $M_c^4 = \pi(Q_0^5)$ in $\mathbb{P}^5$. The compact manifold M_c^4 is sometimes called the compactified Minkowski space or the conformal closure of Minkowski space. In fact, Minkowski space M is embedded as a dense open submanifold of M_c^4 by the map $\psi_1: M \to M_c^4$ defined by

$$\psi_1(x) = \pi(x^0, x^1, x^2, x^3, \tfrac{1}{2}(1 + x^\mu x_\mu), \tfrac{1}{2}(1 - x^\mu x_\mu)).$$

($x \in M$) and the pair $(\psi_1(M), \psi_1^{-1})$ is a local chart of M_c^4. For further details on this embedding, see Ref. [15]. The group $SO_0(4, 2)$ acts naturally in a C^∞ fashion on M_c^4 by the rule $g \cdot \pi(n) = \pi(g\eta)$, where $n \in Q_0^5$ and $g = (G_b^a) 0 \leq a, b \leq 5$ is any element of $SO_0(4, 2)$. The corresponding action on M will then be defined by the formula $g \cdot x = \psi_1^{-1}(g \cdot \psi_1(x))$; the previous equality being valid if $g \cdot \psi_1(x)$ belongs to the domain of the previous open chart, i.e. to $\psi_1(M)$. In coordinates one gets on M:

$$x^\mu \xrightarrow{\varphi(g)} x'^\mu = \omega(g, x)^{-1}[G_\nu^\mu x^\nu + \tfrac{1}{2}(1 + x^2) G_4^\mu + \tfrac{1}{2}(1 - x^2) G_5^\mu] \tag{2.1.3}$$

with

$$\omega(g, x) = \tfrac{1}{2}(G_4^4 + G_5^4 + G_4^5 + G_5^5) + (G_\mu^4 + G_\mu^5) x^\mu + \tfrac{1}{2}(G_4^4 + G_4^5 - G_5^4 - G_5^5) x^2. \tag{2.1.4}$$

The transformation (2.1.3) is defined in an open set O_g whose complement C_g is a light cone or a plane if $G_4^4 + G_4^5 - G_5^4 - G_5^5 = 0$. Formula (2.1.3) gives the desired conformal transformation $\varphi(g)$ and, if one insists on singularities, C_0 is defined as a pseudogroup of analytic diffeomorphism in M [16]. In this article we will be rather 'conservative': we will use Minkowski space M and not M_c^4. In fact, as far as physics is concerned and unless Segal's cosmology is involved, it seems more natural to stay in M [24].

An homomorphism θ of $SU(2, 2)$ onto $SO_0(4, 2)$ is easily obtained with the help of the following device [17]. Denote the hermitian conjugate and the transpose of any matrix A respectively by $A^\dagger$ and ${}^tA$. Define for any $\eta \in \mathbf{R}^6$ the antisymmetric matrix

$$L(\eta) = \begin{pmatrix} 0 & \eta^0 + i\eta^5 & \eta^1 + i\eta^2 & \eta^3 + i\eta^4 \\ -\eta^0 - i\eta^5 & 0 & -\eta^3 + i\eta^4 & \eta^1 - i\eta^2 \\ -\eta^1 - i\eta^2 & \eta^3 - i\eta^4 & 0 & \eta^0 - i\eta^5 \\ -\eta^3 - i\eta^4 & -\eta^1 + i\eta^2 & -\eta^0 + i\eta^5 & 0 \end{pmatrix}.$$

L is obviously an isomorphism of $\mathbf{R}^6$ into a subset E of the set of 4×4 matrices. Let $U\in \mathrm{SU}(2,2)$, i.e. $U^\dagger I_{2,2}U = I_{2,2}$ with

$$I_{2,2}=\begin{pmatrix}1&&&\\&1&&\\&&-1&\\&&&-1\end{pmatrix}.$$

It is then trivial to verify the following facts:

(a) for any $\eta\in\mathbf{R}^6 | L(\eta)^+I_{2,2}L(\eta)I_{2,2}=\eta^a\eta_a I$, where I is the identity matrix;

(b) for any $\eta\in\mathbf{R}^6$ and any $U\in\mathrm{SU}(2,2)$ the matrix $UL(\eta)^tU\in E: UL(\eta)^tU = L(\eta')$;

(c) $\eta'^a\eta'_a=\eta^a\eta_a$.

We therefore obtain for any $U\in\mathrm{SU}(2,2)$, $\eta'=\theta(u)\eta$ with $\theta(u)\in O(4,2)$. The homomorphism θ is continuous (thus analytic) and the differential $d\theta_e$ is easily seen to be an isomorphism of su (2, 2) onto so (4, 2). This allows one to conclude that $\theta(\mathrm{SU}(2,2))=\mathrm{SO}_0(4,2)$ and it is clear that $\theta(Z_4)=Z_2$. The matrix $L(\eta)$ may be written $L(\eta)=\eta^a\Sigma_a$, where the matrices Σ_a are unitary and such that $\mathrm{Tr}(\Sigma_a\Sigma_b^\dagger)=0$ for $a\neq b$. This enables one to compute the matrix coefficients of $\theta(u):\eta'^a=\theta(u)^a_b\eta^b$. With the convention $\Sigma^a=\Sigma^\dagger_a$, $0\leqslant a\leqslant 5$ one has

$$\theta(u)^a_b=\tfrac{1}{4}\mathrm{Tr}(\Sigma^aU\Sigma_b{}^tU).$$

The detailed construction of the homomorphism $\psi:\mathrm{SU}(2,2)\to C_0$ will be omitted, we give only the result. Let us consider the 4×4 matrix

$$J=\begin{pmatrix}0&I\\-I&0\end{pmatrix},$$

and the group $G=\{g\in\mathrm{SL}(4,\mathbf{C})|g^+Jg=J\}$. It is obvious that G is isomorphic to SU (2, 2). For any vector $x\in M$ we define the 2×2 matrix $X=\sigma^\mu x_\mu$, where σ^j are the Pauli matrices and σ^0 the identity matrix. Let

$$g=\begin{pmatrix}g_{11}&g_{12}\\g_{21}&g_{22}\end{pmatrix}\in G,$$

where g_{ij} are 2×2 matrices. Then under $g\in G$, the conformal transformations are realized by the formula

$$X\xrightarrow{\psi(g)}X'=\sigma^\mu x'_\mu=(g_{11}X+g_{12})(g_{21}X+g_{22})^{-1}. \qquad (2.1.5)$$

About the connection between the Weyl groups W, $\bar{W}$ and the homomorphism ψ, one may verify that there exists an injective homomorphism $i : \bar{W} \to G$ such that if we denote by $p : \bar{W} \to W$ the canonical projection one has $p = \psi \circ i$.

Remark. SU (2, 2) is a fourfold covering of C_0 but is not the universal covering group $\bar{C}$ of C_0. In fact SU (2, 2) is homeomorphic to SU (2) × SU (2) × $U(1) \times \mathbf{R}^8$. Consequently $\bar{C}$ is homeomorphic to SU (2) × SU (2) × $\mathbf{R}^9$ and has a discrete but infinite center. A detailed description of $\bar{C}$ exists in Ref. [16] and as shown in Ref. [25] the group $\bar{C}$ may have a role to play in quantum field theory.

2.2. *Causality and conformal transformations*

Causality is the partial ordering defined on M by $x < y$ if $y - x$ is a time-like vector oriented toward the future. A causal automorphism of M may be defined as a bijective map f of M onto M such that both f and f^{-1} preserve the ordering. Then, as shown by Zeeman [18], the group of causal automorphism is the Weyl group W. This result was analyzed critically in Ref. [19] where it was shown that Zeeman's hypotheses were too restrictive. In fact, these hypotheses led to the astonishing result that the classical photon, with which causality is defined, has a covariance group C_0 strictly bigger than W.

An easy computation shows that if $x, y \in O_g$ and $x \overset{g}{\mapsto} x'$, $y \overset{g}{\mapsto} y'$ are given by (2.1.3), then

$$(x' - y')^2 = \omega(g, x)^{-1}\omega(g, y)^{-1}(x - y)^2.$$

This relation shows that to any bounded region $B \subset M$ containing the origin of M there exists a neighborhood V_B of the identity in C_0 such that the action of V_B on B is causal. In other words, the Lie algebra su (2, 2) acts causally on M. This is clearly sufficient for physics where one goes beyond the geometrical point of view.

2.3. *Representations of the conformal group*

(a) *Finite-dimensional representations.* Let us recall a theorem of Weyl: any finite-dimensional representation of a semi-simple connected Lie group is fully reducible. This theorem combined with Cartan's method (take two irreducible representations, make the tensor product of these representations and pick up in the product the highest weight) enables one in principle to get irreducible representations with increasing weights. In fact, as shown by Mack and Salam [20] all finite-dimensional representation of SU (2, 2) can be obtained by reducing out tensor products of the two inequivalent fundamental four-dimensional representation $u \mapsto u$ and $u \mapsto \bar{u}$. Here $u \in \mathrm{SU}(2,2)$ and the bar denotes complex conjugation.

According to Weyl formula the dimension N of the irreducible representation corresponding to the diagram

$$\overset{a_1}{\circ}\!-\!\!-\!\!-\!\overset{a_2}{\circ}\!-\!\!-\!\!-\!\overset{a_3}{\circ} \quad \text{is}$$

$$N = \tfrac{1}{12}(1+a_1)(1+a_2)(1+a_3)(2+a_1+a_2)(2+a_2+a_3) \times (3+a_1+a_2+a_3).$$

As an example we construct explicitly the representations $\overset{k}{\circ}\!-\!\overset{o}{\circ}\!-\!\overset{l}{\circ}$ for any $k, l \in N$. This will be done with the help of twistors which are the tensors of the group SU (2, 2) in the same sense as spinors are tensors of the group SL (2, C). For an introduction to twistor physics, see the illuminating article of Penrose [13]. A contravariant twistor of degree 1 is a four-valued object with four components upon which an element $u = (u^\alpha_\beta) \in \mathrm{SU}(2,2)$ acts according to $T^\alpha \to T'^\alpha = u^\alpha_\beta T^\beta$. A general twistor $T^{\alpha_1\cdots\alpha_K\dot\beta_1\cdots\dot\beta_l}$ with α_i, $\beta_i \in \{1, 2, 3, 4\}$ will cotransform according to

$$T'^{\alpha_1\cdots\alpha_K\dot\beta_1\cdots\dot\beta_l} = u^{\alpha_1}{}_{\gamma_1} \cdots u^{\alpha_K}{}_{\gamma_K} \bar{u}^{\beta_1}{}_{\delta_1} \cdots \bar{u}^{\beta_l}{}_{\delta_l} T^{\gamma_1\cdots\gamma_K\dot\delta_1\cdots\dot\delta_l}. \tag{2.3.1}$$

A metric twistor $\Gamma^{\alpha\dot\beta} = \Gamma_{\alpha\dot\beta} = (I_{2,2})_{\alpha\beta}$ may be introduced with the following obvious cotransformation property $\Gamma'^{\alpha\dot\beta} = \Gamma^{\alpha\dot\beta}$.

Now consider the vector space F of all twistors of valence (k, l) which are:

(i) symmetric with respect to the α_i and to the $\dot\beta_i$ separately;

(ii) traceless in the sense that for any $\alpha_2, \ldots, \alpha_K, \beta_2, \ldots, \beta_l \in \{1, 2, 3, 4\}$ $\Gamma_{\alpha_1\dot\beta_1} T^{\alpha_1\alpha_2\cdots\alpha_K\dot\beta_1\dot\beta_2\cdots\dot\beta_l} = 0$. (This condition is vacuous if $kl = 0$.)

The dimension of F is $\frac{1}{12}(k+1)(l+1)(k+2)(l+2)(k+l+3)$ and it is evident that relation (2.3.1) defines a representation $D^{K;l}$ of SU (2, 2) on the vector space F. The computation of the highest weight of this representation shows that it is equivalent to $\overset{k}{\circ}\!-\!\overset{o}{\circ}\!-\!\overset{l}{\circ}$. For $k = l = 1$ one gets the adjoint representation. For further details on the finite-dimensional representations of SU (2, 2), see Ref. [21].

(b) *Infinite-dimensional unitary representations.* The exhaustive classification of all the unitary strongly continuous irreducible representations of SU (2, 2) or $SO_0(4, 2)$ is a very hard task which has not been done yet. The problem is in fact not even solved for the much simpler group SO (3, 2). The main difficulty is the lack of the multiplicity theorem [23] which enables one to compute exhaustively the unitary irreducible representations of $SO_0(n, 1)$. (In the restriction of any unitary irreducible representation of $SO_0(n, 1)$ to $SO(n)$ the multiplicity of any irreducible of $SO(n)$ is at most one) and especially those of $SO_0(4, 1)$. However, the literature on the subject is enormous (see Refs. [21, 22, 26] and references quoted there).

At a Lie algebra level the most extensive work has been done by Yao who has studied the reduction to the maximal compact subalgebra su (2) ⊕ su (2) ⊕ u (1). Unfortunately, besides the fact that the integrability of these local representations was not checked, completeness was not attained by Yao who has omitted one principal series of representations. Globally the series of papers of Macfadyen [26] are the most comprehensive. In particular, the two principal continuous series and the six principal discrete series of unitary irreducible representations were constructed and then reduced to $\mathscr{P}$. All the representations (principal or degenerate) constructed by Macfadyen are highly reducible when restricted to $\mathscr{P}$.

Concerning the irreducibility when restricted to $\mathscr{P}$ of the irreducible representations of SU (2, 2) one has the following results. The conjecture of Sternheimer [27] that the most degenerate discrete series of representations of SU (2, 2) (ladder series) remain irreducible under $\mathscr{P}$ was proved [28] and it was shown that they correspond to mass-zero and arbitrary integer or half-integer spin j representations. Later all representations of the conformal group which when restricted to $\mathscr{P}$ or W remain irreducible were found [29] and it was proved that the continuous spin, mass-zero, representations of $\mathscr{P}$ have no extension to SU (2, 2), whereas the previous extension in the discrete spin case is unique. Recently [30], the ladder representations have been computed explicitly in terms of integral transforms on the zero-mass shell and if the helicity is half-integer, integer, or odd integer one gets respectively representations of SU (2, 2), $SO_0(4, 2)$, or C_0.

Remark. Gross [31] had already tackled from a slightly different point of view the problem of the extension of mass-zero discrete spin unitary irreducible representation of $\mathscr{P}$. He had shown that, on the Hilbert space of a certain family of solutions of mass-zero relativistic equations, such an extension existed. However, he considered only the inversion $x \mapsto x' = x/x^2$, which together with $\mathscr{P}$ generate C_0, but do not belong to the connected component of the identity C_0.

3. Conformal covariance of field equations

3.1. *Definition of covariance and transformation law of fields*

Poincaré covariance of a field equation $A(\partial)\psi(x) = 0$ is a well-known concept which is taken for granted and which can be extended in the following way [5] when one deals with C_0. Let $g, g' \in C_0$, $x \in M$, $P = (b, \Lambda(A)) \in \mathscr{P}$ with $A \in \mathrm{SL}(2, \mathbb{C})$.

DEFINITION. If $x \mapsto x' = g \cdot x$ is a conformal transformation, we say that the equation $A(\partial)\psi(x) = 0$ is conformal covariant if it goes into $A(\partial')\psi'(x') = 0$ when ψ cotransforms according to $\psi'(x') = S(g, x)\psi(x)$, where $S(g, x)$ is some linear operator depending on the group element g and on the point x in space-time.

Consistency with the group structure of C_0 implies that when both sides are defined:

$$S(g'g, x) = S(g', g \cdot x)S(g, x). \tag{3.1.1}$$

Compatibility with Poincaré covariance requires that if R is the representation of SL (2, C) according to which ψ cotransforms one has

$$S(P, x) = R(A). \tag{3.1.2}$$

Remarks. (a) This notion can be formulated in terms of cocyles (see Ref. [32]); this is a question of language. It can also be formulated with the help of continuation [15, 33] of differential operators (defined on M) to the compact manifold M_c^4 on which $\mathrm{SO}_0(4, 2)$ acts continuously (see Section 2). We shall not use this formalism here. We use Minkowski space M and we are aware of the fact that $x \mapsto g \cdot x$ is defined outside of the closed set $C_g \subset M$.

(b) Condition (3.1.2) and the results of Section 2 show that S is related to C_0 via the covering group SU (2, 2). More precisely relation (3.1.1) holds for $g, g' \in \mathrm{SU}(2, 2)$.

In the case of a finite-dimensional irreducible representation R of

SL(2, C), the conditions (3.1.1) and (3.1.2) determine the matrix S uniquely up to a power of the function ω. In fact if

$$J = \begin{pmatrix} 0 & I \\ -I & 0 \end{pmatrix}$$

is a matrix of order that of S we have [5]:

THEOREM 1. Let R be any finite-dimensional real or complex irreducible representation of SL(2, C) (depending on which linear combinations, real or complex, of the field-components we allow). Denote by C_g the set of singular points of the conformal transformation g. Then any real or complex matrix function $S(g, x)$ satisfying (3.1.1) and (3.1.2), and analytic in g in some neighborhood of $\bar{\mathscr{P}}$ in SU(2, 2) – and thus analytic in x in complex Minkowski space M_c outside of singularities – is given, for special conformal transformations

$$a : x \mapsto x' = \omega(a, x)^{-1}(x + ax^2)$$

by

$$S(a, x) = \omega(a, x)^{n+k\gamma} S_0(a, x), \tag{3.1.3}$$

which determines its values for any $g \in \mathrm{SU}(2, 2)$ by (3.1.1) and (3.1.2). Here, $n \in \mathbf{R}$ is called the conformal degree of S, $\gamma = i$ in the complex case, $\gamma = J$ in the real case, $k \in \mathbf{R}$, and $k = 0$ in the real absolutely irreducible case; $S_0(a, x)$ is the uniquely defined analytic (for x in the connected component of the origin in $M - C_a$, and therefore also outside singularities in M_c) unimodular solution of some matrix differential system

$$\partial_\mu S_0(a, x) - N_\mu(a, x) S_0(a, x) \tag{3.1.4}$$

with initial value $S_0(a, 0) = I$, the identity matrix.

If we write $\mathcal{M}_{\mu\nu} = \mathrm{d}R(M_{\mu\nu})$ for the infinitesimal generators of the Lorentz transformations, the matrix N_μ is given by

$$\begin{aligned} N_\mu(a, x) = 2[&a^\rho \mathcal{M}_{\rho\mu} + (2a_\mu a^\rho \mathcal{M}_{\sigma\rho} + a^2 \mathcal{M}_{\mu\sigma}) \\ &\times \omega(a, x)^{-1}(x^\sigma + a^\sigma x^2)], \end{aligned} \tag{3.1.5}$$

and satisfies the relation $(\partial^\mu \omega(a, x)) N_\mu(a, x) = 0$ (S_0 is constant along lines orthogonal, for the hyperbolic metric, to the hypersurfaces $\omega = \text{const.}$). Besides, the relation (3.1.4) will be shown later to be very useful to prove the covariance of certain field equations.

Now from the theorem the determination of S in the irreducible case follows easily. Let us denote by (j, j') the usual irreducible representation of SL $(2, \mathbf{C})$ of dimension $(2j+1)(2j'+1)$. When $R = (\frac{1}{2}, 0) \oplus (0, \frac{1}{2})$ one knows that if γ_μ are Dirac matrices, one has (up to equivalence) $\mathcal{M}_{\mu\nu} = \frac{1}{2}\gamma_\mu\gamma_\nu$; one then checks easily that $S_0(a, x) = \omega(a, x)^{-1/2}$ $(I + a^\mu x^\nu \gamma_\mu \gamma_\mu)$ satisfies (3.1.4). This gives in the particular representation

$$\gamma_0 = \begin{pmatrix} 0 & I \\ I & 0 \end{pmatrix}, \qquad \gamma_j = \begin{pmatrix} 0 & \sigma^j \\ -\sigma^j & 0 \end{pmatrix}$$

$(\sigma^1, \sigma^2, \sigma^3$ are the Pauli matrices),

$$S_0 = \begin{pmatrix} \Sigma_0 & 0 \\ 0 & \Sigma_0' \end{pmatrix},$$

where the matrices Σ_0 and Σ_0' are given by (see also Ref. [34])

$$\Sigma_0(a, x) = \omega(a, x)^{-1/2}(I + a_\mu x_\nu \bar{\sigma}^\mu \sigma^\nu), \tag{3.1.6}$$

$$\Sigma_0'(a, x) = \omega(a, x)^{-1/2}(I + a_\mu x_\nu \sigma^\mu \bar{\sigma}^\nu), \tag{3.1.6'}$$

with $\sigma^0 = \bar{\sigma}^0 = I$ and $\bar{\sigma}^j = -\sigma^j$ for $j = 1, 2, 3$. These matrices are therefore respectively the solutions of the system (3.1.4) when $R = (\frac{1}{2}, 0)$ and $R = (0, \frac{1}{2})$. In term of spinors, Σ_0 corresponds to the transformation of an undotted spinor which will cotransform according to

$$\varphi_\alpha(x) \mapsto \omega(a, x)^{-1/2}(\delta_\alpha^\beta + a_{\alpha\dot{\alpha}} x^{\beta\dot{\alpha}})\varphi_\beta(x).$$

Σ_0' gives in the same manner the transformation law of a dotted spinor. The representations (j, j') being symmetrized tensor products of $(0, \frac{1}{2})$ and $(\frac{1}{2}, 0)$, one obtains immediately with the help of the theorem in the case where $R = (j, j') = \circledS^{2j}(\frac{1}{2}, 0) \otimes \circledS^{2j'}(0, \frac{1}{2})$:

$$\begin{aligned} S_0^{j,j'}(a, x) &= [\circledS^{2j}\Sigma_0(a, x)] \otimes [\circledS^{2j'}\Sigma_0'(a, x)] \\ &= R(\Sigma_0(a, x)) \end{aligned} \tag{3.1.7}$$

for x in the connected component of the origin in $M - C_a$. Finally, if R is the real representation $(j, j') \oplus (j', j)$ with $j \neq j'$, an expression for S_0 is

$$S_0 = \begin{pmatrix} S_0^{j,j'} & 0 \\ 0 & S_0^{j',j} \end{pmatrix}.$$

Remarks. (a) S_0 depends analytically on the four parameters a^μ and a second-order computation gives [5]

$$S_0(a, x) = I + 2a^\mu x^\nu \mathcal{M}_{\mu\nu} + 4a^\mu a^\nu (\mathcal{M}_{\mu\alpha}\mathcal{M}_{\nu\beta} - g_{\mu\alpha}\mathcal{M}_{\nu\beta})x^\alpha x^\beta + 0(|a|^3).$$

Besides, as shown in Ref. [35], the matrix S_0 may be written (formally):

$$S_0(a, x) = \exp(f(a, x)a^\mu x^\nu \mathcal{M}_{\mu\nu})$$

with

$$f(a, x) = [(ax)^2 - a^2x^2]^{-1/2} \log \frac{1 + ax + [(ax)^2 - a^2x^2]^{1/2}}{1 + ax - [(ax)^2 - a^2x^2]^{1/2}}.$$

(b) The homomorphism $\psi: \mathrm{SU}(2,2) \to C_0$ given in Section 2 allows one to give for any $g \in G \approx \mathrm{SU}(2,2)$ a compact expression for $S(g, x) = \Sigma(g, x)$ in the case where $R = (\frac{1}{2}, 0)$. With the notations of (2.1.5) one has

$$\Sigma(g, x) = [\det(g_{22} + g_{21}X)]^{n-1/2}(g_{22} + g_{21}X).$$

The generalization to any representation $R = (j, j')$ which may be achieved with the previously quoted tensor products is straightforward.

When the field ψ cotransforms according to a finite-dimensional reducible representation of $\mathrm{SL}(2, \mathrm{C})$, the unicity theorem fails for obvious reasons. We shall now derive the expression of $S(g, x)$ in that case. Our result will be substantially the one given in Ref. [20]. However, here global transformations will be considered. According to relation (3.1.1) if $g \in \bar{W}$, $g \mapsto S(g, 0)$ defines a representation of $\bar{W}$ (the little group of $x = 0$). Let $h = (a, \lambda\Lambda(A)) \in W^*$ be the transformation

$$x \mapsto x' = \frac{\lambda\Lambda x + \lambda^2 x^2 a}{\omega(a, \lambda\Lambda x)}$$

and $h \mapsto T(h)$ be any finite-dimensional representation of $\bar{W}$:

$$T(h) = \exp(a^\mu k_\mu)\lambda^{-\Delta}R(A),$$

where Δ (resp. k_μ) is the generator of dilatation (resp. the generators of special conformal transformation) and R a representation of $\mathrm{SL}(2, \mathrm{C})$. The following theorem shows that once a representation T of $\bar{W}$ is given a matrix S satisfying (3.1.1) and (3.1.2) is uniquely determined. The demonstration of the theorem is basically the one of Theorem 1 [58].

THEOREM 1′. Any matrix function $S(g, x)$ satisfying (3.1.1), trivial on the translation, such that $S(h,0) = T(h)$, for all $h \in \bar{W}$ and analytic in g in some neighborhood of $\bar{\mathcal{P}}$ in SU (2, 2) is given, for special conformal transformations by $S(a,x) = \omega(a,x)^{\Delta} T_0(a,x)$; $T_0(a,x)$ is the uniquely defined analytic (for x in the connected component of the origin in $M - C_a$) unimodular solution of some matrix differential system $\partial_\mu T_0(a,x) = M_\mu(a,x) T_0(a,x)$ with initial value $T_0(a,0) = \exp(a^\mu k_\mu)$.

In that case the value of the matrix M_μ is

$$M_\mu(a,x) = N_\mu(a,x) + (a^2 k_\mu - 2a_\mu a^\nu k_\nu)\omega(a,x)^{-1},$$

where N_μ is the matrix given in (3.1.5) computed with the infinitesimal generators of R. The value of $T_0(a,x)$ in the connected component of the origin of $M - C_a$ is easily verified to be

$$T_0(a,x) = R(\Sigma_0(a,x)) \exp\left(\frac{1}{2}\frac{\partial^\mu \omega(a,x)}{\omega(a,x)} k^\mu\right).$$

Let us consider, for example, the case of a field consisting of a real scalar Φ and a real vector A, i.e. $R = (0,0) \oplus (\frac{1}{2}, \frac{1}{2})$. Then it can be shown that if $x \mapsto x' = g \cdot x$ is a conformal transformation, one has (besides an obvious direct sum with two conformal degrees) the following two possibilities and only those:

$$\begin{cases} A'^\mu(x') = \omega(g,x)^{n+1} \partial x'^\mu / \partial x^\nu A^\nu(x) \\ \Phi'(x') = \omega(g,x)^{n+1} \Phi(x) \\ \qquad\qquad + \alpha\omega(g,x)^n (\partial_\mu \omega(g,x)) A^\mu(x), \end{cases} \tag{3.1.8}$$

$$\begin{cases} A'^\mu(x') = \omega(g,x)^{n+1} \partial x'^\mu / \partial x^\nu (A^\nu(x) \\ \qquad\qquad + \alpha\omega(g,x)^{-1} \partial^\nu \omega(g,x) \Phi(x)) \\ \Phi'(x') = \omega(g,x)^{n-1} \Phi(x), \end{cases} \tag{3.1.8'}$$

with $\alpha \in \mathbf{R}^*$ and $n \in \mathbf{R}$. The transformation law (3.1.8) will be met in Subsection 3.2.

Remarks. (a) In the case of a tensor field which is covariant of degree p and contravariant of degree q, it can be shown that the matrix S is given by the coordinate transformation [5, 36]

$$S(g,x) = \omega(g,x)^{n+q-p} \left(\frac{\partial x}{\partial x'}\right)^p \left(\frac{\partial x'}{\partial x}\right)^q. \tag{3.1.9}$$

(b) A generalization of the unicity Theorem 1 to the case of unitary irreducible representations of SL (2, C) is an unsolved problem though it is easy to construct for these representations operators satisfying (3.1.1) and (3.1.2).

(c) An extension of the previous classical considerations to second-quantized theories in which ψ becomes an operator-valued distribution is possible [5].

We end this subsection with a result on the conformal degree n. As explained in Ref. [5] the physical covariance matrices S are given by $S(a, x) = \omega(a, x)^n S_0(a, x)$ with real n and we have:

THEOREM 2. For pure spin representations $(0, j) \oplus (j, 0)$ and $(j/2, j/2)$, the conformal degree n is related to the spin j via the relation $(-1)^{2n} = (-1)^{2j}$ if we suppose that the extended S defined by (3.1.1) and (3.1.2) is real and has analytic extension to complex Minkowski space.

This connection between the spin and the conformal degree is in fact satisfied for all known free fields conformal covariant equations as will be shown in the next subsection.

3.2. *A review of the various conformal covariant equations*

In this subsection we investigate successively the covariance of Lagrangian field equations, of gauge conditions and of non-Lagrangian field equations. In some cases the main lines of the proof of covariance will be given.

Up to now we have not discussed the physical interpretation of the dilatations and special conformal transformations. This point has been extensively studied; see, for example, Refs. [37, 38, 39, 40, 41]. However, no definite interpretation exists for the present time. In this article, conformal transformations are interpreted passively as space-time dependent change of units. This point of view which has been advocated for many years remains to be checked experimentally.

The transformation law of the length element $ds^2 \mapsto ds'^2 = \omega(g, x)^{-2} ds^2$ allows one to give a meaning [5, 36] to the conformal degree n of a physical field. We identify $-n$ with the sum of the physical dimensions in centimeters and seconds of the field. Physical quantities should transform according to their physical dimensions. We shall thus postulate that mass will transform according to $m \mapsto m' = \omega(g, x)m$ when $x \mapsto x' = g \cdot x$. This transformation of mass which was introduced by Haantjes and Schouten [42] enables us to give a sense to conformal covariance of equations for massive particles. The orthodox conception of the conformal covariance of massless field equations is thus enlarged.

To begin with, we state a theorem which gives a necessary and sufficient condition for the conformal invariance of Lagrangian field equations [5].

THEOREM 3. Let $\mathscr{L}(\psi, \partial_\mu\psi, m)$ be a local Lagrangian which is Poincaré invariant, i.e. when $x \mapsto x' = \Lambda(A)x + b$ and $\psi(x) \mapsto \psi'(x') = R(A)\psi(x)$, where R is an irreducible representation of SL (2, C), then $\mathscr{L}(\psi'(x'), \partial'_\mu\psi'(x'), m) = \mathscr{L}(\psi(x), \partial_\mu\psi(x), m)$. Suppose, moreover, that $\mathscr{L}$ is scale-covariant, i.e. there exists an n such that for all $\lambda > 0$ $\mathscr{L}(\lambda^{-n}\psi, \lambda^{-n-1}\partial_\mu\psi, \lambda^{-1}m) = \lambda^{-4}\mathscr{L}(\psi, \partial_\mu\psi, m)$. Then the field equations derived from $\mathscr{L}$ will be conformal covariant with $\psi'(x') = \omega(g, x)^n S_0(g, x)\psi(x)$ for $x' = g \cdot x$, $g \in \mathrm{SU}(2, 2)$ if and only if $\delta \int R_\mu \, d^4x = 0$, where $R_\mu = [(ng_{\mu\nu} + \mathcal{M}_{\mu\nu})\psi]^i \partial\mathscr{L}/\partial(\partial_\nu\psi^i)$.

Moreover, the demonstration of the theorem shows that if $R_\mu = 0$ SU (2, 2) is a symmetry group of the Lagrangian and if R_μ is a gradient SU (2, 2) is a symmetry group of the action integral. This theorem is especially useful, in fact it allows one to prove the covariance of field equations without any manipulation of the cumbersome matrix $S_0(a, x)$. The most important field equations of physics which satisfy the requirements of this theorem will now be given with the corresponding conformal degree n and the cotransformation of the Lagrangian under the special conformal transformation $x' = \omega(a, x)^{-1}(x + ax^2)$.

(1) Dirac equation: $\gamma^\mu\partial_\mu\psi + m\psi = 0$. In that case, $\mathscr{L} = \tilde{\psi}(\gamma^\mu\partial_\mu\psi + m\psi)$, $n = \frac{3}{2}$ and SU (2, 2) is a symmetry group of $\mathscr{L}$:

$$\mathscr{L}(\psi'(x'), \partial'_\mu\psi'(x'), m') = \omega(a, x)^4 \mathscr{L}(\psi(x), \partial_\mu\psi(x), m).$$

(2) Maxwell equations expressed in term of potentials: $\Box A_\mu - \partial_\mu\partial^\nu A_\nu = 0$. Here $\mathscr{L} = -\frac{1}{4}(\partial_\mu A_\nu - \partial_\nu A_\mu)(\partial^\mu A^\nu - \partial^\nu A^\mu)$, $n = 1$, C_0 is a symmetry group of $\mathscr{L}$. The Lorentz condition will be discussed later.

(3) Duffin equation for vector particles [43]: $F^{\mu\nu} = \partial^\mu U^\nu - \partial^\nu U^\mu$, $\partial_\mu F^{\mu\nu} + m^2 U^\nu = 0$. The corresponding Lagrangian is

$$\mathscr{L} = F^{\mu\nu}\partial_\mu U_\nu - \frac{m^2}{2} U_\nu U^\nu - \frac{1}{4}F_{\mu\nu}F^{\mu\nu}.$$

(The representation of SL (2, C) involved here being $(\frac{1}{2},\frac{1}{2})\oplus[(0,1)+(1,0)]$ an obvious generalization of the theorem is needed: in order to obtain R_μ, one has to consider the various fields with their conformal degrees.) Here $F^{\mu\nu}$ is a tensor of degree 2 and U^μ a vector of degree 1. C_0 is a symmetry group of $\mathscr{L}$.

(4) Massless Dirac-Fierz equations: $\partial_{\alpha_1\dot\beta}\psi^{\alpha_1\alpha_2\cdots\alpha_{2j}}=0$, $\partial_{\alpha\dot\beta_1}\varphi^{\dot\beta_1\dot\beta_2\cdots\dot\beta_{2j}}=0$. We consider here the first case ((j, 0) field). We have $2j+1$ functions and $4j$ equations. For integral spin they are equivalent to the first order, zero-mass, tensor equations of Weinberg [44] $\partial_{\mu_1}F_+^{[\mu_1\nu_1]\cdots[\mu_j\nu_j]}=0$ and for any spin they are satisfied by the $2j+1$ quantized fields constructed by Weinberg [45] (among the $4j$ equations, $2j+1$ are equivalent to $(\vec{J}^{(j)}\vec{\nabla}+j\partial_0)\psi=0$). Through a lengthy computation it is possible to prove directly the covariance of these equations [46], the conformal degree of the field being $n=j+1$. However, these equations may be written $D^\mu\partial_\mu\psi=0$ and are derived from the Lagrangian $\mathscr{L}=\varphi^+D^\mu\partial_\mu\psi$. Here D^μ are constant rectangular matrices and φ is a $4j$ components spinor cotransforming under SL (2, C) according to $(\frac{1}{2}, j-\frac{1}{2})$. If the degrees of ψ and φ are respectively $n=j+1$ and $m=2-j$, the hypotheses of the theorem are fulfilled and a straightforward computation shows that $R_\mu=\varphi^+D^\nu(ng_{\mu\nu}+\mathscr{M}_{\mu\nu})\psi=0$.

Remark. The transformation properties under SU (2, 2) of the quantized fields of Weinberg corresponding to these equations have been investigated in Ref. [30].

(5) Klein-Gordon equation: $\Box\varphi+m^2\varphi=0$, $\mathscr{L}=\frac{1}{2}(\partial^\mu\varphi\partial_\mu\varphi-m^2\varphi^2)$, $n=1$ and C_0 is a symmetry group of the action integral:

$$\mathscr{L}(\varphi'(x'),\partial'_\mu\varphi'(x'))=\omega(a,x)^4[\mathscr{L}(\varphi(x),\partial_\mu\varphi(x))$$
$$+\partial_\mu(\omega(a,x)^{-1}(a^\mu+a^2x^\mu)\varphi(x)^2)].$$

(6) Massive Weinberg equation [47] of spin 1: $\gamma^{\mu\nu}\partial_\mu\partial_\nu\psi+m^2\psi=0$, $\mathscr{L}=\partial_\mu\bar\psi\gamma^{\mu\nu}\partial_\nu\psi-m^2\bar\psi\psi$. ψ cotransforms under SL (2, C) according to $(1,0)\oplus(0,1)$, $n=1$ and C_0 is only a symmetry group of the field equations. In that case, $R_\mu=-\partial^\nu(\bar\psi\gamma_{\mu\nu}\psi)$ and one has

$$\mathscr{L}(\psi'(x'),\partial'_\mu\psi'(x'),m')=\omega(a,x)^4[\mathscr{L}(\psi(x),\partial_\mu\psi(x),m)$$
$$-[\partial_\mu\omega(a,x)]\partial_\nu(\omega(a,x)^{-1}\bar\psi(x)\gamma^{\mu\nu}\psi(x))].$$

For further information on the conformal covariance of Lagrangian field equations, see Ref. [32].

Let us now discuss the problem of the conformal covariance of gauge conditions. We begin by stating a result concerning the gauge condition of Gårding [52]

$$\partial_{\alpha_1\dot{\alpha}_1}\psi^{\alpha_1\cdots\alpha_{2j}\dot{\alpha}_1\cdots\dot{\alpha}_{2j'}}(x)=0, \tag{3.2.1}$$

where ψ is a spinor cotransforming under SL(2, C) according to (j, j') and $jj' \neq 0$. Equation (3.2.1) is covariant under SU(2,2) with $n = j+j'+2$. (The proof of this fact is tedious and is a consequence of the covariance of Dirac-Fierz equations.) If $j = j'$, this equation may be written in tensor notations

$$\partial_{\mu_1}\varphi^{\mu_1\mu_2\cdots\mu_{2j}}(x)=0, \tag{3.2.2}$$

where φ is a traceless symmetric tensor of degree $2j$. In that case, $n = 2j+2$ and the proof of the covariance is straightforward with the help of relation (3.1.9).

Equation (3.2.2) contains as a special case the Lorentz condition $\partial_\mu A^\mu = 0$ and here $n = 3$. Therefore the system: Maxwell equations + Lorentz condition is not conformal covariant. The Lorentz condition can be replaced [5] by a nonlinear covariant condition like

$$\partial_\mu(A_\nu A^\nu A^\mu)=0$$

or

$$\partial_\mu(A_\nu(\partial^\mu A^\nu-\partial^\nu A^\mu))=\partial_\mu(A_\nu F^{\mu\nu})=0$$

(the conformal degree of A^μ being $n = 1$). It can also (as suggested by Flato) be replaced by the linear condition

$$\partial_\mu\Box A^\mu=0. \tag{3.2.3}$$

This condition is weaker than the Lorentz condition. However, in the case of a plane monochromatic wave solution of Maxwell equations, it implies the Lorentz condition. Besides this trivial fact, condition (3.2.3) has been met in quantum electrodynamics [48, 49]. Haller and Landowitz [50] have shown that the Gupta-Bleuler condition, $\partial^\mu A_\mu^{(+)}|n\rangle = 0$, is not covariant and fails to define state vectors that remain in the physical subspace. For this reason a generalized Lorentz-gauge formulation was introduced [48]. This formulation which is obtained by supposing the gauge condition (3.2.3), ensures in the presence of interaction between charged particles and photons the existence of an invariant positive-frequency part to $\partial_\mu A^\mu$.

The behavior of condition (3.2.3) under the conformal group has been investigated in Refs. [24, 33, 51]. The result is the following: Equation (3.2.3) + Maxwell equations (expressed as in Paragraph 2 above) do form a system of five equations which is conformal covariant. In fact under a special conformal transformation one has (if $n = 1$)

$$\Box' A'^{\mu}(x') - \partial'^{\mu}\partial'_{\rho}A'^{\rho}(x') = \omega(a,x)^4 \partial_{\nu}x'^{\mu} \times (\Box A^{\nu}(x) - \partial^{\nu}\partial_{\rho}A^{\rho}(x))$$

$$\partial'_{\mu}\Box' A'^{\mu}(x') = \omega(a,x)^4 \partial_{\mu}\Box A^{\mu}(x) - 2\omega(a,x)^3(\partial_{\mu}\omega(a,x)) \times (\Box A^{\mu}(x) - \partial^{\mu}\partial_{\rho}A^{\rho}(x)),$$

i.e. the left-hand side of the equations cotransform under C_0 in an indecomposable manner as in (3.1.8) with $\alpha = -2$.

We shall now give some results concerning the covariance of field equations which are not *a priori* derived from a Lagrangian. We shall not enter in the details except for a certain type of second-order equations and mainly refer to the literature.

(a) *First-order equations.* Spinor equations were treated extensively by McLennan [52]. More recently, Mickelsson and Niederle [53] studied the conformal covariance of massive Gelfand-type equations $L^{\mu}\partial_{\mu}\psi + m\psi = 0$ for finite- as well as infinite-dimensional representations of SL (2, C). Although their definition of covariance is more restrictive than the one given here, they obtain necessary and sufficient conditions for covariance. In particular the conformal degree of the field ψ is always $\frac{3}{2}$. In the finite-dimensional case, the representation R of SL (2, C) corresponding to ψ has always a half-integer spin content. In particular, if the representation R is irreducible one finds that necessarily R is real irreducible and of the type $(j, j+\frac{1}{2}) \oplus (j+\frac{1}{2}, j)$ (interlocking representations). A further analysis then shows that for any $j \in N/2$ there exists a set of four matrices L^{μ} such that the equation is conformally covariant. For $j = 0$ one recovers Dirac's equation.

(b) *Second-order equations.* In this part, we briefly show how to investigate the conditions for a second-order equation to be conformally covariant. We consider an equation of the type

$$A^{\mu\nu}\partial_{\mu}\partial_{\nu}\psi + m^2\psi = 0. \tag{3.2.4}$$

We assume that Equation (3.2.4) is Poincaré covariant and that the square matrices $A^{\mu\nu}$ $(0 \leqslant \mu, \nu \leqslant 3)$ are symmetric with respect to the indices μ and ν. We shall restrict ourselves to the irreducible representations $R = (j, j')$ and $R = (j, j') \oplus (j', j)$ with $j \neq j'$ and real $A^{\mu\nu}$, in the second case. Our results are in agreement with those obtained in Ref. [53] (where a more general situation was studied); however, we do not use the technique of induced representation in certain Banach space of functions and a different method will be followed.

Under a special conformal transformation the cotransformation of the field ψ is given by Theorem 1: $\psi(x) \mapsto \psi'(x') = S(a, x)\psi(x)$ with $S(a, x) = \omega(a, x)^{\Delta} S_0(a, x)$ $(\Delta = n + k\gamma)$, and we have

$$\begin{aligned} &A^{\mu\nu}\partial'_\mu\partial'_\nu\psi'(x') + m'^2\psi'(x') = \\ &= \omega(a, x)^2 S(a, x)[\omega^{-2}S^{-1}A^{\mu\nu}S\partial'_\mu x^\rho\partial'_\nu x^\sigma\partial_\rho\partial_\sigma\psi + m^2\psi \\ &\quad + \omega^2 S^{-1}A^{\mu\nu}(2\partial'_\mu x^\rho\partial'_\nu x^\sigma\partial_\sigma S + \partial'_\nu x^\sigma(\partial_\sigma\partial'_\mu x^\rho)S)\partial_\rho\psi \\ &\quad + \omega^{-2}S^{-1}A^{\mu\nu}\partial'_\mu x^\rho\partial_\rho(\partial'_\nu x^\sigma\partial_\sigma S)\psi]. \end{aligned}$$

Equation (3.2.4) will therefore be conformally covariant if and only if

$$\begin{cases} A^{\mu\nu}\partial'_\mu x^\rho[(\partial_\rho\partial'_\nu x^\sigma)\partial_\sigma S + \partial'_\nu x^\sigma(\partial_\rho\partial_\sigma S)] = 0 & (3.2.5) \\ A^{\mu\nu}\partial'_\mu x^\rho[(\partial_\rho\partial'_\nu x^\sigma)S + 2\partial'_\nu x^\sigma(\partial_\rho S)] = 0 & (3.2.6) \\ \partial'_\mu x^\rho\partial'_\nu x^\sigma A^{\mu\nu}S = \omega^2 S A^{\rho\sigma}. & (3.2.7) \end{cases}$$

$$[\Delta, A^{\mu\nu}] = 0.$$

Poincaré covariance implies that Equation (3.2.7) is satisfied. In fact, $S_0(a, x) = R(\Sigma_0(a, x))$ (see Equation (3.1.7)), $\Lambda^\mu_\nu(\Sigma_0(a, x)) = \omega(a, x)\partial_\nu x'^\mu$ and it is obvious from dilatation covariance that Δ commutes with $A^{\mu\nu}$. A straightforward computation shows that Equations (3.2.5) and (3.2.6) are respectively equivalent to

$$2(\partial_\mu\omega)A^{\mu\nu}S^{-1}\partial_\nu S - (\partial^\mu\omega)A^\nu_\nu S^{-1}\partial_\mu S + \omega A^{\mu\nu}S^{-1}\partial_\mu\partial_\nu S = 0, \tag{3.2.8}$$

$$(\partial_\mu\omega)A^{\mu\nu} - \tfrac{1}{2}(\partial^\nu\omega)A^\mu_\mu + \omega A^{\mu\nu}S^{-1}\partial_\mu S = 0. \tag{3.2.9}$$

The differential equation satisfied by $S(a, x)$ and the equality $(\partial^\mu\omega)\partial_\mu S_0 = 0$ then ensures the equivalence of (3.2.8) and (3.2.9) respectively with

$$2(\partial_\mu\omega)A^{\mu\nu}S^{-1}\partial_\nu S - 4\Delta a^2 A^\mu_\mu + \omega A^{\mu\nu}S^{-1}\partial_\mu\partial_\nu S = 0, \tag{3.2.10}$$

$$(\Delta + I)(\partial_\mu \omega) A^{\mu\nu} - \tfrac{1}{2}(\partial^\nu \omega) A^\mu_\mu + \omega A^{\mu\nu} S_0^{-1} N_\mu(a, x) S_0 = 0, \tag{3.2.11}$$

where N_μ is the matrix introduced in Section 3.1. Equation (3.2.11) may be written with the help of (3.1.5):

$$(\Delta + I)(\partial_\mu \omega) A^{\mu\nu} - \tfrac{1}{2}(\partial^\nu \omega) A^\mu_\mu - \omega \partial_\mu x'^\rho A^{\mu\nu} N_\rho(-a, x') = 0, \tag{3.2.12}$$

and finally the relation $\omega(a, x)\partial_\mu x'^\rho N_\rho(-a, x') = \partial^\rho \omega(a, x) \mathcal{M}_{\mu\rho}$ allows one to transform this equation into

$$[2(\Delta + I)A^{\mu\nu} - g^{\mu\nu} A^\rho_\rho + 2g^{\mu\rho} A^{\nu\sigma} \mathcal{M}_{\rho\sigma}](a_\mu + a^2 x_\mu) = 0.$$

We therefore obtain the following necessary condition for covariance:

$$2(\Delta + I)A^{\mu\nu} - g^{\mu\nu} A^\rho_\rho + 2g^{\mu\rho} A^{\nu\sigma} \mathcal{M}_{\rho\sigma} = 0. \tag{3.2.13}$$

It may then be seen through a lengthy though straightforward calculation that (3.2.13) implies (3.2.10); Equation (3.2.13) together with $[\Delta, A^{\mu\nu}] = 0$ are therefore necessary and sufficient conditions for the conformal covariance of (3.2.4).

Let us now denote by T_1 the first Casimir operator of SL (2, C):

$$T_1 = -\tfrac{1}{4}\mathcal{M}^{\mu\nu}\mathcal{M}_{\mu\nu} = (j(j+1) + j'(j'+1))I.$$

It is easily shown (see Ref. [53]) that (3.2.13) is equivalent to the following system of equations:

$$\begin{cases} (\Delta - I)A^\rho_\rho = 0 & (3.2.14) \\ A^{\mu\rho}\mathcal{M}^\nu_\rho = A^{\nu\rho}\mathcal{M}^\mu_\rho & (3.2.15) \\ [T_1, A^{\mu\nu}] + 2(\Delta - I)A^{\mu\nu} = 0. & (3.2.16) \end{cases}$$

We can now conclude that in both cases $\Delta = I$, i.e. $n = 1$, $k = 0$ is necessary for covariance. Then the only equation to satisfy is (3.2.15). If $R = (j, j')$ two cases are met for the value of the tensor operator $A^{\mu\nu}$.

(1) $jj' = 0$ then $A^{\mu\nu} = a g^{\mu\nu} I$ with $a \in \mathbb{C}$ and (3.2.15) is satisfied if and only if $j = j' = 0$.

(2) $jj' \neq 0$ then $A^{\mu\nu} = aT_1 g^{\mu\nu} + b(g_{\rho\sigma}\mathcal{M}^{\mu\rho}\mathcal{M}^{\nu\sigma} + \mathcal{M}^{\mu\nu})$ with $a, b \in \mathbb{C}$ and it is easy to prove that (3.2.15) is satisfied if and only if $j = j'$ and $a = 2b$.

If $R = (j, j') \oplus (j', j)(j \neq j')$ an analogous method shows that (3.2.15) is satisfied if and only if $j' = j + 1$ (or $j = j' + 1$). The matrices $A^{\mu\nu}$ are traceless: $A^\mu_\mu = 0$ and may be written (in a base such that $R(A)$ is decomposed into conjugate blocks:

$$R(A) = \begin{pmatrix} S(A) & 0 \\ 0 & \overline{S(A)} \end{pmatrix},$$

$$A = \begin{pmatrix} 0 & ab^{\mu\nu} \\ \overline{ab^{\mu\nu}} & 0 \end{pmatrix}.$$

Here the $b^{\mu\nu}$ are $(2j+1)(2j+3)$ square matrices uniquely determined by condition (3.2.15) and Poincaré covariance up to the factor $a \in \mathbb{C}$ (the bar denotes complex conjugation). We summarize the previous analysis by the following propositions:

PROPOSITION. Let R be a real or complex finite dimensional irreducible representation of $SL(2,\mathbb{C})$ and ψ a field cotransforming under $SL(2,\mathbb{C})$ according to R. A Poincaré covariant equation $A^{\mu\nu}\partial_\mu\partial_\nu\psi + m^2\psi = 0$ is conformally covariant if and only if

$$R = (j, j) \quad \text{and} \quad A^{\mu\nu} = a(2T_1 g^{\mu\nu} + g_{\rho\sigma}\mathcal{M}^{\mu\rho}\mathcal{M}^{\nu\sigma} + \mathcal{M}^{\mu\nu})$$

or

$$R = (j, j+1) \oplus (j+1, j) \quad \text{and}$$

$$A^{\mu\nu} = \begin{pmatrix} 0 & ab^{\mu\nu} \\ \bar{a}\bar{b}^{\mu\nu} & 0 \end{pmatrix}$$

when R is decomposed in two conjugate blocks. The $b^{\mu\nu}$ are uniquely determined and $a \in \mathbb{C}$ is arbitrary. In both cases the conformal degree of the field is $n = 1$ and $j \in N/2$.

Remark. With our covariance principle we have shown that the field ψ cotransforms under C_0 in a physically covariant manner: $\Delta = n$, i.e. $k = 0$. For the representation $(0,0)$ and $(\frac{1}{2}, \frac{1}{2})$, we recover Klein-Gordon and Duffin equations. For the representation $(0,1) \oplus (1,0)$ we have the Weinberg equation $\gamma^{\mu\nu}\partial_\mu\partial_\nu\psi + m^2\psi = 0$ which was studied above.

(c) *Higher-order equations.* Few things are known about the conformal covariance of such equations. We give two results. The massive Weinberg equations [47] of order $2j$ with $2(2j+1)$ components $\gamma^{\mu_1\cdots\mu_{2j}}\partial_{\mu_1\cdots\mu_{2j}}\psi + m^{2j}\psi = 0$ were found [54] to be conformally covariant with a conformal degree $n = 2 - j$. The iterated Klein-Gordon equation for a scalar field $\Box^K\varphi + m^{2K}\varphi = 0$ ($\Box^K = \Box(\Box^{K-1})$ for $K \geq 1$) is easily verified to be conformally covariant with $n = 2 - K$.

4. Outlook

We conclude this article with two suggestions and a remark:

(1) In the problem of the localization of a massless particle, the conformal group may have a role to play. Up to now this problem has been treated [55, 56] in the context of unitary irreducible representations of $\bar{\mathscr{P}}$. However, as already mentioned for finite spin, these representations may be extended to unitary representations of SU (2, 2). One has therefore a possibility of redefining the problem of the localization and to give a new interpretation to the four parameters corresponding to accelerations. For further details, see Ref. [56].

(2) Let $A(\partial)\psi = 0$ be a linear conformally covariant field equation and θ a particular solution of this equation. According to the definition given at the beginning of Section 3 it is obvious that if $\psi'(x') = S(g, x)\psi(x) + \theta(x')$, $A(\partial)\psi = 0$ implies $A(\partial')\psi'(x') = 0$. The previous cotransformation of the field ψ is an element of a (generalized) covariance group G_S of the equation $A(\partial)\psi = 0$. G_S is a semi-direct product of SU (2, 2) by the vector space of solutions. The problems connected with the structure of groups like G_S are unsolved. In particular, most of the unitary irreducible representations of G_S are unknown. In the case where SU (2, 2) is replaced by $\bar{\mathscr{P}}$ interesting results [57] have been obtained in neutrino physics with the equation $\gamma^\mu \partial_\mu \psi = 0$. Thus the study of the unitary irreducible representations of these groups seems promising.

Finally, a natural question to ask is [9]: does there exist a connection between the space-time conformal group and the dynamical conformal group of the Kepler problem? This is far from being clear.

Physique Mathématique
Université de Dijon

REFERENCES

[1] Conformal invariance was introduced in physics by Cunningham, E., *Proc. London Math. Soc.* **8** (1909), 77, and Bateman, H., *Ibid.* **8** (1910), 223 who proved that Maxwell equations are covariant under the conformal group.

[2] Bjorken, J. D., *Phys. Rev.* **179** (1969), 1547.

[3] Gross, D. J. and Wess, J., Preprint TH 1076 (CERN) (1969); and also Boulware, D. G., Brown, L. S., and Peccei, R. D., *Phys. Rev. D.* **2** (1970), 293, where the case of the Compton amplitude for a spin-zero target was investigated and gave the same result.

[4] Todorov, I. T., 'Strong Interaction Physics', *Lecture Notes in Physics*, Vol. 17, Springer Verlag, 1972.
[5] Flato, M., Simon, J., and Sternheimer, D., *Ann. of Phys. (N.Y.)* **61** (1970), 78.
[6] Segal, I., *Astron. & Astrophys.* **18** (1972), 143.
[7] Flato, M. and Sternheimer, D. *J. Math. Phys.* **7** (1966), 1932.
[8] Flato, M. and Sternheimer, D., *Phys. Rev. Lett.* **16** (1966), 1185; and *Commun. Math. Phys.* **12** (1969), 296.
[9] Wybourne, B. G., *Classical Groups for Physicists*, Wiley, New York, 1974.
[10] Barut, A. O., in *Springer Tracts In Modern Physics*, Vol. 50, Springer Verlag, Berlin, 1969.
[11] See, for example, the solution of the relativistic Kepler problem in Dirac and Klein-Gordon forms for particles having both electric and magnetic charge given by Barut, A. O. and Bornzin, G. L., *J. Math. Phys.* **12** (1971), 841.
[12] Nambu, Y., *Progr. Theor. Physics*, Suppls. 37, 38 (1966), 368.
[13] Penrose, R. and MacCallum, M. A. H., *Physics Reports*, **6**C(4) (1973).
[14] Borisov, A. B. and Ogievetsky, V. I., *Acad. Nauk S.S.S.R. Teoret. i Matemat. Physica* **21**(3) (1974), 329.
[15] Go, T. H. and Mayer, D. H., *Rep. Math. Phys.* **5** (1974), 43.
[16] Go, T. H., *Comm. Math. Phys.* **41** (1975), 157.
[17] Barut, A. O., 'De Sitter and Conformal Groups and Their Applications', *Lectures in Theoretical Physics*, Vol. 13, Colorado Associated University Press, 1971.
[18] Zeeman, E. C., *J. Math. Phys.* **5** (1964), 490.
[19] Flato, M. and Sternheimer, D., *C.R. Acad. Sci. Paris* **263** (1966), 935.
[20] Mack, G. and Salam, A., *Ann. Phys. (N.Y.)* **53** (1969), 174.
[21] Yao, T., *J. Math. Phys.* **8** (1967), 1931.
[22] Yao, T., *J. Math. Phys.* **9** (1968), 1615; **12** (1971), 315.
[23] Dixmier, J., *Bull. Soc. Math. France* **89** (1961), 9.
[24] Bayen, F. and Flato, M., *J. Math. Phys.* **17** (1976), 1112.
[25] Lüscher, L. and Mack, G., *Comm. Math. Phys.* **41** (1975), 203.
[26] Macfadyen, N. W., *J. Math. Phys.* **12** (1971), 1436; **14** (1973), 57, 638.
[27] Sternheimer, D., *J. Math. Pures Appl.* **47** (1969), 289.
[28] Mack, G. and Todorov, I., *J. Math. Phys.* **10** (1969), 2078.
[29] Mickelsson, J. and Niederle, J., *J. Math. Phys.* **13** (1972), 23.
[30] Post, G., Preprint RWTH Aachen (1974).
[31] Gross, L., *J. Math. Phys.* **5** (1964), 687.
[32] Kotecký, R. and Niederle, J., *Czech. J. Phys.* **B25** (1975), 123.
[33] Mayer, D. H., *J. Math. Phys.* **16** (1975), 884.
[34] Wess, J., *Nuovo Cimento* **18** (1960), 1086.
[35] Rosen, G., *Ann. Phys. (N.Y.)* **77** (1973), 452.
[36] Flato, M., Hillion, P., and Simon, J., *C.R. Acad. Sci. Paris* **268** (1969), 1065.
[37] Page, L., *Phys. Rev.* **49** (1936), 254, 468, 946.
[38] Kastrup, H. A., *Annalen der Physik* **7** (1962), 388.
[39] Fulton, T., Rohrlich, F., and Witten, L., *Rev. Mod. Phys.* **34** (1962), 442.
[40] Go, T. H., Kastrup, H. A., and Mayer, D. H., Properties of dilatations and conformal transformations in Minkowski space. Preprint, München (1971).
[41] Barut, A. O. and Haugen, R. B., *Ann. Phys. (N.Y.)* **71** (1972), 519.
[42] Haantjes, J. and Schouten, J. A., *Koninkl. Ned. Akad. Wetenshap. Proc.* **43** (1940), 1288.
[43] Gelfand, I. M., Minlos, R. A., and Shapiro, Z. Ya., Representations of the Rotation and Lorentz Groups and Their Applications, Pergamon Press, 1963.
[44] Weinberg, S., *Phys. Rev.* **138** (1965), 988.
[45] Weinberg, S., *Phys. Rev.* **134** (1964), 882.

[46] Bayen, F., *C.R. Acad. Sci. Paris* **269** (1969), 546; *Nuovo Cimento* **68A** (1970), 39.
[47] Weinberg, S., *Phys. Rev.* **133** (1964), 1318.
[48] Haller, K. and Tomczak, S. P., *Nuovo Cimento* **8B** (1972), 1.
[49] Haller, K., *Phys. Rev. D.* **10** (1974), 4299.
[50] Haller, K. and Landowitz, L. F., *Phys. Rev.* **171** (1968), 1749.
[51] Bayen, F., *C.R. Acad. Sci. Paris* **280** (1975), 1083.
[52] McLennan, J. A., Jr., *Nuovo Cimento* **3** (1956), 1360; *Ibid.* **5** (1957), 640.
[53] Mickelsson, J. and Niederle, J., *Ann. I.H.P.* **23** (1975), 277.
[54] Flato, M. and Simon, J., *C.R. Acad. Sci. Paris* **269** (1969), 546.
[55] Wightman, A. S., *Rev. Mod. Phys.* **34** (1962), 845.
[56] Angelopoulos, E., Bayen, F., and Flato, M., *Physica Scripta* **9** (1974), 173.
[57] Flato, M. and Piard, A., *Jour. Math. Phys.* **15** (1974), 1288, and references quoted there.
[58] Bayen, F., *C.R. Acad. Sci. Paris* **282** (1976), 605.

LUIS BEL

QUANTUM MECHANICS OF PREDICTIVE POINCARÉ INVARIANT SYSTEMS

1. Introduction

The classical predictive relativistic mechanics of systems of interacting point-like structureless particles has been developed in two main steps. The first step has led to the study of predictive Poincaré invariant systems (P.I.S.) [1–5]. No exact P.I.S. is known to correspond to any interaction which is interesting from the physical point of view, but many results have been obtained in the framework of perturbation theory [6–9].

The second step has been the study of possible Hamiltonian-like formalisms. This has been difficult in the sense that it has been impossible to maintain the classical notion of a Hamiltonian system and that several attempts have been needed to find a fruitful generalization. Here again, most of the interesting results have been obtained in the framework of perturbation theory [10].

In this paper we present an attempt to quantize P.I.S. There are many obscurities in non-relativisitic quantum mechanics and very often we reach only an illusion of comprehension because of the extreme simplicity of the systems under study. Quantum mechanics of P.I.S. inherits very crucially the general difficulties of quantum mechanics because of technical complications which are present, even at the first order of perturbation theory, when we consider P.I.S. which are interesting from the physical point of view, and also because of the necessity of working with a generalized notion of Hamiltonian systems. We only presume here to give a few definitions and to set down a few principles which look promising to us and which are justified by the first results which we have been able to obtain from them. Now and then difficult problems of rigor appear. We have not tried to solve them and only formal correctness is claimed.

Cahen & Flato (eds.), Differential Geometry and Relativity, 197–214.

2. Predictive Poincaré invariant systems

Let M_4 be Minkowski space-time and let TM_4 be the manifold of pairs* (x^α, π^β) where $x^\alpha \in M_4$, and π^β is a time-like $(\pi^\beta\pi_\beta < 0)$ future oriented $(\pi^0 > 0)$ tangent vector to M_4 at the point x^α. In predictive relativistic mechanics the phase space of two point-like structureless particles is $(TM_4)^2$ whose points will be designated by† $(x^\alpha_a, \pi^\beta_b)$. The Poincaré group acts on $(TM_4)^2$ as the derivative group of the trivial extension to $(M_4)^2$. The generators of the space-time translations and of the Lorentz subgroup are respectively‡

$$\begin{aligned} \mathbf{P}_\lambda &= -\varepsilon_a \delta^\alpha_\lambda \frac{\partial}{\partial x^\alpha_a}, \\ \mathbf{J}_{\lambda\mu} &= (\delta^\alpha_\lambda \eta_{\mu\beta} - \delta^\alpha_\mu \eta_{\lambda\beta}) \left(x^\beta_a \frac{\partial}{\partial x^\alpha_a} + \pi^\beta_a \frac{\partial}{\partial \pi^\alpha_a} \right), \end{aligned} \tag{2.1}$$

where $\eta_{\lambda\mu}$ is the metric tensor of M_4. The Lie brackets of these vector fields satisfy the commutation relations of the Lie algebra $\underline{P}$ of the Poincaré group:

$$\begin{aligned} &[\mathbf{P}_\lambda, \mathbf{P}_\mu] = 0, \\ &[\mathbf{P}_\lambda, \mathbf{J}_{\mu\nu}] = \eta_{\lambda\nu}\mathbf{P}_\mu - \eta_{\lambda\mu}\mathbf{P}_\nu, \\ &[\mathbf{J}_{\mu\nu}, \mathbf{J}_{\rho\sigma}] = \eta_{\mu\rho}\mathbf{J}_{\nu\sigma} + \eta_{\nu\sigma}\mathbf{J}_{\mu\rho} - \eta_{\nu\rho}\mathbf{J}_{\mu\sigma} - \eta_{\mu\sigma}\mathbf{J}_{\nu\rho}. \end{aligned} \tag{2.2}$$

By definition the interaction of the two point-like structureless particles is described by a predictive Poincaré invariant system (P.I.S.) if these particles obey a second-order ordinary system of differential equations

$$\frac{dx^\alpha_a}{d\tau} = \pi^\alpha_a, \qquad \frac{d\pi^\alpha_a}{d\tau} = \theta^\alpha_a(x^\beta_b, \pi^\gamma_c); \tag{2.3}$$

such that, $\mathbf{H}_a$ being the vector fields of $(TM_4)^2$

$$\mathbf{H}_a = \delta_{ab}\left(\pi^\alpha_b \frac{\partial}{\partial x^\alpha_b} + \theta^\alpha_b \frac{\partial}{\partial \pi^\alpha_b} \right), \tag{2.4}$$

we have

$$\mathscr{L}(\mathbf{H}_a)(\pi^\alpha_a \pi_{a\alpha}) = 0, \tag{2.5}$$

* $\alpha, \beta, \gamma, \ldots = 0, 1, 2, 3$. We use the signature +2 of M_4.
† $a, b, c, \ldots = 1, 2$. We use these indices indistinctly in an up or down position.
‡ $\varepsilon_a = 1$. We use the summation convention both for the Greek and Latin indices.

where $\mathscr{L}$ is the Lie derivative operator, and

$$[\mathbf{P}_\mu, \mathbf{H}_a] = 0, \qquad [\mathbf{J}_{\lambda\mu}, \mathbf{H}_a] = 0, \tag{2.6}$$

$$[\mathbf{H}_1, \mathbf{H}_2] = 0. \tag{2.7}$$

These last equations, when developed, yield*

$$\pi^\rho_{a'} \frac{\partial \theta^\alpha_a}{\partial x^{a'\rho}} + \theta^\rho_{a'} \frac{\partial \theta^\alpha_a}{\partial \pi^{a'\rho}} = 0. \tag{2.8}$$

From (2.4) and (2.5) it follows that $\pi^2_a = -\pi^\alpha_a \pi_{a\alpha}$ are first integrals of the system of differential equations (2.3), which are interpreted as the squares of the masses of the particles†.

$$\pi^2_a = m^2_a. \tag{2.9}$$

On the other hand we see, from the consideration of (2.2), (2.6), and (2.7), that to each P.I.S. there corresponds a trivial extension of $\underline{P}$ by an abelian Lie algebra $\underline{A}_2$ of dimension two:

$$\underline{G} = \underline{P} \oplus \underline{A}_2. \tag{2.10}$$

We shall call the group G of transformations of $(TM_4)^2$ generated by $\underline{G}$ the complete symmetry group of the P.I.S. which is being considered.

Let $f(x^\alpha_a, \pi^\beta_b)$ be a scalar or tensor function. And let us define

$$R_a(\lambda) f(x^\alpha_a, x^\alpha_{a'}, \pi^\beta_b) = f(x^\alpha_a + \lambda \pi^\alpha_a, x^\alpha_{a'}, \pi^\beta_b). \tag{2.11}$$

We shall say that f tends to zero at the infinite past (resp. future) and we shall write

$$\lim_{x^2 \to \infty_{p(\text{resp. } f)}} f = 0 \tag{2.12}$$

if we have

$$\lim_{\mu \to -\infty \,(\text{resp. } +\infty)} R_1(h_1\mu) R_2(h_2\mu) f = 0 \tag{2.13}$$

* By convention $a' \neq a$.

† We use units for which the speed of light in vacuum is 1.

for each values of $h_a > 0$ such that

$$k - \Lambda < \frac{\pi^2_a h_a}{h_{a'}} < k + \Lambda \quad \text{where} \quad k = -\pi^\alpha_1 \pi_{2\alpha},$$

$$\Lambda^2 = k^2 - \pi^2_1 \pi^2_2 \tag{2.14}$$

and every configuration* with $\Lambda^2 \neq 0$.

We say that a P.I.S. is separable if there exists a real number $p > 0$ for which

$$\lim_{x^2 \to \infty_{\text{p and f}}} x^p \theta^\alpha_a = 0 \quad \text{where} \quad x = (x^\alpha x_\alpha)^{1/2},$$

$$x^\alpha = x^\alpha_1 - x^\alpha_2. \tag{2.15}$$

We say that s is the separability index of a given P.I.S. if s is the greatest number p for which (2.15) hold. From now on, unless otherwise stated, we shall assume $s > 2$.

We shall assume also that the two particles involved in the P.I.S. that we consider are of the same type and correspondingly we shall assume the S_2 symmetry

$$\theta^\alpha_a = \theta^\alpha(x^\beta_a, x^\beta_{a'}, \pi^\gamma_a, \pi^\gamma_{a'}), \quad \theta^\alpha_{a'} = \theta^\alpha(x^\beta_{a'}, x^\beta_a, \pi^\gamma_{a'}, \pi^\gamma_a). \tag{2.16}$$

Finally, we shall assume that the functions defining the P.I.S. that we consider are smooth enough in the neighborhood of every space-like configuration ($x^2 > 0$).

3. Hamiltonian Forms and Conservative Systems

Let us consider a P.I.S. which satisfies the assumptions of the preceding section. We shall say that $\Omega_{\text{p (resp. f)}}$ is a Hamiltonian form if $\Omega_{\text{p (resp. f)}}$ is a symplectic 2-form of $(TM_4)^2$ which is invariant, in an obvious sense, by S_2 and which fulfills the following two sets of conditions:

$$\mathscr{L}(\Lambda)\Omega_{\text{p (resp. f)}} = 0, \tag{3.1}$$

where Λ is any of the generators of the complete group, and

$$\lim_{x^2 \to \infty_{\text{p (resp. f)}}} (\Omega - \Omega^{(0)}) = 0 \quad \text{with} \quad \Omega^{(0)} = \mathrm{d}x^\alpha_a \wedge \mathrm{d}\pi^a_\alpha. \tag{3.2}$$

* A configuration is a point of $(TM_4)^2$.

This equation means that each tensor component of $\Omega_{p\,(\text{resp. f})} - \Omega^{(0)}$ with respect to the co-basis $(dx^\alpha_a, d\pi^\beta_b)$ tends to zero at the infinite past (resp. future). We shall assume that Ω_p and Ω_f exist and are unique*. Furthermore, we shall assume that the P.I.S.'s under consideration are conservative, i.e. are such that $\Omega_p = \Omega_f = \Omega$.

Let Λ be the generating function of the infinitesimal canonical transformation generated by an element $\mathbf{\Lambda}$ of $\underset{\sim}{G}$, i.e. the function which is defined up to an additive constant by the formula

$$i(\mathbf{\Lambda})\Omega = -d\Lambda, \tag{3.3}$$

where $i(\mathbf{\Lambda})$ is the interior product operator. We shall assume that*

$$H_a = +\tfrac{1}{2}\pi^2_a. \tag{3.4}$$

Let $[\Lambda_1, \Lambda_2]$ be the Poisson bracket, in the sense of Ω, of two functions Λ_1 and Λ_2. From the structure of G and the invariance of Ω by S_2 it follows that

$$[P_\lambda, P_\mu] = 0, \quad [P_\mu, H_a] = 0, \quad [J_{\lambda\mu}, H_a] = 0, \quad [H_1, H_2] = 0, \tag{3.5}$$

$$\begin{aligned} [P_\lambda, J_{\mu\nu}] &= \eta_{\lambda\nu}P_\mu - \eta_{\lambda\mu}P_\nu, \\ [J_{\mu\nu}, J_{\rho\sigma}] &= \eta_{\mu\rho}J_{\nu\sigma} + \eta_{\nu\sigma}J_{\mu\rho} - \eta_{\mu\sigma}J_{\nu\rho} - \eta_{\nu\rho}J_{\mu\sigma}. \end{aligned} \tag{3.6}$$

In fact, (3.3) guarantees (3.6) only up to some additive constants. But it is known that P_α and $J_{\lambda\mu}$ can always be chosen such that (3.6) are satisfied. These functions are interpreted as the energy-momentum and the generalized† angular momentum of the P.I.S.

We shall say that a canonical coordinate system, i.e. a system of coordinates (q^α_a, p^b_β) of $(TM_4)^2$ for which Ω takes the form

$$\Omega = dq^\alpha_a \wedge dp^a_\alpha, \tag{3.7}$$

is a coordinate system adapted to Ω, if it is symmetric with respect to S_2 and if we have on one hand‡

$$\begin{aligned} &\mathcal{L}(\mathbf{P}_\mu)(q^\alpha_a - x^\alpha_a) = 0, \qquad \mathcal{L}(\mathbf{J}_{\lambda\mu})q^\alpha_a = \delta^\alpha_\lambda q_{a\mu} - \delta^\alpha_\mu q_{a\lambda}, \\ &\mathcal{L}(\mathbf{P}_\mu)P^b_\beta = 0, \qquad \mathcal{L}(\mathbf{J}_{\lambda\mu})p^b_\beta = \eta_{\lambda\beta}p^b_\mu - \eta_{\mu\beta}p^b_\lambda, \end{aligned} \tag{3.8}$$

* This assumption has been proved to be true at any order of perturbation theory in Ref. [10].

† We say generalized because it includes the center of mass formula.

‡ Up to now the existence of adapted coordinate systems has been proved only in the framework of first-order perturbation theory.

and on the other hand

$$\lim_{x^2 \to \infty_{p \text{ and } f}} (q_a^\alpha - x_a^\alpha) = 0, \qquad \lim_{x^2 \to \infty_{p \text{ and } f}} (p_\beta^b - \pi_\beta^b) = 0. \tag{3.9}$$

For each canonical coordinate system satisfying (3.8), P_α and $J_{\lambda\mu}$ take the form

$$P_\alpha = \varepsilon_a p_\alpha^a, \qquad J_{\lambda\mu} = q_{a\lambda} p_\mu^a - q_{a\mu} p_\lambda^a.$$

4. Quantization of P.I.S.

We shall consider only spinless particles and we shall assume that the quantum states of the P.I.S. are defined by functions $\Psi(x_a) = \Psi(x_1^\alpha, x_2^\alpha)$, which have a Fourier transform $\underline{\Psi}(k_b) = \underline{\Psi}(k_1^\beta, k_2^\beta)$:

$$\Psi(x_a) = \frac{1}{(2\pi\hbar)^4} \int_{R^4 \times R^4} \eta(k_b) E(x_c, k_d) \underline{\Psi}(k_e), \tag{4.1}$$

$h = 2\pi\hbar$ being the Planck constant, and

$$\eta(k_b) = dk_1^0 \wedge \cdots \wedge dk_1^3 \wedge dk_2^0 \wedge \cdots \wedge dk_2^3, \tag{4.2}$$

$$E(x_e, k_d) = \exp\left(\frac{i}{\hbar} x_a^\alpha k_\alpha^a\right).$$

Let $\hat{F}$ be a linear operator acting on the space of the functions Ψ and let $\tilde{F}(x_a, k_b)$ be the function defined by

$$\hat{F} E(x_a, k_b) = \tilde{F}(x_c, k_d) E(x_a, k_b). \tag{4.3}$$

We call this function the indicial function of $\hat{F}$. This operator being linear we have

$$\hat{F}\Psi(x_a) = \frac{1}{(2\pi\hbar)^4} \int_{R^4 \times R^4} \eta(k_b) E(x_c, k_d) \underline{\Psi}(k_e) \tilde{F}(x_f, k_g), \tag{4.4}$$

and therefore the knowledge of the indicial function of $\hat{F}$ is equivalent to the knowledge of $\hat{F}$ itself. We shall write

$$\tilde{F} = \theta(\hat{F}), \qquad \hat{F} = \theta^{-1}(\tilde{F}). \tag{4.5}$$

Two linear operators $\hat{F}$ and $\hat{G}$ being given, we have

$$\tilde{F} \circ \tilde{G} \equiv \theta(\hat{F}\hat{G}) = \frac{1}{(2\pi\hbar)^8} \int_{R^4 \times R^4} \eta(\lambda_a) \int_{R^4 \times R^4} \times \eta(k'_b) E(\lambda_c, k_d - k'_d) \tilde{F}(x_e, k'_f) \tilde{G}(x_g + \lambda_g, k_h). \tag{4.6}$$

The composition law thus defined, which we call the quantum composition of two functions, is linear and associative. Other simple properties are

$$x_a^\alpha \circ \tilde{G} = x_a^\alpha \tilde{G} = \tilde{G} \circ x_a^\alpha + i\hbar \frac{\partial \tilde{G}}{\partial k_\alpha^a},$$
$$\tilde{F} \circ k_\alpha^a = \tilde{F} k_\alpha^a = k_\alpha^a \circ \tilde{F} + i\hbar \frac{\partial \tilde{F}}{\partial x_a^\alpha}; \tag{4.7}$$

and if $\tilde{F}$ and $\tilde{G}$ are independent of $\hbar$,

$$\lim_{\hbar \to 0} \tilde{F} \circ \tilde{G} = \tilde{F}\tilde{G}. \tag{4.8}$$

The quantum bracket of two functions is by definition the expression

$$\{\tilde{F}, \tilde{G}\} = \tilde{F} \circ \tilde{G} - \tilde{G} \circ \tilde{F}. \tag{4.9}$$

This bracket defines on the set of functions $\tilde{F}$ a structure of Lie algebra which is distributive with respect to the quantum composition of two functions, i.e.

$$\{\tilde{F}, \tilde{G} \circ \tilde{H}\} = \tilde{G} \circ \{\tilde{F}, \tilde{H}\} + \{\tilde{F}, \tilde{G}\} \circ \tilde{H}. \tag{4.10}$$

Other simple properties are

$$\{x_a^\alpha, \tilde{G}\} = i\hbar \frac{\partial \tilde{G}}{\partial k_\alpha^a}, \qquad \{\tilde{F}, k_\alpha^a\} = i\hbar \frac{\partial \tilde{F}}{\partial x_a^\alpha}; \tag{4.11}$$

and if $\tilde{F}$ and $\tilde{G}$ are independent of $\hbar$,

$$\lim_{\hbar \to 0} \frac{1}{i\hbar} \{\tilde{F}, \tilde{G}\} = \frac{\partial \tilde{F}}{\partial x_a^\alpha} \frac{\partial \tilde{G}}{\partial k_\alpha^a} - \frac{\partial \tilde{G}}{\partial x_a^\alpha} \frac{\partial \tilde{F}}{\partial k_\alpha^a} = [\tilde{F}, \tilde{G}]_{\Omega^{(0)}}. \tag{4.12}$$

Moreover, it is clear that

$$\theta([\hat{F}, \hat{G}]) = \{\tilde{F}, \tilde{G}\}, \tag{4.13}$$

where $[\hat{F}, \hat{G}]$ is the commutator of the two operators.

We shall say that a quantum bracket is exact if

$$\{\tilde{F}, \tilde{G}\} = i\hbar [\tilde{F}, \tilde{G}]_{\Omega^{(0)}}. \tag{4.14}$$

When $\tilde{F}$, or $\tilde{G}$, is a linear combination of the variables x_a^α and k_α^a the quantum bracket of $\tilde{F}$ and $\tilde{G}$ is exact. When $\tilde{F}$ and $\tilde{G}$ are linear combinations of the variables x_a^α, or k_α^a, the quantum bracket of $\tilde{F}$ and $\tilde{G}$ is also exact.

Let us consider the following set of functions which we assume to be inversible for each space-like configuration:

$$\tilde{x}^{\alpha}_{a} = \tilde{x}^{\alpha}_{a}(x_b, k_c), \quad \tilde{\pi}^{\alpha}_{a} = \tilde{\pi}^{\alpha}_{a}(x_b, k_c);$$
$$x^{\alpha}_{a} = q^{\alpha}_{a}(\tilde{x}_b, \tilde{\pi}_c), \quad k^{\alpha}_{a} = p^{\alpha}_{a}(\tilde{x}_b, \tilde{\pi}_c); \tag{4.15}$$

and let us assume also that they are symmetric with respect to S_2, that they have the symmetry properties*

$$\mathscr{L}(\mathbf{P}_{\mu})(\tilde{x}^{\alpha}_{a} - x^{\alpha}_{a}) = 0, \qquad \mathscr{L}(\mathbf{J}_{\lambda\mu})\tilde{x}^{\alpha}_{a} = \delta^{\alpha}_{\lambda}\tilde{x}_{a\mu} - \delta^{\alpha}_{\mu}\tilde{x}_{a\lambda},$$
$$\mathscr{L}(\mathbf{P}_{\mu})\tilde{\pi}^{b}_{\beta} = 0, \qquad \mathscr{L}(\mathbf{J}_{\lambda\mu})\tilde{\pi}^{b}_{\beta} = \eta_{\lambda\beta}\tilde{\pi}^{b}_{\mu} - \eta_{\mu\beta}\tilde{\pi}^{b}_{\mu}, \tag{4.16}$$

and that they satisfy the limit conditions

$$\lim_{x^2 \to \infty_{\mathrm{p\ and\ f}}} (\tilde{x}^{\alpha}_{a} - x^{\alpha}_{a}) = 0, \qquad \lim_{x^2 \to \infty_{\mathrm{p\ and\ f}}} (\tilde{\pi}^{\alpha}_{a} - k^{\alpha}_{a}) = 0. \tag{4.17}$$

To any such set of functions we shall associate the mapping

$$\rho : F(x_a, \pi_b) \to \tilde{F}(x_a, k_b) = F[\tilde{x}_a(x_c, k_d), \tilde{\pi}_b(x_e, k_f)]. \tag{4.18}$$

Let us consider a conservative P.I.S. and let Ω be the corresponding Hamiltonian form. We shall say that ρ is an adapted canonical isomorphism of the system if for each pair of functions F and G we have

$$\rho([F, G]) = [\tilde{F}, \tilde{G}]_{\Omega^{(0)}}. \tag{4.19}$$

The mapping ρ associated with the functions (4.15) is an adapted canonical isomorphism if and only if the functions q^{α}_{a} and p^{b}_{β}, considered as functions of x^{α}_{a} and π^{β}_{b}, are an adapted canonical coordinate system of Ω.

To each adapted canonical isomorphism we shall associate the mapping

$$\varphi : F \to \hat{F} = \varphi(F) = (\theta^{-1} \circ \rho)(F). \tag{4.20}$$

We shall say that a commutator is exact (with respect to φ) if

$$[\hat{F}, \hat{G}] = i\hbar\varphi([F, G]). \tag{4.21}$$

From (4.13), (4.19), and (4.20) it follows that this condition is equivalent to (4.14).

* The left-hand terms of these equations have the same meaning as if the Lie derivatives acted on functions of $(x^{\alpha}_{a}, \pi^{\beta}_{b})$.

Finally, we shall say that φ is a quantizer of the P.I.S. that we consider if all the commutators $[\hat{\Lambda}_1, \hat{\Lambda}_2]$, where Λ_1 and Λ_2 are any two of the functions H_a, P_α, or $J_{\lambda\mu}$, are exact. To each quantizer there corresponds a quantum representation of the Lie algebra [(3.5) and (3.6)].

From the assumptions we made and from (3.10), which leads to

$$\tilde{P}_\alpha = \varepsilon_a k^a_\alpha, \qquad \tilde{J}_{\lambda\mu} = x_{a\lambda} k^a_\mu - x_{a\mu} k^a_\lambda, \tag{4.22}$$

it is easy to see that φ is a quantizer if and only if

$$\{\tilde{H}_1 \cdot \tilde{H}_2\} = 0. \tag{4.23}$$

Equations (2.9), which are, using (3.4), equivalent to

$$2H_a = m^2_a, \tag{4.24}$$

are constraints which we have to impose to the solutions of the differential system (2.3) to take into account the masses of the particles. Whenever a P.I.S. has been quantized, i.e. a quantizer has been found, these constraints lead very naturally to the evolution equations of the quantum states

$$2\hat{H}_a \Psi = m^2_a \Psi. \tag{4.25}$$

Non-relativistic quantum mechanics suggests adding to these equations the boundary condition

$$\{r\Psi(x^\alpha_1, x^\alpha_2 = x^\alpha_1 + rn^\alpha)\}_{r=0} = 0 \tag{4.26}$$

for each space-like unit vector n^α. This condition is a restriction on the P.I.S. and a restriction on the quantizer.

5. Free particles

The only exact P.I.S. which is known and which has an obvious physical meaning is the free particle system: $\theta^\alpha_a = 0$. The Hamiltonian form is $\Omega = \Omega^{(0)}$, and the simplest adapted canonical isomorphism is the trivial isomorphism

$$\rho : \tilde{x}^\alpha_a = x^\alpha_a, \qquad \tilde{\pi}^\alpha_a = k^\alpha_a, \tag{5.1}$$

which leads to the evolution equations

$$\left(\Box_a - \frac{m^2_a}{\hbar^2}\right)\Psi = 0$$

with

$$\Box_a = \Delta_a - \frac{\partial^2}{\partial x^{a\circ 2}}, \qquad \Delta_a = \delta^{ij}\frac{\partial^2}{\partial x^{ai}\,\partial x^{aj}}. \tag{5.2}$$

Let $\Psi_{(1)}$ and $\Psi_{(2)}$ be two solutions of the Klein-Gordon system (5.2). To this pair of solutions we may associate the bi-current

$$J_{\alpha_1\alpha_2} = \partial_{\alpha_1}\Psi^*_{(1)}\partial_{\alpha_2}\Psi_{(2)} + \partial_{\alpha_1}\Psi_{(2)}\partial_{\alpha_2}\Psi^*_{(1)} - \Psi^*_{(1)}\partial_{\alpha_1\alpha_2}\Psi_{(2)} - \Psi_{(2)}\partial_{\alpha_1\alpha_2}\Psi^*_{(1)}, \tag{5.3}$$

where $\partial_{\alpha_a} = \partial/\partial x^\alpha_a$. This bi-current is conservative with respect to x^α_1 and x^α_2:

$$\partial_{\alpha_1}J^{\alpha_1\alpha_2} = 0, \qquad \partial_{\alpha_2}J^{\alpha_1\alpha_2} = 0. \tag{5.4}$$

Let us write, with notations which are easy to understand

$$\langle \Psi_{(1)}, \Psi_{(2)}\rangle = \int_{x^0_1=x^0_2=0} \mathrm{d}^3\mathbf{x}_1\,\mathrm{d}^3\mathbf{x}_2 J_{0_10_2}. \tag{5.5}$$

The value of the right-hand term is independent of the hyperplane $x^0_1 = x^0_2 = 0$ which is chosen to calculate the integral. Therefore, this formula defines on the set of solutions of system (5.2) a symmetric sesquilinear form which is invariant by the Poincaré group.

6. Linearized P.I.S.

Let us consider functions θ^α_a which can be represented by power series of a coupling constant g of the following form:

$$\theta^\alpha_a = g\theta^{(1)\alpha}_a + g^2\theta^{(2)\alpha}_a + \cdots. \tag{6.1}$$

These functions define a formal P.I.S. if the coefficients of these developments are solutions of (2.6) and (2.7) considered order by order. In particular, this implies

$$D_{a'}\theta^{(1)\alpha}_a = 0 \quad \text{with} \quad D_a = \pi^\alpha_a \frac{\partial}{\partial x^{a\alpha}}. \tag{6.2}$$

From now on we shall neglect all terms which are of order higher than g, except those which are meaningful, i.e. independent of the consideration of terms of order higher than g in (6.1). To this approximation any canonical coordinate system of $\Omega = \Omega^{(0)} + g\Omega^{(1)}$, which can be written as

$$p_a^\alpha = \pi_a^\alpha + g p_a^{(1)\alpha}, \qquad q_a^\alpha = x_a^\alpha + g q_a^{(1)\alpha}, \tag{6.3}$$

can be obtained by integrating the differential equations

$$D_a p_{a\alpha}^{(1)} = -\theta_a^{(1)\alpha} + \frac{\partial \Phi_a^{(1)}}{\partial x^{a\alpha}}, \qquad D_{a'} p_{a\alpha}^{(1)} = \frac{\partial \Phi_{a'}^{(1)}}{\partial x^{a\alpha}}, \tag{6.4}$$

where $\Phi_a^{(1)} = \pi_a^\alpha p_{a\alpha}^{(1)}$,

$$D_a q_a^{(1)\alpha} = p_a^{(1)\alpha} - \frac{\partial \Phi_a^{(1)}}{\partial \pi_\alpha^a}, \qquad D_{a'} q_a^{(1)\alpha} = -\frac{\partial \Phi_{a'}^{(1)}}{\partial \pi_\alpha^a}.$$

From these equations and from

$$\frac{\mathrm{d}}{\mathrm{d}\lambda} R_a(\lambda) = R_a(\lambda) D_a = D_a R_a(\lambda), \tag{6.5}$$

where $R_a(\lambda)$ is the operator defined by (2.11), it can be seen that the more general expressions providing an adapted canonical coordinate system of Ω are

$$\begin{aligned} p_a^\alpha &= \pi_a^\alpha + g \int_{-\infty}^0 \mathrm{d}\lambda R_1(\lambda) R_2(\lambda) \left(\frac{\partial \Phi^{(1)}}{\partial x_\alpha^a} - \theta_a^{(1)\alpha} \right), \\ q_a^\alpha &= x_a^\alpha - g \int_{-\infty}^0 \mathrm{d}\lambda R_1(\lambda) R_2(\lambda) \left(\frac{\partial \Phi^{(1)}}{\partial \pi_\alpha^a} - p_a^{(1)\alpha} \right), \end{aligned} \tag{6.6}$$

$\Phi^{(1)}$ being any invariant under S_2 function which satisfies the equations

$$\mathscr{L}(\mathbf{P}_\alpha)\Phi^{(1)} = 0, \qquad \mathscr{L}(\mathbf{J}_{\lambda\mu})\Phi^{(1)} = 0, \tag{6.7}$$

and which is such that the limit conditions (3.9) are verified. Or equivalently, such that

$$\begin{aligned} \int_{-\infty}^{+\infty} \mathrm{d}\lambda R_1(\lambda) R_2(\lambda) \left(\frac{\partial \Phi^{(1)}}{\partial x_\alpha^a} - \theta_a^{(1)\alpha} \right) &= 0, \\ \int_{-\infty}^{+\infty} \mathrm{d}\lambda R_1(\lambda) R_2(\lambda) \left(\frac{\partial \Phi^{(1)}}{\partial \pi_\alpha^a} - p_a^{(1)\alpha} \right) &= 0. \end{aligned} \tag{6.8}$$

To the adapted canonical coordinate system (6.6) there corresponds the adapted canonical isomorphism

$$\rho : \tilde{x}_a^\alpha = x_a^\alpha - g q_a^{(1)\alpha}(x_b, k_c), \qquad \tilde{\pi}_a^\alpha = k_a^\alpha - g p_a^{(1)\alpha}(x_b, k_c). \tag{6.9}$$

The corresponding functions $\tilde{H}_a$ are

$$\tilde{H}_a = -\tfrac{1}{2} k_a^\alpha k_{a\alpha} + g \tilde{\Phi}_a^{(1)} \equiv \tilde{H}_a^{(0)} + g \tilde{H}_a^{(1)}, \tag{6.10}$$

where

$$\tilde{\Phi}_a^{(1)} = \int_{-\infty}^{0} d\lambda R_1(\lambda) R_2(\lambda) D_a \tilde{\Phi}^{(1)}$$

with (6.11)

$$\tilde{\Phi}^{(1)}(x_a, k_b) = \Phi^{(1)}(x_a, k_b).$$

The mapping φ defined by (4.20) is a quantizer of the P.I.S. at the first order if and only if*

$$\Box(\tilde{\Phi}_1^{(1)} - \tilde{\Phi}_2^{(1)}) = 0. \tag{6.12}$$

And the evolution equations of the quantum states are

$$\left(\Box_a + \frac{2g}{\hbar^2}\Phi_a^{(1)} - \frac{m_a^2}{\hbar^2}\right)\Psi = 0. \tag{6.13}$$

To each transformation

$$\Phi^{(1)\#} = \Phi^{(1)} + V^{(1)}, \tag{6.14}$$

which preserves (6.7) and (6.8), and the invariance by S_2, there corresponds the transformation of adapted canonical coordinate systems

$$q_a^{\alpha\#} = q_a^\alpha - g\frac{\partial T^{(1)}}{\partial \pi_\alpha^a}, \qquad p_\alpha^{a\#} = p_\alpha^a + g\frac{\partial T^{(1)}}{\partial x_\alpha^a}, \tag{6.15}$$

where $T^{(1)}$ is the S_2 invariant function which satisfies (6.7) and which is defined by

$$T^{(1)} = \int_{-\infty}^{0} d\lambda R_1(\lambda) R_2(\lambda) V^{(1)}. \tag{6.16}$$

Conversely, to each canonical transformation (6.15) which preserves the invariance under S_2 and (3.8) and (3.9) there corresponds a transformation (6.14) with

$$V^{(1)} = (D_1 + D_2)T^{(1)}. \tag{6.17}$$

By an adapted canonical transformation we have

$$\tilde{\Phi}_a^{(1)\#} = \tilde{\Phi}_a^{(1)} + D_a\tilde{T}^{(1)} \quad \text{where} \quad \tilde{T}^{(1)}(x_a, k_b) = T^{(1)}(x_a, k_b), \tag{6.18}$$

* When acting on a function which is invariant by space-time translations $\Box_1 = \Box_2 \equiv \Box$.

and therefore from (6.12) it follows that the new adapted canonical coordinate system provides a second quantizer if and only if

$$(D_1 - D_2)\square T^{(1)} = 0. \tag{6.19}$$

The functions $\tilde{H}_a$ are transformed according to

$$\tilde{H}_a^{\#} = \tilde{H}_a + g[\tilde{H}_a^{(0)}, \tilde{T}^{(1)}]_{\Omega^{(0)}} = \tilde{H}_a + gD_a\tilde{T}^{(1)}. \tag{6.20}$$

But condition (6.19) guarantees the existence of an S_2 invariant function $\tilde{F}^{(1)}$ which satisfies (6.17) and which is such that

$$D_a\tilde{T}^{(1)} = -\frac{i}{\hbar}\{\tilde{H}_a^{(0)}, \tilde{F}^{(1)}\} = D_a\tilde{F}^{(1)} - \frac{i\hbar}{2}\square\tilde{F}^{(1)}. \tag{6.21}$$

From (6.20) it then follows for the operators $\hat{H}_a$ the transformation law

$$\hat{H}_a^{\#} = \hat{H}_a - \frac{ig}{\hbar}[\hat{H}_a^{(0)}, \hat{F}^{(1)}]. \tag{6.22}$$

Therefore, if

$$\Psi = \Psi^{(0)} + g\Psi^{(1)} \tag{6.23}$$

is a linearized solution of the evolution equations (6.13), then

$$\Psi^{\#} = \Psi + \frac{ig}{\hbar}\hat{F}^{(1)}\Psi^{(0)} \tag{6.24}$$

is a linearized solution of the evolution equations corresponding to the second quantizer.

The integration of (6.21) with the boundary condition

$$\lim_{x^2\to\infty_p} (\tilde{F}^{(1)} - \tilde{T}^{(1)}) - 0 \tag{6.25}$$

gives

$$\tilde{F}^{(1)} = \tilde{T}^{(1)} + i\hbar\int_{-\infty}^{0} \mathrm{d}\lambda R_1(\lambda)R_2(\lambda)\square\tilde{F}^{(1)}. \tag{6.26}$$

Let $\tilde{F}^{(1)}$ be the analytical solution of this equation which coincides with $\tilde{T}^{(1)}$ when $\hbar = 0$:

$$\begin{aligned}\tilde{F}^{(1)} &= \tilde{T}^{(1)} + i\hbar\tilde{F}_1^{(1)} + \cdots + (i\hbar)^n\tilde{F}_n^{(1)} + \cdots;\\ \tilde{F}_n^{(1)} &= \int_{-\infty}^{0} \mathrm{d}\lambda R_1(\lambda)R_2(\lambda)\square\tilde{F}_{n-1}^{(1)}.\end{aligned} \tag{6.27}$$

Let us consider the linear mapping $\alpha : \Psi \to \Psi^{\#}$, where $\Psi^{\#}$ is the function defined by (6.24) and the function (6.27). If α preserves the boundary condition (4.26), α is a mapping of the space of states corresponding to the first quantizer onto the space of states corresponding to the second quantizer.

Let us consider a quantizer defined by a function $\Phi^{(1)}$. Let $S^{(1)}$ be the function defined by (6.16) with $V^{(1)} = -\Phi^{(1)}$ and let $G^{(1)}$ be the function defined by the formula (6.27) with $\tilde{T}^{(1)} = \tilde{S}^{(1)}$. Let us now consider the mapping $\beta : \Psi \to \Psi_0^{\#}$, where Ψ is a quantum state corresponding to the quantizer associated with $\Phi^{(1)}$ and $\Psi_0^{\#}$ is the function defined by (6.24) with $\tilde{F}^{(1)} = \tilde{G}^{(1)}$. We have $\beta \circ \alpha = \beta$ and therefore $\Psi_0^{\#}$ may be considered as a representative of the equivalence class of all quantum states corresponding to all possible quantizers and which define the same physical state. The functions $\Psi_0^{\#}$ are not, in general, quantum states corresponding to a particular quantizer because (6.8) cannot hold, in general, when $\Phi^{(1)} = 0$.

Since these functions $\Psi_0^{\#}$ are solutions of the Klein-Gordon system (5.2), if we write

$$\langle \Psi_{(1)}, \Psi_{(2)} \rangle = \langle \Psi_{0(1)}^{\#}, \Psi_{0(2)}^{\#} \rangle,$$

the right-hand term being given by (5.5) the symmetric sesquilinear form thus defined is invariant by the mapping α and by the Poincaré group.

7. The Scalar and Vector Interactions

As a partial application of the general results presented in the preceding section we consider here the problem of two interacting spinless scalar or vector 'charges' e_a. The linearized P.I.S. corresponding to these interactions are defined by the functions

$$\theta_a^{\alpha} = g r_a^{-3}(1 + \mu r_a) \exp(-\mu r_a)[\eta_a A_a x^{\alpha} - (A_a + k L_{aa'}) z_a \pi_a^{\alpha} + (A_a z_{a'} + \pi_a^2 L_{aa'} z_a) \pi_{a'}^{\alpha}], \tag{7.1}$$

where* $\eta_a = (-1)^{a+1}$, $g = e_1 e_2$,

$$\begin{aligned} r_a &= +[x^2 + \pi_{a'}^{-2}(x\pi_{a'})^2]^{1/2}, \qquad (x\pi_a) = x_\alpha \pi_a^{\alpha}, \\ z_a &= \eta_a \Lambda^{-2}[\pi_{a'}^2(x\pi_a) - k(x\pi_{a'})]; \end{aligned} \tag{7.2}$$

* The remaining quantities have been defined in Section 2.

and

$$A_a = -\pi_a, \qquad L_{aa'} = \pi_a^{-1}\pi_{a'}^{-2}k \tag{7.3}$$

for the scalar interaction, and for the vector interaction

$$A_a = \pi_{a'}^{-1}k, \qquad L_{aa'} = -\pi_{a'}^{-1}. \tag{7.4}$$

$\mu(\geqslant 0)$ is the inverse of the range of the interaction.

Let us consider first the case with $\mu > 0$. The separability index of these systems is infinite. It is not difficult to find two particular functions $\Phi^{(1)}$, invariant by S_2 and satisfying (6.7) and (6.8). These functions are

$$\Phi^{(1)} = -2B\Pi^{-1}R^{-1}\exp(-\mu R), \tag{7.5}$$

where

$$R = [x^2 + \Pi^{-2}(x\Pi)^2]^{1/2}, \qquad \Pi^\alpha = \varepsilon^a\pi_a^\alpha,$$
$$\Pi^2 = -\Pi^\alpha\Pi_\alpha, \qquad (x\Pi) = x^\alpha\Pi_\alpha \tag{7.6}$$

and where

$$B = -\pi_a\pi_{a'} \quad \text{or} \quad B = k, \tag{7.7}$$

depending on whether we consider the scalar or vector interaction; and

$$\Phi^{(1)} = -\tfrac{1}{2}B[\pi_{a'}^{-1}r_a^{-1}\exp(-\mu r_a) + \pi_a^{-1}r_{a'}^{-1}\exp(-\mu r_{a'})]. \tag{7.8}$$

Let us consider the quantizer associated with function (7.5). The evolution equations of the linearized quantum states (6.23) are for this quantization

$$\left(\Box_a - \frac{m_a^2}{\hbar^2}\right)\Psi^{(0)} = 0, \qquad \left(\Box_a - \frac{m_a^2}{\hbar^2}\right)\Psi^{(1)} = \frac{-1}{\hbar^2}\hat{\Phi}^{(1)}\Psi^{(0)} \tag{7.9}$$

Let us assume that

$$\Psi^{(0)} = \exp\left(\frac{i}{\hbar}x_a^\alpha k_\alpha^a\right), \qquad \hat{P}_\alpha\Psi = P_\alpha\Psi \tag{7.10}$$

with $P_\alpha = \varepsilon_a k_\alpha^a$, and $-k_a^\alpha k_{a\alpha} = m_a^2$. $\Psi^{(0)}$ is then a solution of the first system of (7.9), and adding and substracting the two equations of the second system it is easily seen that for any Galilean frame of reference for which $P_i = 0$ $(i = 1, 2, 3)$, $\Psi^{(1)}$ takes the form

$$\Psi^{(1)} = \psi^{(1)}(x^i) \exp\left[\frac{-i}{2P\hbar}(m_1^2 - m_2^2)x^0\right] \exp\left[\frac{-iP}{2\hbar}(x_1^0 + x_2^0)\right]; \qquad P = P^0, \tag{7.11}$$

where $\psi^{(1)}$ is a solution of the equation

$$(\hbar^2\Delta + |\mathbf{k}|^2)\psi^{(1)} = 2BP^{-1}r^{-1}\exp(-\mu r)\exp\left(\frac{i}{\hbar}\mathbf{kx}\right); \tag{7.12}$$

and where $\mathbf{x} = (x^i)$, $r = |\mathbf{x}|$

$$\mathbf{k} = \mathbf{k}_1 = -\mathbf{k}_2,$$
$$|\mathbf{k}|^2 = \frac{1}{4P^2}[P^2 - (m_1 - m_2)^2][P^2 - (m_1 + m_2)^2] \tag{7.13}$$

and

$$B = -m_1 m_2 \quad \text{or} \quad B = \tfrac{1}{2}(P^2 - m_1^2 - m_2^2), \tag{7.14}$$

depending on whether we consider the scalar or vector interaction.

Note that (7.12) is formally identical to the correspondingly linearized Schrödinger equation for a particle of mass m moving in a Yukawa potential of strength $gBP^{-1}m^{-1}$. From this remark we may conclude immediately, using available information, that, in particular, for a scattering problem the asymptotic behavior of $\psi^{(1)}$ is

$$g\psi^{(1)} \underset{r\to\infty}{\sim} f(\theta; P)r^{-1}\exp\left(\frac{i}{\hbar}|\mathbf{k}|r\right), \tag{7.15}$$

where θ is the scattering angle and where, using standard notations,

$$f(\theta; P) = \frac{\hbar}{|\mathbf{k}|}\sum_{l=0}^{\infty}(2l+1)e^{i\delta_l}\sin\delta_l P_l(\cos\theta) \tag{7.16}$$

with

$$\delta_l = -\frac{gB|\mathbf{k}|}{P\hbar^3}\int_0^\infty rj_l^2\left(\frac{|\mathbf{k}|r}{\hbar}\right)\exp(-\mu r)\,dr. \tag{7.17}$$

Let $\Psi_0^{\#} = \beta(\Psi)$ be the image of a scattering state Ψ by the mapping β defined in the preceding section. The x_a^0 dependence of $\Psi_0^{\#}$ is identical to that of Ψ and the x^i dependence of the factor, using the direction of $\mathbf{k}$ as the direction of the positive $z = x^3$ axis, is

$$\psi_0^{\#} = \exp\left(\frac{i}{\hbar}|\mathbf{k}|z\right)\left\{1 + \frac{igB}{\hbar|\mathbf{k}|P}\exp\left(\frac{i\hbar\mu^2}{2|\mathbf{k}|}z\right)\right.$$

$$\left.\cdot\int_{-\infty}^{z}\frac{\exp\left[-\mu(r^2-z^2+u^2)^{1/2}-\frac{i\hbar\mu^2}{|\mathbf{k}|}u\right]}{(r^2-z^2+u^2)^{1/2}}\,\mathrm{d}u\right\} + g\psi^{(1)}. \tag{7.18}$$

The r^{-1} asymptotic behavior of $\psi_0^{\#}$, when $r - z$ tends to infinite is the same as that of $g\psi^{(1)}$ because the integral in (7.18) does not contribute to this behavior. Therefore, Equations (7.15) (with $\psi^{(1)} = \psi_0^{\#(1)}$), (7.16), and (7.17) have an intrinsic meaning, i.e. they are independent of the quantizer we have used to perform the calculations.

Let us finally say a few words about the case $\mu = 0$ (the vector interaction is then the electromagnetic interaction). When $\mu = 0$ the separability index of the corresponding P.I.S. is $s = 2$ and this forces to insert a few modifications. For instance, we have to substitute the first condition (3.9) by the less restrictive one:

$$\lim_{x^2\to\infty_{\mathrm{p\ and\ f}}} x^{-1}(q_a^\alpha - x_a^\alpha) = 0 \tag{7.19}$$

in the definition of an adapted canonical coordinate system. The general expressions of the adapted canonical coordinates p_a^α is still given by the first set of (6.6), but we cannot use the second one to obtain the canonical mates q_a^α, which have to be found with a different integration procedure of the second set of (6.4). The function $\Phi^{(1)}$ has to satisfy the first set of (6.8), but not the second. The definition of a quantizer has to be correspondingly modified and the function $S^{(1)}$ used at the end of the preceding section to define the mapping β and the function $\Psi_0^{\#}$ has to be found as the solution of the equations

$$D_a S^{(1)} = -\Phi_a^{(1)}, \tag{7.20}$$

which satisfies the asymptotic conditions

$$\lim_{x^2\to\infty_{\mathrm{p}}} \frac{\partial S^{(1)}}{\partial x_a^\alpha} = 0. \tag{7.21}$$

With $\mu = 0$, the function (7.5) still defines a quantizer. Using the same notations as before, the scattering state corresponding to the first quantizer is

$$\psi = \psi_0^{\#} + \frac{igB}{\hbar P|\mathbf{k}|} \ln\left[\frac{|\mathbf{k}|}{\hbar}(r-z)\right] \exp\left(\frac{i}{\hbar}|\mathbf{k}|z\right), \tag{7.22}$$

where $\psi_0^{\#}$, the x^i dependent factor of $\Psi_0^{\#} = \beta(\Psi)$ is

$$\psi_0^{\#} = \exp\left(\frac{i}{\hbar}|\mathbf{k}|z\right) - \frac{igB}{\hbar P|\mathbf{k}|}\int_{\infty}^{r-z} v^{-1} \exp\left(\frac{i}{\hbar}|\mathbf{k}|v\right) \mathrm{d}v. \tag{7.23}$$

The asymptotic behavior of this function when $r - z$ tends to infinite is

$$\psi_0^{\#} \underset{r-z\to\infty}{\sim} \exp\left(\frac{i}{\hbar}|\mathbf{k}|z\right) - \frac{gB}{P|\mathbf{k}|^2}\frac{\exp\left(\frac{i}{\hbar}|\mathbf{k}|r\right)}{r-z}. \tag{7.24}$$

Identifying the factor of the second exponential as the scattering amplitude, using for instance the value (7.14) of B for the electromagnetic interaction and the value of $|\mathbf{k}|$ given by (7.13), we obtain the differential scattering cross section

$$\sigma(\theta, P) = \left\{\frac{gP(P^2 - m_1^2 - m_2^2)}{[P^2-(m_1-m_2)^2][P^2-(m_1+m_2)^2]}\right\}^2 \cdot \frac{1}{\sin^4\frac{\theta}{2}}. \tag{7.25}$$

At the limit of low-energy Equation (7.25) reduces to Rutherford's formula and when one of the masses becomes infinite Equation (7.25) reduces to Mott's formula for a spinless particle. Since σ does not depend on $\hbar$ it must have also a classical origin. As a matter of fact it is possible to obtain (7.25) by a direct integration of the equations of motion (2.3) using the appropriate functions (7.1).

Université Pierre et Marie Curie, Paris

REFERENCES

[1] Droz-Vincent, Ph., *Phys. Scripta* **2** (1970), 129.
[2] Bel, L., *Ann. Inst. H. Poincaré* **14** (1971), 189.
[3] Arens, R., *Arch. for Rat. Mech. and Analysis* **47** (1972), 255.
[4] Wray, J. G., *Phys. Rev. D.* **1** (1970), 2212.
[5] Künzle, H. P., *J. Math. Phys.* **15** (1974), 1033.
[6] Bel, L., Salas, A., and Sanchez-Ron, J. M., *Phys. Rev. D.* **7** (1973), 1099.
[7] Bel, L. and Martin, J., *Phys. Rev. D.* **8** (1973), 4347.
[8] Bel, L. and Martin, J., *Phys. Rev. D.* **9** (1974), 2760.
[9] Hirondel, D., *J. Math. Phys.* **15** (1974), 1689.
[10] Bel, L. and Martin, J., *Ann. Inst. H. Poincaré* **22** (1975), 173.

R. BUDIC and R. K. SACHS*

SCALAR TIME FUNCTIONS: DIFFERENTIABILITY

1. Introduction

It is a pleasure to contribute this paper in honor of Professor Lichnerowicz, whose pioneering efforts in applying global differential geometry to general relativity have had such a rich harvest.

We shall analyze, in an appropriate context, how smoothly the Lorentz volume of the past of a point in a general relativistic space-time changes as the point moves. It turns out that the key questions concern the 'past, null cut locus' – roughly, the points where different past lightlike geodesics from the given point first cross. While our methods are in part applicable to more general space-times, we here work exclusively with a special kind of globally hyperbolic space-time, called deterministic.

Briefly, in a deterministic space-time, each observer finds his future determined entirely by data which he has, at least in principle, been able to observe. We find, essentially, that the volume of the past of a point defines a time function which is C^1. The gradient of the time function is unique, and might be taken to define absolute rest.

The notion of absolute rest can be given meaning in space-times with sufficient symmetry (i.e. those with irrotational killing vector fields, or spatially homogeneous space-times), but generally disappears with the assumed symmetry. It is perhaps surprising that a concept of absolute rest can be resurrected in deterministic space-times.

Definitions are given in Section 2. Section 3 includes basic results on deterministic space-times that are needed in the proof of our main result – the existence of a C^1 global time function. The theorem is stated in Section 4, and proof appears in Sections 5 and 6.

2. Definitions

2.1. Throughout (M, g) will be a smooth, connected, Hausdorff manifold with a smooth Lorentz metric, g. (M, g) is assumed to be

* Partially supported by the National Science Foundation under Grant No. GP29257.

Cahen & Flato (eds.), Differential Geometry and Relativity, 215–224.

time- and space-oriented, and the volume element is denoted ω [1, 2].

A smooth curve with nonzero tangent is said to be timelike if its tangent is everywhere timelike and causal if the tangent is timelike or null. A causal curve, γ, joins p and q if p and q are, respectively, the future and past endpoints of γ [1, 2].

We define the past of p, $I^-p : I^-p = \{q \in M | p$ and q are joined by a timelike curve$\}$. If p and q are joined by a piecewise timelike curve then they are joined by a timelike curve; i.e. $q \in I^-q'$ and $p \in I^+q'$ implies $q \in I^-p$. I^-p is open and its closure is denoted $\overline{I^-p}$ [1, 2].

The past of a set, S, I^-S, is the set of all those events which are in the past of at least one point of S. Dual definitions, such as I^+S, will be taken for granted.

2.2. We assume M satisfies a distinguishing condition [3]: $I^-p = I^-q$ or $I^+p = I^+q$ implies $p = q$. A causal curve is said to be past inextendible iff it has no past endpoint (cf. [2]); we use the future of past inextendible curves to define points on the 'edge' or 'boundary' of M [1, 4].

If γ has a past endpoint, p, then $I^+\gamma = I^+p$ and the set I^+p may be taken to represent p. This is well defined since M satisfies the distinguishing condition and no two points have the same future; i.e. $I^+p = I^+q$ implies $p = q$. In addition, if γ has no past endpoint, then for each $p \in M$, $I^+\gamma \neq I^+p$; $I^+\gamma$ is called a *past ideal point*, and may be interpreted as the past endpoint of γ. Two curves γ and γ' are assigned the same past endpoint iff $I^+\gamma = I^+\gamma'$ [4].

3. Deterministic space-times

Various definitions of deterministic space-times are possible, and we take the following in terms of ideal points: M is *deterministic* iff M has a unique past ideal point. In [5, 6], it is shown that this is equivalent to the intuitive definition given in Section 1.

A number of results follow directly from the definition of determinism. Since these are basic to what follows, we discuss them briefly and refer the reader to Refs. [5, 6] for a fuller treatment.

PROPOSITION 3.1. Deterministic space-times are globally hyperbolic.

The proof is given in [6]. □

Recall that any causal curve joining two points, q and q', which is not a null geodesic can be deformed to yield a timelike curve with endpoints q and q' [1, 2]. If any two points in the boundary of the past of a point, $\partial I^- p$, are joined by a causal curve then it must be a null geodesic. $\lambda \subseteq \partial I^- p$ is called a generator iff λ is a null geodesic and for any null geodesic, λ', $\lambda \subseteq \lambda' \subseteq \partial I^- p$ implies $\lambda = \lambda'$; i.e. λ is maximal. $\partial I^- p$ is the union of its generators [1, 2, 7].

By Proposition 3.1, it follows that in a deterministic space-time each generator of $\partial I^- p$ has p as a future endpoint [1, 2, 7]. A generator may also have a past endpoint; if λ and λ' are distinct generators of $\partial I^- p$, for example, and gravity is sufficiently strong that λ and λ' intersect at $q \in \partial I^- p$, then q must be the past endpoint of both λ and λ'. Every extension of λ to the past of q contains points of $I^- p$ [1]. (See Figure 1.)

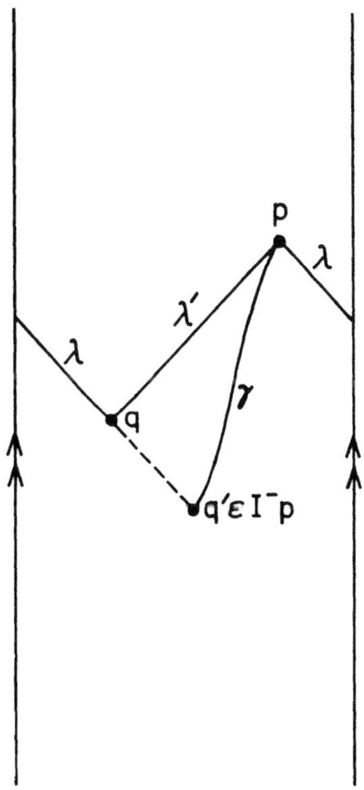

Fig. 1. Here M is a two-dimensional analog of the Einstein static universe. (↟) indicate identifications. q is in the intersection of the generators λ and λ' of $\partial I^- p$. q' is contained in an extension of λ and is joined to p by a timelike curve γ.

In a deterministic space-time, this behavior occurs with a vengeance:

PROPOSITION 3.2. In a deterministic space-time, each generator of $\partial I^- p$ has a past endpoint in $\partial I^- p$.

Proof. For each generator of $\partial I^- p$, $\lambda \cap I^- p = \emptyset$ and, so, $p \notin I^+\lambda$. But p is the future endpoint of some past inextendible timelike curve, γ, and $p \in I^+\gamma$, a past ideal point. If λ has no past endpoint, then $I^+\lambda$ and $I^+\gamma$ are distinct past ideal points, a contradiction. Therefore, each generator of $\partial I^- p$ has a past endpoint, q, and since the boundary is closed we have $q \in \partial I^- p$. □

Such endpoints are called the *cut locus* of p. The proposition may be roughly interpreted as saying that there are no 'points at infinity' or 'singular points' on any observers past 'light cone'.

PROPOSITION 3.3. For any finite set $\{p_n\}$ in a deterministic space-time, $\bigcap_{n=1}^{N} I^- p_n$ is nonempty.

Proof. The proof for $N = 2$ is similar to Proposition 3.2. If $I^- p_1 \cap I^- p_2 = \emptyset$, past inextendible timelike curves γ_1 and γ_2 in $I^- p_1$ and $I^- p_2$, respectively, define distinct past ideal points, a contradiction. The proof for any N proceeds by induction. □

4. Global Time Functions

4.1. Define $\Delta(p, q) = \overline{I^- p} - (I^- p \cap I^- q)$, and $V(p, q) = \int_{\Delta(p,q)} \omega - \int \Delta(q, p)\omega$. Fix q arbitrarily and define $T: M \to R$ where $T(p) = V(p, q)$.

THEOREM 4.1. Let M be a deterministic space-time.

(A) T is finite and is differentiable for each $p \in M$.

(B) The constant time surfaces (T = const.) and $\mathrm{d}T$ are unique (i.e. independent of the choice of q).

(C) $\mathrm{d}T$ is a timelike vector field, and the constant time surfaces are properly imbedded spacelike 3-submanifolds – T is a *global time function.*

4.2. The proof is rather long, so we first make a number of observations.

(i) Recall that a continuous function, T, is differentiable at p iff for each vector v at p, vT exists, and for some form, $\mathrm{d}T$, $vT = \mathrm{d}T(v)$. It is sufficient to show that in some coordinate patch T is C^1 (i.e. T, i is continuous).

(ii) $I^-p \cap I^-q$ is nonempty by Proposition 3.3, and, for any q' in the intersection, $\Delta(p, q) = \Delta(p, q') - \Delta(q, q')$. (See Figure 2.) Therefore, $V(p, q) = V(p, q') - V(q, q')$; $V(p, q)$ will be finite and depend differentiably on p if it is finite and differentiable when $q \in I^-p$.

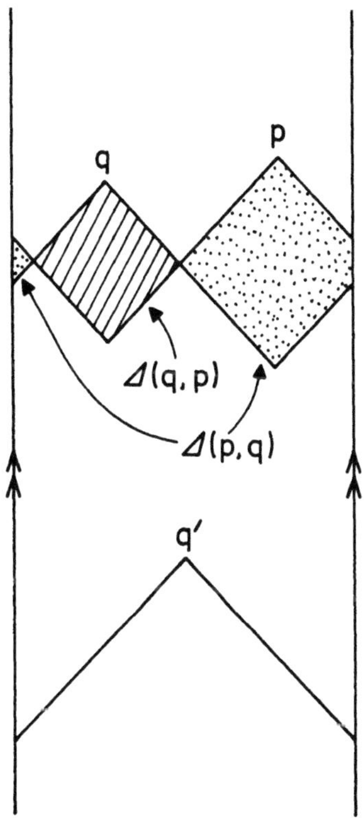

Fig. 2. (↟) indicate identifications. For $q' \in I^-p \cap I^-q$, $\Delta(p, q) = \Delta(p, q') - \Delta(q, q')$ and $\Delta(q, p) = \Delta(q, q') - \Delta(p, q')$. $V(p, q) = V(p, q') - V(q, q')$.

(iii) When $q \in I^-p$, $\Delta(p, q) = \overline{I^-p} - I^-q = \cap_{p' \in \gamma} I^-p'$ where γ is any timelike curve joining p and q. $\Delta(p, q)$ is a union of compact generators over the compact set of null directions along γ, and is found in Section 6 to be compact. The compactness of $\Delta(p, q)$ insures that $V(p, q)$ is finite.

(iv) If $q \in I^-p$, and p is moved slightly to the future along a timelike curve, γ, to a point p', $V(p', q) - V(p, q)$ is essentially an 'integral of the extent of the past light cones' along γ. The partial derivative of V along the timelike curve will exist and be continuous if the 'integrand' is continuous. The proof that this is the case appears in Section 6 and depends primarily on the result, discussed in Section 5, that the cut locus of a point varies continuously as the point is moved.

We indicate in Section 6 how one obtains partial derivatives along spacelike coordinate lines in terms of derivatives along timelike lines.

Finally we discuss parts (B) and (C).

5. The Cut Locus

The cut locus in the case of Riemannian manifolds is discussed in [8] for example. Here we examine the (past, null) cut locus, $C(p)$, in Lorentzian manifolds. Restating our definition: $q \in C(p)$ iff p and q are the future and past endpoints of a generator, λ, of ∂I^-p. q will be in the cut locus of p if q is either conjugate to p along λ, or $q \in \lambda \cap \lambda'$ for a generator $\lambda \neq \lambda'$ [1]. Every extension of λ to the past of q must contain points which can be joined to p by a timelike curve and, in fact, by a timelike geodesic since deterministic space-times are globally hyperbolic [1, 2].

Any smooth future-pointing timelike vector field on M, E_4, can be used to define a distance to the cut locus as follows. Let $T(M)$ be the tangent bundle with projection π, and let $\hat{O} \subseteq T(M)$ be the following subset of past directed null vectors: $\hat{O} = \{(p, u) | u \cdot u = 0$ and $u \cdot E_4 = 1\}$. Each (p, u) defines a null direction, and if for α_o, $\exp_p \alpha_o u \in C(p)$, we call α_o the distance to the cut locus in the direction u. By Proposition 3.2, α_o is defined on all of $\hat{O}$ in a deterministic space-time.

We define $O = \{(p, \alpha u) | (p, u) \in \hat{O}$ and $0 \leq \alpha \leq \alpha_0(p, u)\}$. $\hat{O}(W) = \pi^{-1}(W) \cap \hat{O}$, and $O(W) = \pi^{-1}(W) \cap O$ for any subset $W \subseteq M$. $\hat{O}(W)$ is simply the set of all null directions, (p, u), where p is restricted to W. If W is compact, then $\hat{O}(W)$ is compact.

LEMMA 5.1. *If M is deterministic and W is compact, then $O(W)$ is compact.*

Proof. Any sequence, $\{(p_n, v_n)\} \subseteq O(W)$, defines a sequence, $\{(p_n, u_n)\} \subseteq \hat{O}(W)$, where $v_n = \alpha_n u_n$ for some $\alpha_n \leq \alpha_o(p_n, u_n)$. Since

$\hat{O}(W)$ is compact, we assume $(p_n, u_n) \to (p, u)$. $\alpha_o(p, u)$ is finite since M is deterministic, and we want to show that α_n has a limit $\beta \leqslant \alpha_o(p, u)$; for then, $(p, \beta u) \in O(W)$ is a limit of $\{(p_n, v_n)\}$. If α_n has no limit in the interval $[0, \alpha_o(p, u)]$, one can find $\beta > \alpha_o(p, u)$ and a subsequence $\{\alpha_{n'}\}$ where $\alpha_{n'} \geqslant \beta$ for each n'. But one finds $\bar{p} \equiv \exp_p \beta u \in I^- p$, since $\beta > \alpha_0(p, u)$. For any $q' \in I^- p \cap I^+ \bar{p}$, $I^+ q'$ and $I^- q'$ are neighborhoods of p and $\bar{p}$. For sufficiently large n', $p_{n'} \in I^+ q'$ and $\exp_{p_{n'}} v_{n'} \in I^- q'$; thus $\exp_{p_{n'}} v_{n'} \in I^- p_{n'}$, contradicting $\alpha_{n'} \leqslant \alpha_o\ (p_{n'}, u_{n'})$. □

From Lemma 5.1 we have that α_o is upper semi-continuous (cf. [1], p. 214) on $\hat{O}$. We will need that α_o is in fact continuous on $\hat{O}$, and therefore uniformly continuous on any $\hat{O}(W)$, for W with compact closure.

LEMMA 5.2. In a deterministic space-time, α_o is continuous on $\hat{O}$.

Proof. Let $\{(p_n, u_n)\}$ be a sequence in $\hat{O}$ which converges to (p, u) and let W be a neighborhood of p with compact closure. We define $v_n = \alpha_o(p_n, u_n) u_n$. Since (p_n, u_n) is eventually in $\hat{O}(W)$, (p_n, v_n) is eventually in $O(W)$, and will, by Lemma 5.1, have a limit $(p, v) \in O(\bar{W})$. Thus $v = \beta u$ for $\beta \leqslant \alpha_o(p, u)$; we will show that $\beta = \alpha_o(p, u)$.

Suppose $\{(p_{n'}, v_{n'})\}$ is a subsequence which converges to (p, v) and $\beta < \alpha_o(p, u)$. Then $\bar{p} \equiv \exp_p v$ is joined to p by a null geodesic, λ, without conjugate points. One can therefore find a neighborhood, $U_{(p,v)} \subseteq T(M)$, of (p, v), such that $\pi(U_{(p,v)}) = U_p$ is a neighborhood of p, and, for each $p' \in U_p$, $\pi^{-1}(p') \cap U_{(p,v)}$ is mapped one to one and onto a neighborhood of $\bar{p}$, $U_{\bar{p}}$.

For sufficiently large n', $(p_{n'}, (1 + \frac{1}{n'}) v_{n'}) \in U_{(p,v)}$, $\bar{p}_{n'} \equiv \exp_{p_{n'}} (1 + \frac{1}{n'}) v_{n'}$ converges to $\bar{p}$, and $\bar{p}_{n'} \in I^- p_{n'}$. It follows that $\bar{p}_{n'}$ is joined to $p_{n'}$ not only by a null geodesic, $\lambda_{n'}$, but also by a timelike geodesic, $\gamma_{n'}$, with tangent $(p_{n'}, \tilde{v}_{n'})$ contained in the complement of $U_{(p,v)}$, $\sim U_{(p,v)}$. Global hyperbolicity guarantees that the $\gamma_{n'}$ will have a limit, $\tilde{\gamma}$, joining p and $\bar{p}$ [2]. $\tilde{\gamma}$ will be a null geodesic, and its tangent, a limit of $\{(p_{n'}, \tilde{v}_{n'})\}$, will therefore be in $\sim U_{(p,v)}$. λ and $\tilde{\gamma}$ are distinct geodesics joining p and $\bar{p}$. Any extension of λ to the past of $\bar{p}$ must enter $I^- p$; thus $\bar{p} \in C(p)$, contradicting $\beta < \alpha_o(p, u)$. □

6. Proof of Theorem

To prove the theorem, we first show that for $q \in I^- p$, $V(p, q)$ is finite and depends differentiably on p.

6.1. That $V(p,q)$ is finite follows immediately from Lemma 5.1. We define $E: T(M) \to M$, where $E(p,v) \equiv \exp_p v$, and choose γ to be a timelike curve joining p and q. E maps the compact set $O(\gamma)$ continuously onto $\Delta(p,q)$. $\Delta(p,q)$ is therefore compact [9]; $V(p,q)$ is finite.

6.2. We choose a coordinate patch (U_p, x^a) such that $U_p \ni p$ has compact closure contained in I^+q and $x^a: \bar{U}_p \to N \times [a,b]$, where N is some compact subset of R^3 and $[a,b]$ is some closed interval of R^1. (Latin indices run from 1 to 4 and Greek indices run from 1 to 3.) One may choose the coordinate patch so that $x^\alpha =$ const. defines a three-parameter family of timelike curves $\gamma(x^\alpha; t)$. For $q \in I^-p$, $V(p,q) = V(\gamma(x_o^\alpha; t_o), \gamma(x_o^\alpha; a)) +$ const. for some (x_o^α, t_o);

$$\frac{\partial}{\partial t} V(\gamma(x_o^\alpha; t), q)|_{t=t_o} = \frac{\partial}{\partial t} V(\gamma(x_o^\alpha; t), \gamma(x_o^\alpha; a))|_{t=t_o}$$

if the derivative exists. Hereafter, $\Delta(\gamma(x^\alpha; t), \gamma(x^\alpha; a))$ is denoted $\Delta(x^\alpha, t)$ and the volume of this region is denoted $V(x^\alpha, t)$.

To show that the derivative does exist and is continuous, choose the vector field E_4 to be tangent to the timelike curves $x^\alpha =$ const. on U_p, and let E_α be a triad chosen smoothly on U_p such that (E_α, E_4) is an orthonormal basis. We use this basis and the coordinates on U_p to define coordinates (x^a, θ, φ) on $\hat{O}(U_p)$ and $(x^a, \alpha, \theta, \varphi)$ on $O(U_p)$. Coordinates on the slice $x^\alpha = x_o^\alpha$ yield coordinates $(t, \alpha, \theta, \varphi) \equiv (y^a)$ on a subset of the past directed null vectors over the timelike curve $\gamma(x_o^\alpha; t)$. We define a map

$$F_{x_o}: O(\gamma(x_o^\alpha; t)) \to \Delta(x_o^\alpha, t),$$

where

$$F_{x_o} \circ y^a = E|_{O(\gamma(x_o^\alpha; t))}.$$

F_{x_o} is a diffeomorphism everywhere except $\alpha = 0$ and $\alpha = \alpha_o(x_o^\alpha, t, \theta, \varphi)$. But by Lemma 5.2, $\alpha_o(x_o^\alpha, t, \theta, \varphi)$ is a C^0 surface; $\alpha = 0$ and $\alpha = \alpha_o(x_o^\alpha, t, \theta, \varphi)$ both have measure zero as does their image, γ and $\cup_{p' \in \gamma} C(p')$, under the smooth map F_{x_o} [10]. $F_{x_o}^{-1}$ therefore yields coordinates everywhere on $\Delta(x_o^\alpha, t)$ except a set of measure zero.

We use these coordinates to define

$$A_\delta(x_o^\alpha, t) = \int_0^\pi \int_0^{2\pi} \int_\delta^{\alpha_o - \delta} \mathrm{d}\theta\, \mathrm{d}\varphi\, \mathrm{d}\alpha \sqrt{-\det \delta F_{x_o} g}$$

and

$$V_\delta(x_o^\alpha, t) = \int A_\delta(x_o^\alpha, t)\,\mathrm{d}t.$$

(Here δF_{x_o} is the dual map.) Since a basis for the neighborhood system of $\alpha = 0$ and $\alpha = \alpha_o(x_o^\alpha, t, \theta, \varphi)$ is mapped by F_{x_o} into a basis for γ and its cut locus, we see that $V(x_o^\alpha, t) = \lim_{\delta\to 0} V_\delta(x_o^\alpha, t)$.

One can see that $V(x_o^\alpha, t)$ is continuous; a short calculation shows that $(\partial/\partial t)V(x_o^\alpha, t)$ exists if $A(x_o^\alpha, t) \equiv \lim_{\delta\to 0} A_\delta(x_o^\alpha, t)$ depends continuously on t. In that case $(\partial/\partial t)V = A$.

One finds that A depends continuously on x^α as well. For, by Lemma 5.2, α_o which defines the range of integration is continuous on a compact set: thus for (x', t') near (x, t), $\alpha_o(x', t', \theta, \varphi)$ is near $\alpha_o(x, t, \theta, \varphi)$ for all θ and φ. Similarly, the integrand, $\sqrt{-\det \delta F_x g(y^a)} \equiv f(x^\alpha, y^a)$, is continuous on a compact set (by Lemma 5.1) and so, for (x', t') near (x, t), $f(x', t', \alpha, \theta, \varphi)$ is near $f(x, t, \alpha, \theta, \varphi)$ for all θ and φ and almost all α. Then $A(x', t')$ is near $A(x, t)$; the partial derivatives along timelike coordinate lines exist and are continuous.

6.3. To determine the partial derivatives along a family of spacelike coordinate lines, we choose the timelike coordinate lines so that $E_4 + E_\alpha = \bar{E}_4$ is timelike on U_p for each α. Choosing coordinates, $\bar{x}^a$, such that $\bar{E}_4$ is tangent to the lines $\bar{x}^\alpha = \text{const.}$, one finds that $(\partial/\partial x^\alpha)V$ exists and $(\partial/\partial x^\alpha)V = (\partial/\partial t)V - (\partial/\partial \bar{x}^\alpha)V$. The partial derivatives of V are continuous; $V(p, q)$ depends differentiably on p.

6.4. The remainder of the proof is straightforward. To prove (B), we need only show that $V(p, \bar{q}) - V(p, q)$ is a constant that depends only on q and $\bar{q}$. By Proposition 3.3, we can choose $q' \subset I^- p \cap I^- q \cap I^- \bar{q}$. Then,

$$\begin{aligned} V(p, \bar{q}) - V(p, q) &= [V(p, q') - V(\bar{q}, q')] \\ &\quad - [V(p, q') - V(q, q')] \\ &= V(q, \bar{q}). \end{aligned} \tag{6.1}$$

(Cf. Section 4.2(ii).)

To prove (C), let $\bar{p}$ be any point on a causal curve with past endpoint p. A rearrangement of (6.1) shows that $T(\bar{p}) - T(p) = V(\bar{p}, p) > 0$. T must therefore increase along any causal curve, and $H_{T_o} \equiv \{p \mid T(p) = T_o\}$ contains no causally related points (i.e. H_{T_o} is an acausal set).

Next, recall that edge (H_{T_o}) is the set of all points p' where given any neighborhood of p', $U_{p'}$, one can find $q' \in I^-p' \cap U_{p'}$ and $q'' \in I^+p' \cap U_{p'}$ and q' and q'' are joined by a timelike curve γ_1 which meets H_{T_o} and γ_2 which does not [1].

But if γ_1 meets H_{T_o}, so must γ_2. For $\gamma_1 \cap H_{T_o} \neq \emptyset$ implies $T(q'') \geqslant T_o \geqslant T(q')$; T must increase continuously along γ_2 from $T(q')$ to $T(q'')$ and thus take on the value T_o. It follows that edge $(H_{T_o}) = \emptyset$ and H_{T_o} is an acausal, properly imbedded C^1 submanifold; dT is everywhere timelike [1].

University of California

REFERENCES

[1] Hawking, S. W. and Ellis, G. F. R., *The Large Scale Structure of Space-Time*, Cambridge Univ. Press, Cambridge, 1973.
[2] Penrose, R., *Techniques of Differential Topology in Relativity*, N.S.F. Conference Board of the Mathematical Sciences, Regional Conference Series in Applied Mathematics #7, SIAM, Philadelphia, 1972.
[3] Kronheimer, E. H. and Penrose, R., *Proc. Camb. Phil. Soc.* **63**, (1967), 481.
[4] Geroch, R., Kronheimer, E. H., and Penrose, R., *Proc. Roy. Soc. (Lond.)* **A327** (1972), 545.
[5] Budic, R. and Sachs, R. K., *Gen. Rel. Grav.*, 1975 (to appear).
[6] Budic, R. and Sachs, R. K., to appear.
[7] Hawking, S. W., in *Black Holes* (C. DeWitt and B. DeWitt, eds.), Gordon and Breach, New York, 1973.
[8] Bishop, R. L. and Crittenden, R. J., *Geometry of Manifolds*, Academic Press, New York, 1964.
[9] Kelley, J. L., *General Topology*, Van Nostrand, Princeton, 1955.
[10] Sternberg, S., *Lectures on Differential Geometry*, Prentice-Hall, Englewood Cliffs, N.J., 1964.

YVONNE CHOQUET-BRUHAT

THE PROBLEM OF CONSTRAINTS IN GENERAL RELATIVITY: SOLUTION OF THE LICHNEROWICZ EQUATION

This paper is dedicated to André Lichnerowicz. The splitting property through conformal methods has been discovered by him, as well as the master equation whose solutions on a manifold give admissible initial data sets for Einstein's equations. In his fundamental paper of 1944 ('Journal de Mathématiques pures et appliquées) he uses the master equation, which will now be called the Lichnerowicz equation, to construct the first rigorous general solutions of the n-body problem in general relativity.

It is a great pleasure for me to have the opportunity to say how much I have admired during the years the papers of André Lichnerowicz, so beautifully written and full of neat and basic results. I would also like to say how deeply I have appreciated, during those same years, his illuminating comments and his always friendly encouragement.

Introduction

It has been known for a long time that all information about a space-time is contained in its Cauchy data. This statement, proved in the analytic case by Cartan, Darmois, Lichnerowicz [1] and in the differentiable case by Choquet [Fourès]-Bruhat, has attained its more precise formulation in Choquet-Bruhat and Geroch [8, 9]. An initial data set (M, g, K) is a three-dimensional manifold M, endowed with a properly Riemannian metric g and a field K of symmetric 2-tensors. It is proved in Ref. [9] that to each initial data set corresponds a space-time $(V, {}^{(4)}g)$, with V four-dimensional manifold and ${}^{(4)}g$ Lorentzian metric on V, which admits a submanifold S diffeomorphic to M, such that the metric induced on S by ${}^{(4)}g$ and the second fundamental form of S as submanifold of $(V, {}^{(4)}g)$ are respectively the images by the diffeomorphism $M \to S$ of g and K. The space-time satisfies Einstein empty space equations

$$\mathrm{Ricc}\,({}^4g) = 0, \qquad \mathrm{Ricc}\,({}^4g)\text{: Ricci tensor of } {}^{(4)}g \tag{1}$$

Cahen & Flato (eds.), Differential Geometry and Relativity, 225–235.

if and only if the initial data set (M, g, K) satisfies the constraint equations which read

$$R(g) - K \cdot K + (\operatorname{tr} K)^2 = 0 \tag{2}$$

[$R(g)$ scalar Riemannian curvature of g, $K \cdot K = K_i^h K_h^i$, $\operatorname{tr} K = K_i^i$] – Equation (2) is often called the Hamiltonian constraint – and

$$\delta(K - g \operatorname{tr} K) = 0 \tag{3}$$

(i.e. $\nabla_i K^{ij} - g^{ij} \nabla_i K_l^l = 0$, ∇_i covariant derivative in the metric g).

It is proved in Refs. [8, 9] that there exists one and only one, up to isometries, maximal (the ordering being defined by embedding of one space-time into another) Einsteinian, globally hyperbolic* space-time $(V, {}^{(4)}g)$ corresponding to a given initial data set (M, g, K) satisfying the constraints. Such a space-time is called the Cauchy development of (M, g, K).

The minimum regularity of (M, g, K) required in this theorem is†

$$g \in H_4^{\text{loc}}(M), \qquad K \in H_3^{\text{loc}}(M), \qquad M \in H_5^{\text{loc}}.$$

It is also true in C^∞ (Choquet-Bruhat [11]).

An analogous theorem is valid for Einstein equations with sources

$$\operatorname{Ricc}({}^{(4)}g) - \tfrac{1}{2}{}^{(4)}g R({}^{(4)}g) = T \tag{1}$$

when the conservation law $\delta T = 0$ for the stress energy tensor T gives rise to hyperbolic equations (perfect fluid, electromagnetic field with zero or infinite conductivity, . . .). The constraints equations then read

$$R(g) - K \cdot K + (\operatorname{tr} K)^2 = Q, \tag{2}$$

$$\delta(K - g \operatorname{tr} K) = s, \tag{3}$$

where Q is a positive scalar and s a vector on M. If, in the Cauchy development, the local equation of S is $x^0 = 0$, then

$$Q = 2T^{00}, \qquad s_i = S_i^0.$$

This paper is devoted to the study of the constraint set (2) and (3) by the conformal method initiated by A. Lichnerowicz.

* Cf. Leray [13] and Choquet-Bruhat [5].

† The existence in H_4^{loc}, and uniqueness in H_5^{loc} is proved in Choquet-Bruhat [6]. The fact that uniqueness is also valid in H_4 was noticed by Fisher and Marsden in Ref. [16] and Hawking (cf. [15]).

1. Conformal methods

The conformal method

If the (positive definite) metric g is given up to a conformal factor, the scalar curvature $R(g)$ is an elliptic operator for this conformal factor, with principal part the Laplace operator. In particular, if one sets (Lichnerowicz [2])

$$g = \varphi^4 \gamma,$$

then the Christoffel symbol Γ^k_{ij} of g and $\bar{\Gamma}^k_{ij}$ of γ are related by

$$\Gamma^j_{ih} = \bar{\Gamma}^j_{ij} + 2\varphi^{-1}(\delta^j_i \partial_h \varphi + \delta^j_h \partial_i \varphi - \gamma^{jl}\gamma_{ih}\partial_l \varphi)$$

and the scalar curvatures of g and γ by

$$R(g) = \varphi^{-4}R(\gamma) - 8\varphi^{-5}\Delta_\gamma \varphi.$$

On the other hand, if A is a mixed 2-tensor, then

$$\delta_g(\varphi^{-6}A) = \varphi^{-6}\delta_\gamma A - 2\varphi^{-1}\nabla\varphi \operatorname{tr} A \qquad (\nabla\varphi \equiv \operatorname{grad} \varphi).$$

Therefore, if we define the tensor A with zero trace by

$$A = \varphi^6(K - \tfrac{1}{3}I \operatorname{tr} K), \qquad [(I)^k_i = g^k_i = \delta^k_i]$$

and denote

$$\tau = \operatorname{tr} K,$$

the system (2) and (3) of constraints read*

$$8\Delta_\gamma \varphi - R(\gamma)\varphi + A \cdot A\varphi^{-7} - \tfrac{2}{3}\tau^2\varphi^5 + Q\varphi^5 = 0, \tag{4}$$

$$\delta_\gamma A - \tfrac{2}{3}\varphi^6 I \nabla\tau = s\varphi^6. \tag{5}$$

Equation (4) is the Lichnerowicz equations.

In the case $s = 0$, which is met in many physical situations, the system admits a particular family of solutions, obtained by replacing the system (4) and (5) by†

$$\nabla\tau = 0, \qquad \text{i.e.} \quad \tau = cte, \tag{6}$$

$$\delta_\gamma A = 0, \tag{7a}$$

* Note also that the tensor densities $|\det \gamma|^{1/2}A$ and $|\det g|^{1/2}(K - \tfrac{1}{3}I \operatorname{tr} K)$ are equal.
† The extension from the case $\tau = 0$ to the case $\tau = cte$ is due to O'Murchada and York [17].

$$\operatorname{tr} A = 0. \tag{7b}$$

From A and τ satisfying these equations, and $\varphi > 0$ solution of the corresponding Lichnerowicz equation one deduces a solution of the constraints

$$K = \varphi^{-6} A + \tfrac{1}{3} I \tau, \qquad g = \varphi^4 \gamma.$$

The solutions thus obtained are characterized by the geometric property $\operatorname{tr} K = cte$, they correspond to initial manifolds of constant mean extrinsic curvature, to maximal (with respect to area) submanifolds of the space-time when this constant is zero.

Solution of the non-Hamiltonian constraint, geometric interpretation

The solutions A of (7a) and (7b) are a linear subspace of the space of 2-tensors on M. The operator δ_γ is the dual to the Lie derivative operator $\mathscr{L}_\gamma$, and the equation $\delta_\gamma A = 0$ expresses* that A is orthogonal at γ to the orbit of γ through the diffeomorphisms of M.

Solution of the Lichnerowicz equation on a closed manifold

The Lichnerowicz equation is of the form

$$\Delta\varphi - c\varphi + a\varphi^{-7} + b\varphi^5 = 0 \tag{L}$$

with

$$c = \tfrac{1}{8} R, \qquad a = \tfrac{1}{8} A \cdot A \geqslant 0, \qquad b = \frac{1}{4}\left(S^{00} - \frac{\tau^2}{3}\right).$$

If instead of considering the sources as *a priori* given on the manifold M one supposes that they are of the type

$$S^{00} = \varphi^{-12} \bar{S}^{00},$$

$\bar{S}^{00}$ given on M, satisfying the energy condition, then

$$a = \tfrac{1}{8} A \cdot A + \tfrac{1}{4} \bar{S}^{00} \geqslant 0, \qquad b = -\frac{\tau^2}{12} \leqslant 0.$$

This point of view, adopted by O'Murchada and York [17] is legitimate if the sources are themselves fields, known by their tensor densities, and if the stress energy tensor is quadratic in these fields: indeed if $\mathscr{B}$ is

* Cf. Wheeler [19] and Deser [18] and, for a proof in H_s spaces or a compact manifold, Berger and Ebin [24].

a given tensor density, the corresponding tensors in the metrics g and γ are respectively $\bar{B} = \mathscr{B}\, |\det \gamma|^{-1/2}$ and $B = \mathscr{B}\, |\det g|^{-1/2} = \bar{B}\varphi^{-6}$.

I have proved in Ref. [10] global theorems for the Lichnerowicz equation (existence, nonexistence, uniqueness, and stability) in the Hölder spaces $C^{k,\alpha}$ on a closed manifold, by using a general method to solve nonlinear elliptic equations recently developed by J. Leray and myself (cf. [12]). We shall here, following the same general ideas, present such a study in the Sobolev spaces H_s, where the Cauchy development is known to exist*.

Regularity hypothesis on (M, γ)

The manifold M is always taken to be C^∞. On the given metric γ we make the following assumption:

(1) γ is positive definite at each point of M,

(2) γ belongs to the Banach† space H_4.

Note that, M being three-dimensional, H_2 is an algebra and $\gamma \in H_4$ implies $\gamma \in C^2$.

Let us recall some known lemmas:

Maximum principle. If $u \in C^2(M)$ verifies at a point $x \in M$

$$\Delta u < 0, \qquad (\text{resp. } \Delta u > 0),$$

then u cannot reach a minimum (resp. a maximum) at this point.

Under the regularity hypothesis made on γ, we have (cf. Refs. [21–23]):

Lemma for linear equations. On the closed manifold M the linear equation

$$\Delta u - u = f, \qquad f \in H_2(M)$$

has one, and only one, solution $u \in H_4(M)$, which satisfies (C_1 and C_2 constant depending on M and γ):

$$\|u\|_{H_4} \leq C_1 \|f\|_{H_2}$$

and

$$\|u\|_{H_2} \leq C_2 \|f\|_{L_2}.$$

* Note that, for $n = 3$, on a closed manifold, $C^{2,\alpha} \subset H_4 \subset C^{4,\alpha}$. The inclusions being strict, our present result is not included in our previous one.

† In a given finite atlas (and thus in all such atlases) each component of γ, in every chart (U, φ) belongs to $H_4(\omega)$, with $\omega = \varphi(U)$ bounded open ball of $\mathbf{R}^3$.

These two inequalities and the maximum principle will furnish the *a priori* estimates necessary for the application of the Leray-Schauder theory of degree to the Lichnerowicz equation.

THEOREM. On the closed manifold $M(C^\infty)$, with a positive Riemannian metric $\gamma \in H_4$, the Lichnerowicz equation

$$\Delta\varphi - c\varphi + a\varphi^{-7} + b\varphi^5 = 0 \tag{L}$$

with c, a, $b \in H_2$, has a solution

$$\varphi > 0, \qquad \varphi \in H_4$$

if the following conditions are satisfied:

(1) $a \geqslant 0$ for all $x \in M$;
(2) at the points $x \in M$, where $a = 0$, then $b < 0$;
(3) if α_1 denotes the first positive root of $P(z)$ (which exists under the above assumption) and if α_2 denotes either the second positive root of $P(z)$, or $+\infty$ if $P(z)$ has only one positive root, then, on M

$$\sup_{x \in M} \alpha_1 < \inf_{x \in M} \alpha_2.$$

This last condition is always satisfied if $b < 0$ on M (then $\alpha_2 \equiv +\infty$).

Proof. We denote by Ω the bounded open set of the space of functions $H_2(M)$ defined by*

$$\Omega : \|v\|_{H_2} < c, \qquad 0 < l < v < m,$$

where c, l, m are positive numbers to be determined later.

The closure $\bar{\Omega}$ of Ω is

$$\|v\|_{H_2} \leqslant c, \qquad 0 < l \leqslant v \leqslant m.$$

The solutions of Equation (L) are the fixed points of the mapping

$$\mathscr{F} : \Omega \to H_2, \qquad v \mapsto u = \mathscr{F}(v),$$

where u is the (unique) solution of the linear equation on M:

$$\Delta u - u = cv - av^{-7} - bv^5 - v. \tag{E}$$

* Since M is three-dimensional, $C^0(M) \subset H_2(M)$.

Due to the hypothesis made on c, a, b

$$cv - av^{-7} - bv^5 - v \in H_2$$

when $v \in \Omega$; therefore, Equation (E) has indeed a unique solution u in H_2, which is even in H_4 by the linear lemma. The embedding of H_4 into H_2 being compact (M is a compact manifold), the mapping $\mathscr{F}$ itself is a compact mapping. The mapping $\mathscr{F}$ has a fixed point in Ω if the mapping

$$Id - \mathscr{F} : \Omega \to H_2$$

has nonzero degree at the point $0 \in \Omega$.

We denote by $\mathscr{F}_0 : \Omega \to H_2$ a compact mapping such that $Id - \mathscr{F}_0$ has degree 1 at 0; for instance, we take for $\mathscr{F}_0$ the constant map defined by resolution of the linear problem

$$\Delta u - u = -\frac{1+m}{2}. \tag{E$_0$}$$

The maps $Id - \mathscr{F}$ and $Id - \mathscr{F}_0$ have the same degree (cf. Leray and Schauder [14]) if the mapping $\mathscr{F}_t = t\mathscr{F} + (1-t)\mathscr{F}_0$ has no fixed point on the boundary $\partial\Omega = \bar{\Omega} - \bar{\Omega}$ for $0 \leqslant t \leqslant 1$. The theorem will be proved if we show that, under its hypothesis, all possible solutions $u \in \bar{\Omega}$ of

$$\Delta u = t(cu - au^{-7} + bu^5) + (1-t)\left(u - \frac{1+m}{2}\right) \tag{E$_t$}$$

are in fact in Ω, for every t, $0 \leqslant t \leqslant 1$.

We first show that we can choose l and m, $0 < l < m$ such that

$$l \leqslant u \leqslant m \quad \text{implies} \quad l < u < m.$$

Indeed, at a point $\bar{x} \in M$ where $u(\bar{x}) = l$ [resp. $u(\bar{x}) = m$] we must have, since $l \leqslant u \leqslant m$ on M, $(\Delta u)(\bar{x}) \geqslant 0$ [resp. $\Delta u(\bar{x}) \leqslant 0$]. Therefore, if we can choose l and m such that the right-hand side of Equation (E$_t$) is negative [resp. positive] whenever $u = l$ [resp. $u = m$], we shall have proved $l < u < m$, for this choice of l and m. Clearly, it is necessary and sufficient to verify

$$cl - al^{-1} - bl^5 < 0, \quad cm - am^{-7} - bm^5 > 0, \quad \forall x \in M. \tag{I}$$

Let us denote by $P(z)$ the polynomial

$$P(z) = -bz^3 + cz^2 - a.$$

Its variation (with respect to z) depends on the signs of the coefficients, a, b, and c, which are functions of $x \in M$. According to these signs we have Figures 1(a) to (e).

It is clear from Figures 1(a) to (e) that one can choose l and m satisfying the inequalities (I), with $0 < l < m$, at each point $x \in M$, where $a > 0$, $b < 0$, or $a = 0$, $b < 0$, $c < 0$, or $a > 0$, $b \geqslant 0$, $c > 0$,

$$\frac{4P^3}{27Q^2} > 0.$$

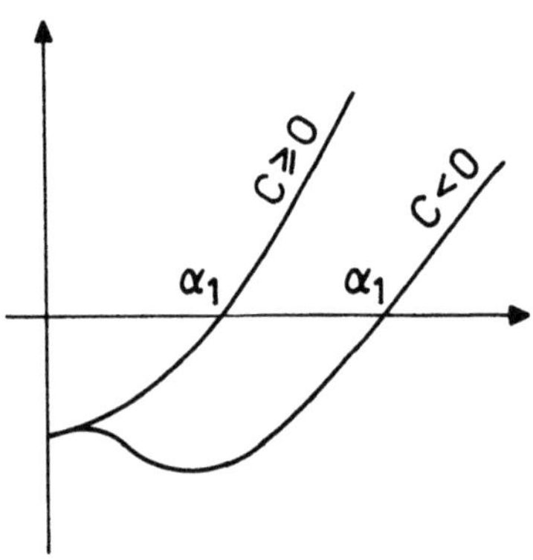

a>0, b<0

(a)

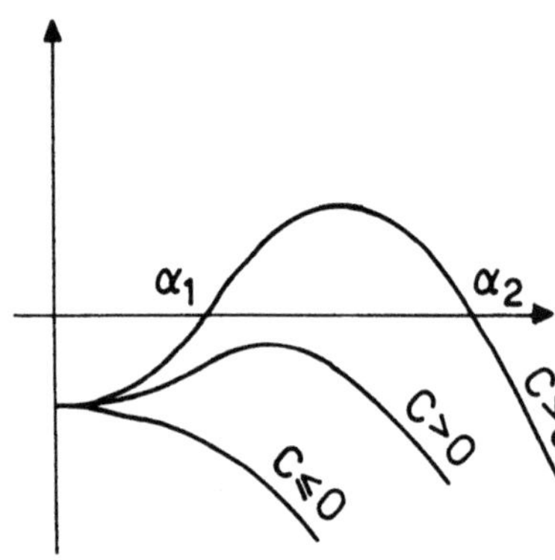

a>0, b>0

(b)

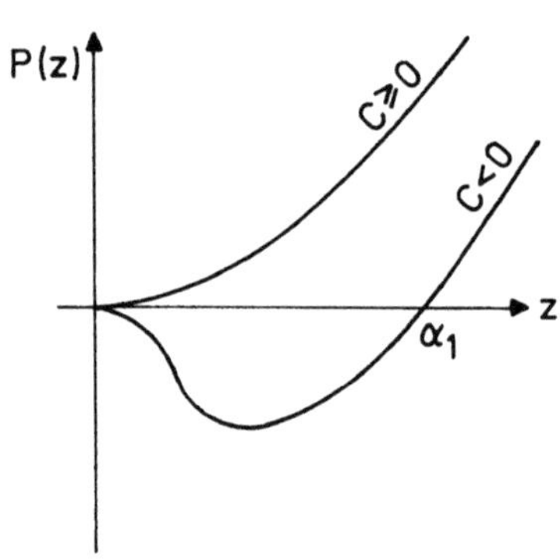

a = 0, b<0

(c)

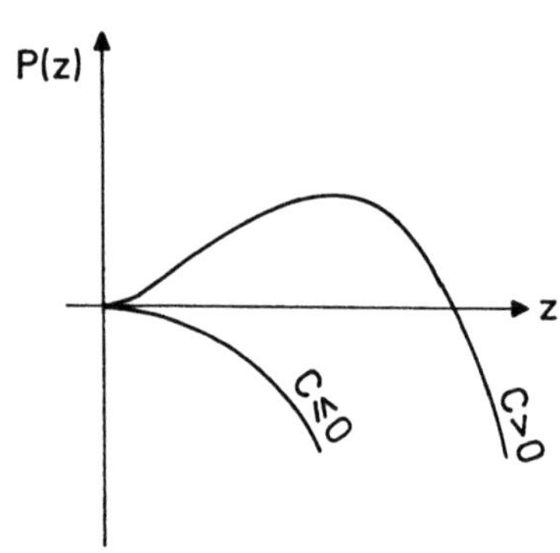

a = 0, b≥0

(d)

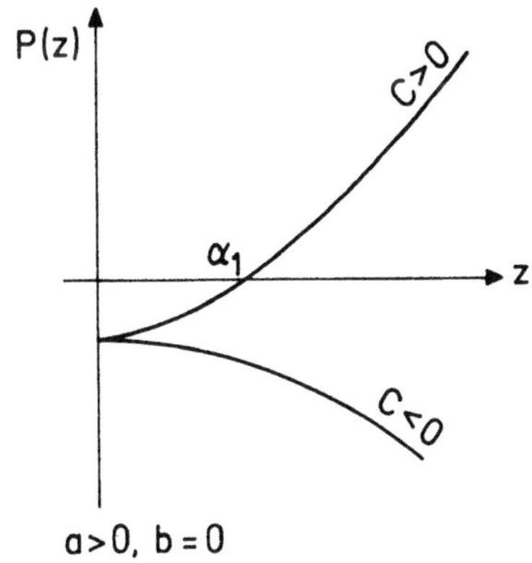

(e)

Fig. 1.

The necessary and sufficient condition for the existence of constant l and m on M (i.e. choice independent of x) is

$$\operatorname*{Sup}_{x\in M} \alpha_1 < \operatorname*{Inf}_{x\in M} \alpha_2,$$

where α_1 denotes, as indicated in Figures 1(a) to (e), the first positive root of $P(z)$, and α_2 either the second or $+\infty$ if $P(z)$ has no second positive root.

In particular, if M is compact, the condition is always satisfied when $a > 0$, $b > 0$ on M.

We now show that we can choose K such that the solutions of Equation (E_t) – equivalent (convexity) to the solutions of Equations (E) and (E_0) – verify

$$\|u\|_{H_2} < K.$$

If M is compact, u continuous, and $0 < l < u < m$, then $cu - au^{-7} + bu^5 - u$ has an L_2 norm bounded by a constant c depending only on M, a, b, c, l, m. Therefore, if we choose

$$K > C_2\left\{\max\left(C, \left\|\frac{l+m}{2}\right\|_{L_2}\right)\right\},$$

we will have the required inequality.

Corollaries

If we apply the theorem to the Lichnerowicz equation corresponding to the Hamiltonian constraint in empty space:

$$8\Delta\varphi - R\varphi + A \cdot A\varphi^{-7} - \tfrac{2}{3}\tau^2\varphi^5 = 0 \tag{L}$$

(τ is a constant and $A \cdot A \geqslant 0$), we obtain:

COROLLARY 1. The Lichnerowicz equation in empty space has a solution $\varphi > 0$ on an initial closed manifold of constant mean extrinsic curvature $\tau \neq 0$ if either

(a) $A \cdot A \neq 0$ on M,
(b) $R > 0$ at the points $x \in M$, where $A \cdot A = 0$.

The formulation of the result in the presence of sources depend on their scaling. In the case of scaled sources, and $\tau = 0$, the equation reads

$$8\Delta\varphi - R\varphi + (A \cdot A + 2\bar{S}^{00})\varphi^{-7} = 0.$$

The existence theorem then gives:

COROLLARY 2. The Lichnerowicz equation with scaled sources on a maximal closed initial manifold has a solution $\varphi > 0$ if $A \cdot A + 2\bar{S}^{00} > 0$ and $R > 0$ on M.

Uniqueness and nonexistence theorems. They are the same as that in the $C^{k,\alpha}$ case (cf. Choquet-Bruhat [10]). A uniqueness theorem corresponding to the case $\tau = 0$ had already been given by Lichnerowicz [2].

Université de Paris

REFERENCES

[1] Lichnerowicz, A., *Problèmes globaux en Mécanique Relativiste*, Hermann, 1939.
[2] Lichnerowicz, A., *J. Math. Pures Appl.* **23** (1944), 37–63.
[3] Lichnerowicz, A., *Théories relativistes de la gravitation*, Masson, 1955.
[4] Choquet-Bruhat, Y., *Acta Matematica* (1952).
[5] Choquet-Bruhat, Y., in *Batelle Rencontres*, (J. A. Wheeler and C. DeWitt, eds.), Gordon and Breach, 1967.
[6] Choquet-Bruhat, Y., *Ann. I.H.P.* **8**(4) (1968), 327–338.
[7] Choquet-Bruhat, Y., *Jour. G.R.G.* (1971).
[8] Choquet-Bruhat, Y. and Geroch, R., *C.R. Acad. Sci.* **268** (1969).
[9] Choquet-Bruhat, Y. and Geroch, R., *Comm. Maths. Phys.* **21** (1971), 211–218.
[10] Choquet-Bruhat, Y., *C.R. Acad. Sci.* (1972) et Istituto di Acta Mat. (1973).

[11] Choquet-Bruhat, Y., *J.G.R.G.* **5** (1974), 47–64.
[12] Choquet-Bruhat, Y. and Leray, J., *C.R. Acad. Sci.* **274** (1972), 81–85.
[13] Leray, J., *Hyperbolic Differential Equations*, I.A.S., Princeton, 1952.
[14] Leray, J. and Schauder, J., *Ann. Ec. N.S.* (1934).
[15] Hawking, S. and Ellis, G., Global Structure of Space Time', 1973.
[16] Fischer, A. and Marsden, J., *Comm. Maths. Phys.* (1973).
[17] O'Murchada, N. and York, J., *Phys. Rev. D.* **10** (1974).
[18] Deser, S., *Ann. I.H.P.* (1968).
[19] Wheeler, J. A., *Geometrodynamics*, Academic Press, New York, 1962.
[20] Misner, C., Thorne, K., and Wheeler, J. A., *Gravitation*, Freeman, 1974.
[21] Lions, J. L. and Magenès, E., *Problèmes aux limites elliptiques non homogènes*, Vol. 1, Dunod, 1968.
[22] Palais, S., Seminar on the Atiyah Singer Index Theorem', Ann. of Math. Stud, Princeton, 1965.
[23] Morrey, J., *Multiple Integrals in the Calculus of Variations*, Springer Verlag, Berlin, 1966.
[24] Berger, M. and Ebin, D., *J. Diff. Geom.* **3** (1969), 379–392.

PH. DROZ-VINCENT

HAMILTONIAN CONSTRUCTION OF PREDICTIVE SYSTEMS

1. Introduction

This paper is devoted to the explicit construction of relativistic two-particle systems with *finitely predictive interactions*. We mean that the equations of motion are a *differential system* of *finite* order, in contrast with the generic case of action-at-a-distance theories where these equations are usually more complicated (integrodifferential, difference-differential, etc.).

Actually we always consider second-order differential systems. Finitely predictive dynamics has widely developed during the last years, first in the three-dimensional formalism, then in a manifestly covariant way [1, 2]. But the explicit construction of models has for a long time been a challenging task, mainly because, in contrast with the classical case, the relativistic accelerations for N bodies ($N \geqq 2$) *cannot* be chosen arbitrarily.

Besides the well-known fact that the accelerations must be orthogonal to the velocities, N-particle dynamics requires a strong *nonlinear* condition on the relativistic forces: world-line condition in the three-dimensional approach, *predictivity condition* in the covariant formulation [1, 2]. With the help of Hamiltonian formalism we have indicated how to get many solutions of this condition [3]. Once commuting Hamiltonians (in the sense of Poisson bracket) have been found [4], the remaining problem is only to solve a set of linear equations

$$\{H, x^{\alpha'}\} = 0, \qquad \{H', x^{\alpha}\} = 0, \tag{1.1}$$

that we may call *position equations*. If the Hamiltonians have enough symmetry the appropriate technique is the conventional use of first integrals (completed with the concept of 'partial integral' which is more specially adapted to the many-time formalism).

This recipe leads to explicit results, that is to say the right-hand side of the equations of motion can be finally constructed in terms of positions and velocities. But the orthogonality of accelerations to velocities is generally not satisfied.

Cahen & Flato (eds.), Differential Geometry and Relativity, 237–248.

In Ref. [4] we have indicated how mass constancy can be obtained in principle by suitably choosing the temporal parts of the positions. This *ad hoc* procedure gives no explicit results, so we prefer to abandon it. Rather than accepting particles with varying masses we shall enlarge slightly the axiomatics of predictive dynamics, as follows: the same mathematical structure which could be considered as a system with varying masses may as well be interpreted as a system of constant masses described with the 'wrong' parameter. The correspondence between this nonaffine parameter and the proper-times is fixed because we make the simplest assumption about the Hamiltonians (which are constant of the motion, anyway). We postulate that these Hamiltonians are just the halves of the squared masses.

Finally, for any system with nonconstant masses, we explicitly show which change of parameters is to be performed in order to recover constant masses.

In Section 3 we write down Hamiltonians and apply the method.

Recall that relativistic Hamiltonians are basically scalar generating functions, the framework being that of a symplectic phase space. The positions *are not* canonical variables, except in the free case [5].

Hamiltonization of a given system (and the puzzling question of unicity) is not our present problem, since we follow exactly the reverse way. Doing so, the meaning of the canonical variables q, p, q', p', is not *a priori* fixed.

But, for each model, i.e. for each solution of the position equations, the ordinary variables (positions and velocities) are differently related to the canonical ones.

Starting from oscillator-like Hamiltonians is very attractive. In this case, the equations of evolution can be integrated. Moreover, the general solution of the position equations is known [4, 6].

Unfortunately, we have failed to construct a solution which reduces to

$$x = q, \qquad x' = q'$$

when the coupling constant is made to vanish [7].

For this reason the canonical variables cannot receive a clear interpretation, and the analogy with harmonic oscillator remains formal. That is why, in the present article, we shall consider only free particle-like Hamiltonians.

Though they are formally trivial, they do not forbid interaction at all.

In fact, interaction does appear when the position equations are solved.

We get easily the most general solution of these equations.

We can manage to have $x \to q$, $x' \to q'$ when $k \to 0$, and the behavior for large spatial distances can be discussed. The canonical variables are trivially submitted to linear evolution equations. Actually they form a system of Hamilton-Jacobi coordinates [8].

Notations

In space-time M_4 the positions are x, x'. Whenever it is possible, Greek indices are omitted, e.g. we write x for x^α, etc. The scalar product is written in compact form, for instance: $v \cdot v'$ stands for $v^\alpha v'_\alpha$. We write $v_{\alpha'}$ for v'_α, etc., and v^2 for $v \cdot v$, etc. The four-vectors v, v' are related to the unit velocities through $v = mu$, $v' = m'u'$. Phase space $T(M_4) \times T(M_4)$, i.e. the space of quadruplets x, u, x', u'. Ordinary coordinates $x^\alpha, u^\alpha, x^{\alpha'}, u^{\alpha'}$. The indices are always moved by the Minkowski metric $\eta_{\mu\nu}$

$$\partial_\alpha = \partial / \partial x^\alpha, \qquad \partial_{\alpha'} = \partial / \partial x^{\alpha'}$$

According to the traditions of differential geometry, we identify contravariant vector fields with differential operators. Thus, the Lie derivative of a *scalar* function f by a vector field X can be noted Xf as well as $L_X f = i_X\, df$. More generally Xv^α will stand for the result of applying the differential operator X to the component v^α of v. Lie bracket [ ]; Poisson bracket { }.

In the case of constant masses, the parameters τ, τ' are related to the proper-times by $s = m\tau$, $s' = m'\tau'$. The speed of light is taken as unity. Duality of skew-symmetric tensors of M_4 is noted $*$. If a, b, c are four-vectors we write

$$(a \wedge b \wedge c)^\alpha = \eta^{\alpha}_{\ \mu\nu\rho} a^\mu b^\nu c^\rho,$$

where $\eta_{\alpha\mu\nu\rho}$ is the Levi-Civita tensor. We set

$$z = q - q', \qquad y = \tfrac{1}{2}(p - p'), \qquad r = x - x'.$$

2. Masses and Parametrization

In the *a priori* Hamiltonian approach we construct two-particle systems as follows [4]:

Consider some sixteen-dimensional space endowed with a symplectic form

$$\Omega = \mathrm{d}q^{\alpha} \wedge \mathrm{d}p_{\alpha} + \mathrm{d}q^{\alpha'} \wedge \mathrm{d}p_{\alpha'}. \tag{2.1}$$

Write down, in terms of the canonical variables q, p, q', p' the Hamiltonians H and H' satisfying

$$\{H, H'\} = 0. \tag{2.2}$$

Let X (resp. X') be the vector field generated by H (resp. H'), that is to say

$$i_X\Omega = \mathrm{d}H, \qquad i_{X'}\Omega = \mathrm{d}H'. \tag{2.3}$$

Note that (2.2) implies the predictivity condition

$$[X, X'] = 0. \tag{2.4}$$

Now, let us find a solution of (1.1), say x^{α}, $x^{\beta'}$.
We have shown [4] that, defining

$$v = Xx, \qquad v' = X'x', \tag{2.5}$$

then X and X' have the standard form

$$X = v \cdot \partial + \xi \cdot \partial/\partial v, \qquad X' = v' \cdot \partial' + \xi' \cdot \partial/\partial v' \tag{2.6}$$

in terms of the new coordinates x, v, x', v', provided, of course, these variables can be taken as coordinates, which is actually a condition about x and x' as solutions of (1.1).

In this procedure, starting from the Hamiltonians is the simplest way to construct X and X' such that (2.4) holds. But, as we can see in Ref. [4, §IV], the role of H and H' is not essential here. In order to derive the standard from (2.6) it is not necessary that X and X' are Hamiltonian vector fields. All we need actually is (2.4) and

$$Xx' = X'x = 0, \tag{2.7}$$

which is a more general version of (1.1). More briefly:

PROPOSITION 2.1. When (2.4) and (2.7) hold, X and X' take the standard form (2.6) in terms of x, v, x', v', where v and v' are defined by (2.5), provided $D(x, v, x', v')/D(q, p, q', p')$ is finite and does not vanish.

According to (2.6) the trajectories of X and X' are solution of a second-order system

$$d^2x/d\tau^2 = \xi, \qquad d^2x'/d\tau'^2 = \xi'. \tag{2.8}$$

So far, we have always made the identification

$$\begin{aligned} v^2 &= m^2, & v'^2 &= m'^2, \\ \tau &= s/m, & \tau' &= s'/m', \end{aligned} \tag{2.9}$$

where m, m' are the masses and s, s' are the proper-times. This interpretation is sound when v^2 and v'^2 are constants of the motion.

But, with our present construction, v^2 and v'^2 are generally not constant, except if we are very lucky in solving (2.7).

Hence the question: suppose we know explicitly X, X', $x^\alpha, x^{\beta'}$ satisfying (2.4) and (2.7), how do we alter this solution in order to have a system with constant masses?

The way we have indicated in Ref. [4] consists implicitly in changing a little bit x and x'. It does not lead to explicit formulas [9]. We now propose a quite different procedure, namely changing conformally the vector fields X, X' and keeping the same x, x'. Since the change in X and X' is conformal their integral curves are preserved. But they will be described by new parameters.

It is essential to assume that we start from a Hamiltonian system. We mean that X and X' are respectively generated by H and H' satisfying (2.2) [10]. We consider the general case where v^2 and v'^2 are not first integrals.

Define

$$\lambda = \sqrt{2H/v^2}, \qquad \lambda' = \sqrt{2H'/v'^2}. \tag{2.10}$$

Consider the new vector fields

$$Y = \lambda X, \qquad Y' = \lambda' X'. \tag{2.11}$$

From (2.4) and (2.7) we have

$$X' v^2 = X v'^2 = 0.$$

Thus

$$X'\lambda = X\lambda' = 0. \tag{2.12}$$

Thus

$$[Y, Y'] = 0. \tag{2.13}$$

From (2.7) and (2.11) we get obviously

$$Yx' = Y'x = 0. \tag{2.14}$$

Comparison of (2.13) and (2.14) with (2.4) and (2.7) shows that the change of vector fields (2.11) preserves the form of (2.4) and (2.7). Replacing X, X' by Y, Y' in (2.5) we shall now introduce

$$w = Yx, \qquad w' = Y'x' \tag{2.15}$$

and apply Proposition 2.1 to the structure defined by Y, Y', x, x'.

The new vector fields have the standard form [11] in terms of the new coordinates x, w, x', w'.

$$Y = w \cdot \partial + \eta \cdot \partial / \partial w, \qquad Y' = w' \cdot \partial' + \eta' \cdot \partial / \partial w', \tag{2.16}$$

where we define

$$\eta = Yw, \qquad \eta' = Y'w'. \tag{2.17}$$

The integral curves are again given by a second-order system:

$$\mathrm{d}^2x/\mathrm{d}\sigma^2 = \eta, \qquad \mathrm{d}^2x'/\mathrm{d}\sigma'^2 = \eta'. \tag{2.18}$$

They are the same curves as in (2.8), but they are described by new parameters σ, σ'.

From (2.5) and (2.15) we have

$$w = \lambda v, \qquad w' = \lambda' v'. \tag{2.19}$$

Then (2.10) yields automatically

$$w^2 = 2H, \qquad w'^2 = 2H', \tag{2.20}$$

which ensures that, in contrast with v^2 and v'^2, w^2 and w'^2 are constants of the motion.

Since $v = \mathrm{d}x/\mathrm{d}\tau$, $w = \mathrm{d}x/\mathrm{d}\sigma$, etc., we have

$$\lambda = \mathrm{d}\tau/\mathrm{d}\sigma, \qquad \lambda' = \mathrm{d}\tau'/\mathrm{d}\sigma'. \tag{2.21}$$

Note that σ, σ' can be considered as affine parameters. We shall now reject (2.9) and replace it by

$$\begin{aligned} w^2 &= m^2, \qquad & w'^2 &= m'^2, \\ \sigma &= s/m, \qquad & \sigma' &= s'/m'. \end{aligned} \tag{2.22}$$

It is clear that the choice of λ, λ' in (2.10) is not so arbitrary: it is deeply motivated by the need of having (2.20), which gives to H and H' the simplest kinematical interpretation.

The force law is given by (2.17). Its explicit form in terms of q, p, q', p' is given by applying twice Y (resp. Y') to x (resp. x'). Then one has to make a change of variables which is known since x, x', λ, λ' are known explicitly. When ξ and ξ' are $\neq 0$, then η and η' will not vanish, except maybe for some exceptional configurations requiring that ξ is a four-vector parallel to v (resp. ξ', v'). Thus, starting from a Hamiltonian system where v^2 and v'^2 are allowed to vary, the procedure we have just described provides an interacting system with constant masses. This transformation will be referred to as the *mass-parameter correction.*

So the remaining task is the construction of Hamiltonian systems, up to the constancy of masses.

3. Construction of models

We consider simply the free particle-like Hamiltonians,

$$H = \tfrac{1}{2}p^2, \qquad H' = \tfrac{1}{2}p'^2. \tag{3.1}$$

Naturally, $\{H, H'\}$ vanishes, since q, p, q', p' are submitted to the standard Poisson bracket relations [12].

The Hamiltonian equations of motion are trivial and immediately integrated, which yields

$$q = p\tau + q_{(0)}, \qquad q' = p'\tau' + q'_{(0)}, \tag{3.2}$$

where $q_{(0)}$ and $q'_{(0)}$ are initial conditions.

But (3.2) makes sense only after (1.1) is solved.

Finding a solution of (1.1) gives the correspondence between x, v, x', v' and the canonical variables, since, once x and x' are obtained, applying X and X' respectively to x and x' provides the canonical expressions of v and v'.

Then inserting the change of variables

$$q, p, q', p' \mapsto x, v, x', v'$$

into (3.2) provides the trajectories

$$x = x(\tau), \qquad x' = x'(\tau').$$

Note that (1.1) states simply that $x^{\alpha'}$ (resp. x^{β}) are integrals of the vector field X (resp. X') generated by H (resp. H'). Instead of the trivial solution, $x = q$, $x' = q'$, we shall consider more generally

$$x = q + F, \qquad x' = q' + F', \tag{3.3}$$

where the extra terms F, F' include interaction and a coupling constant k.

DEFINITIONS. A *first integral* or constant of the motion is an integral of both vector fields X, X'.

A *partial integral* relative to X (resp. X') is a phase-space function which is invariant by X (resp. X') but not by X' (resp. X).

Its Lie derivative by X (resp. X') and Poisson bracket by H (resp. H') vanish.

The question of Poincaré invariance is settled according to a general discussion included in Ref. [4].

Recall that, in the case of a $\mathscr{P}$-invariant Hamiltonian formulation, z, p, p' are vector invariants of $\mathscr{P}$ while the six canonical quadratic scalars

$$z^2, p^2, p'^2, z \cdot p, z \cdot p', p \cdot p'$$

are scalar invariants of $\mathscr{P}$.

The most general *scalar* invariant is a function of these six scalars.

The most general *vector* invariant is a linear combination of z, p, p', and $z \wedge p \wedge p'$ *with scalar invariant coefficients.*

The Hamiltonian formulation is $\mathscr{P}$-invariant provided $x - q$, $x' - q'$, $p - v$, $p' - v'$ are vector invariants. But, since H and H' are obviously $\mathscr{P}$-invariants in (3.1), requiring that $x - q$ and $x' - q'$ are vector invariants is actually sufficient.

Let us introduce the following notations:

$$Q = \tfrac{1}{2}(q + q'), \qquad P = p + p',$$
$$M = Q \wedge P + z \wedge y,$$
$$|P| = (P^2)^{1/2}, \qquad L = P \cdot *M/|P|.$$

We consider only the cases where $P^2, p^2, p'^2 > 0$. Thus the following projectors are well defined

$$\Pi = \delta - P \otimes P/P^2,$$
$$\bar{\omega} = \delta - p \otimes p/p^2,$$
$$\bar{\omega}' = \delta - p' \otimes p'/p'^2.$$

The application of Π is represented by $\sim$; for example, $\tilde{y} = \Pi y$. Note that $P, \tilde{y}, L$ are three mutually orthogonal vector invariants of $\mathscr{P}$. Thus

$$I = P \wedge L \wedge \tilde{y}$$

is a vector invariant. We have an orthogonal (not normalized!) tetrad $P, \tilde{y}, L, I$.

In the generic case, z, p, p' and therefore any vector invariant of $\mathscr{P}$, can be expressed as a linear combination of $P, \tilde{y}, L, I$ with scalar invariant coefficients.

But with our choice (3.1), it turns out that $P, \tilde{y}, L$ and therefore I, are constants of the motion [13]. So, developing vector invariants on this basis will bring out some simplifications.

The most general possible form of F and F' in (3.3) is

$$\begin{aligned} F &= AP + B\tilde{y} + CL + DI, \\ F' &= A'P - B'\tilde{y} + C'L - D'I, \end{aligned} \tag{3.4}$$

where A, B, C, D (resp. A', B', C', D') must be scalar invariant integrals of X' (resp. X) in order to satisfy (1.1).

Using now the fact that H and H' have the simple expression (3.1) we get:

PROPOSITION 3.1. p, p', q', $q \wedge p$ (resp. p, p', q, $q' \wedge p'$) are a complete set of integrals of X (resp. X').

Proof. p, p', q' are obviously twelve independent quantities. Knowing the components of $q \wedge p$ is, generally, equivalent to the knowledge of three more independent quantities (e.g. the q^i in a frame adapted to p). The whole set is thus equivalent to fifteen independent integrals, which is the maximum, since phase space has sixteen dimensions.

It is convenient to introduce

$$l = z \wedge p, \qquad l' = -z \wedge p', \tag{3.5}$$

and to consider alternatively the *complete* set of integrals

$$p, p', q', l \qquad \text{(resp. } p, p', q, l'\text{)}.$$

It is then obvious that

PROPOSITION 3.2. The most general $\mathscr{P}$*-invariant* integral of X (resp. X') is a function of p, p' and l (resp. p, p', l').

PROPOSITION 3.3. *A, B, C, D* (resp. *A', B', C', D'*) must be scalars formed only from p, p', l (resp. p, p', l').

Remark. In particular, any $\mathscr{P}$-invariant constant of the motion can be expressed in terms of either p, p', l or p, p', l' as well.

Practically, we shall write the constants of the motion more compactly, without the details of this dependence.

Example. P^2 instead of $p^2+p'^2+2p \cdot p'$.

Finally, all the predictive systems deducible from the Hamiltonians (3.1) are explicitly defined by (3.3), (3.4), and Proposition 3.3.

However, in order to construct physically realistic systems one is compelled to go into more details.

For instance one wishes to discuss the behavior of the interaction at large spatial distance.

In the framework of relativity this concept is far from being clear.

Fortunately, some suitable $\mathscr{P}$-invariant scalars seem to generalize the familiar squared distance of Newtonian mechanics.

Using l and l' defined by (3.5) we set

$$\rho^2 = -\tfrac{1}{2}l^{\alpha\beta}l_{\alpha\beta}/p^2, \qquad \rho'^2 = -\tfrac{1}{2}l'^{\alpha\beta}l'_{\alpha\beta}/p'^2. \tag{3.6}$$

It turns out immediately that

$$\rho^2 = -(\bar{\omega}z)^2, \qquad \rho'^2 = -(\bar{\omega}'z)^2. \tag{3.7}$$

Since $\bar{\omega}$ and $\bar{\omega}'$ are projectors onto spacelike hyperplanes, ρ^2 and ρ'^2 are $\geqslant 0$.

In so far as the canonical coordinates can be confused with x, v, x', v' at very large spatial distances (this has to be checked *a posteriori* for each model), ρ^2 and ρ'^2 are approximations of

$$R^2 = -r^2+(r \cdot v)^2/v^2, \qquad R'^2 = -r^2+(r \cdot v')^2/v'^2, \tag{3.8}$$

which are the squared distances measured in the three-dimensional proper-space of each particle. Note that, in the present case, ρ^2 and ρ'^2 are partial integrals.

This suggests strongly that ρ^2 (resp. ρ'^2) occurs in F' (resp. F) and that F' (resp. F) vanishes when ρ^2 (resp. ρ'^2)$\to\infty$.

Since $Xz = p$ and $X'z = -p'$ we get from (3.7)

$$X'\rho^2 = 2\bar{\omega}z \cdot p', \tag{3.9}$$

$$X'^2\rho^2 = -2(\bar{\omega}p')^2, \tag{3.10}$$

and (by symmetry under particle exchange) analogous formulas for $X\rho'^2$ and $X^2\rho'^2$. These relations are useful when computing v, v', ξ, ξ'.

As an example consider the model [14]

$$F = kI \exp(-\rho'^2/a^2), \qquad F' = -kI \exp(-\rho^2/a^2), \tag{3.11}$$

where k is a coupling constant and a is some length characteristic of the range of interaction.

Applying X' to F' we get

$$v' = p' + 2kIa^{-2}(\bar{\omega}z \cdot p') \exp(-\rho^2/a^2), \tag{3.12}$$

and, applying X' again

$$\xi' = -kIa^{-2}[2(\bar{\omega}p')^2 + 4a^{-2}(\bar{\omega}z \cdot p')^2] \exp(-\rho^2/a^2). \tag{3.13}$$

Symmetry under particle exchange would give v and ξ. Note that

$$|\bar{\omega}z \cdot p'| < |\bar{\omega}z||\bar{\omega}p'| = \rho|\bar{\omega}p'|.$$

Thus, for p, p', and I fixed, we have

$$x' \to q', \qquad v' \to p',$$

and $\xi' \to 0$ when $\rho \to \infty$ (resp. $x \to q$, etc. when $\rho' \to \infty$). Note also that I is orthogonal to p'. Thus $v'^2 = p'^2 + 0(k^2)$. The mass-parameter correction is of second order in k.

Conclusions

The mass-parameter correction explained in Section 2 permits, systematically, to perform the Hamiltonian construction, without care about the constancy of mass.

Of course, it involves some alteration of the Hamiltonian formulation, since the new vector fields Y, Y' are generally no longer generators of infinitesimal symplectic transformations.

Now, the affine parameters σ, σ' become different from the parameters τ, τ' occurring in Hamilton's equations of motion.

Classically, this departure from conventional habits is not a difficulty, since the orbits, first integrals and partial integrals are preserved by this change.

Perhaps, some trouble could result at the quantum mechanical level. Anyway we should not use Hamilton-Jacobi coordinates for quantization. But these Hamilton-Jacobi coordinates are very efficient for the construction of systems, as we have shown in Section 3.

Now we can explicitly construct all the interactions admitting such coordinates.

The problem is: does any predictive system admit Hamilton-Jacobi coordinates, in a $\mathscr{P}$-invariant way?

This question is still open, although the answer is yes at the first order in perturbation theory [8].

Collège de France

NOTES AND REFERENCES

[1] Droz-Vincent, Ph., *Lett. Nuovo Cim.* **1** (1969), 839; *Physica Scripta* **2** (1970), 129; *Ann. I.H.P.* **XX**(3) (1974), 269.

[2] Wray, J. G., *Phys. Rev. D.* **1**(8) (1970), 2212. See the 'Time Symmetric' formalism in Bel, L., *Ann. I.H.P.* **XIV**(3) (1971), 189.

[3] Droz-Vincent, Ph., *Lett. Nuovo Cim.* **7**(6) (1973), 206; *C.R. Acad. Sci. Paris* **278A** (1974), 807.

[4] Droz-Vincent, Ph., *Reports on Math. Phys.* **8**(1) (1975), 91.

[5] From Currie's theorem we could not expect that they are canonical.

[6] Droz-Vincent, Ph., *C.R. Acad. Sci. Paris* **280A** (1975), 1169.

[7] When $k = 0$, the solution we have obtained reduces to free motion; with a peculiar Hamiltonian formulation where the canonical variables are not x, x'. The difficulty is that our first integrals do not remain sufficiently independent when $k = 0$.

[8] The interest of such coordinates has been recently stressed by Bel, L. and Martin, J., *Ann. I.H.P.* **XXII**(3) (1975), 173.

[9] Changing the time component of $x - q$ and $x' - q'$ did not violate covariance because P defines the rest frame of the system. But the change was defined only by Cauchy data on some hypersurface.

[10] Our procedure would fail for $\{H, H'\} = \text{const.} \neq 0$. But this never happens if we have symmetry under particle interchange.

[11] In (2.16) it is understood that ∂ and ∂' act on functions of the independent variables x, w, x', w'. In principle the regularity of the change $x, v, x', v' \mapsto x, w, x', w'$ has to be checked. In most usual cases λ and λ' will differ from unity by first-order corrections at most, so there will be no trouble.

[12] $\{q, p\} = \{q', p'\} = \delta, \qquad \{q, q'\} = \{p, p'\} = 0.$

[13] More precisely L is only a pseudo vector invariant. $\tilde{y}$ and I undergo a change of sign under particle exchange. The minus in (3.4) insures the symmetry of all the coefficients.

[14] There is no compelling reason to have (3.13) linear in k. But, as a result of the choice (3.1), linearity of F and F' implies linearity of ξ, ξ'.

G.-M. MARLE

SYMPLECTIC MANIFOLDS, DYNAMICAL GROUPS, AND HAMILTONIAN MECHANICS

SUMMARY

Let (M, ω) be a symplectic manifold and G a Lie group acting on M by symplectic diffeomorphisms (i.e. $g^*\omega = \omega$ for all $g \in G$). Souriau has defined the moment of this group action as a suitable map $J : M \to \mathcal{G}^*$ (dual of the Lie algebra $\mathcal{G}$ of G). Some well-known results are first briefly outlined in Part I: there exists an action of G on $\mathcal{G}^*$ for which J is equivariant, whose orbits ϑ_ξ are symplectic manifolds (Kirillov-Souriau-Kostant's theorem); if ξ is a regular value of $J, J^{-1}(\xi)$ quotiented by the action of the isotropy subgroup of ξ is, under suitable assumptions, a symplectic manifold (Meyer's theorem).

Part II is devoted to a synthesis of these two theorems. It is shown that $J^{-1}(\vartheta_\xi)$, quotiented by a suitable equivalence relation, is a symplectic manifold, which has a local product structure (Theorem 3). Examples of Hamiltonian systems which may be reduced by this theorem are given (Theorem 4). Then Part III deals with groups of symplectic similarities, for which the property of equivariance of J is extended (Theorem 5). Finally, Part IV shows how these results may be applied to mechanical systems with time-dependent Hamiltonians: Lie groups of canonical transformations are defined and related to dynamical groups of a product symplectic manifold. A way for constructing groups of canonical transformations from groups of isometries of the base manifold is indicated; in particular, it explains how the Galilean group arises in mechanics from the group of space isometries.

All manifolds and maps considered, unless otherwise specified, are assumed C^∞. If M is a manifold, $\mathcal{D}(M)$, $\mathcal{T}(M)$, $\Lambda^p(M)$ are the sets of C^∞ real functions, vector fields, differential p-forms on M, respectively.

Cahen & Flato (eds.), Differential Geometry and Relativity, 249–269.

1. Outline of classical definitions and well-known facts

1.1. *Symplectic manifolds and Hamiltonian vectors fields* (Refs. [1, 4, 7, 9, 12, 20])

Let (M, ω) be a connected symplectic manifold, i.e. ω is a closed 2-form on M, with maximal rank $2n = \dim(M)$. ω defines an isomorphism between $\mathcal{T}(M)$ and $\Lambda^1(M)$:

$$X \mapsto i(X)\omega = {}^bX, \qquad X \in \mathcal{T}(M), \qquad {}^bX \in \Lambda^1(M), \tag{1.1}$$

whose inverse is noted

$$\beta \mapsto {}^{\#}\beta, \qquad \beta \in \Lambda^1(M), \qquad {}^{\#}\beta \in \mathcal{T}(M). \tag{1.2}$$

With this isomorphism and the usual bracket operation on $\mathcal{T}(M) \times \mathcal{T}(M)$, $\Lambda^1(M)$ becomes a Lie algebra, isomorphic with $\mathcal{T}(M)$:

$$[\alpha, \beta] = i([{}^{\#}\alpha, {}^{\#}\beta])\omega, \qquad \alpha \text{ and } \beta \in \Lambda^1(M). \tag{1.3}$$

We also define Poisson brackets, with which $\mathcal{D}(M)$ becomes a Lie algebra

$$\{f, g\} = i({}^{\#}\mathrm{d}f)\,\mathrm{d}g = -i({}^{\#}\mathrm{d}g)\,\mathrm{d}f = \omega({}^{\#}\mathrm{d}g, {}^{\#}\mathrm{d}f), \qquad f \text{ and } g \in \mathcal{D}(M). \tag{1.4}$$

Note that the exterior derivative $\mathrm{d}: \mathcal{D}(M) \to \Lambda^1(M)$ is a Lie algebra homomorphism: $\mathrm{d}(\{f, g\}) = [\mathrm{d}f, \mathrm{d}g]$.

A vector field $X \in \mathcal{T}(M)$ is said to be locally Hamiltonian if $\mathcal{L}(X)\omega = 0$ ($\mathcal{L}$ being the Lie derivative) or, equivalently, if $i(X)\omega$ is closed. X is said to be globally Hamiltonian, associated to the Hamiltonian $H \in \mathcal{D}(M)$ if

$$i(X)\omega = \mathrm{d}H, \quad \text{or} \quad X = {}^{\#}\mathrm{d}H. \tag{1.5}$$

Note that if X and Y are locally Hamiltonian, $[X, Y]$ is Hamiltonian: $i([X, Y])\omega = \mathrm{d}(\omega(Y, X))$.

1.2. *Dynamical groups and moments* (Ref. [19])

Let G be a connected Lie group, acting effectively on the left on a connected manifold M, by

$$(g, m) \in G \times M \mapsto g \cdot m \in M. \tag{1.6}$$

Let $X \in \mathcal{G}$, the Lie algebra of G. X_M will indicate the vector field on

M, defined by

$$X_M(m) = \frac{\mathrm{d}}{\mathrm{d}t}(\exp(-tX)\cdot m)\big|_{t=0}. \tag{1.7}$$

The minus sign ensures that $X \mapsto X_M$ is a Lie algebra homomorphism of $\mathscr{G}$ into $\mathscr{T}(M)$: $[X, Y]_M = [X_M, Y_M]$.

G is said to be a dynamical group of the symplectic manifold (M, ω) if, for all $g \in G$,

$$g^*\omega = \omega. \tag{1.8}$$

This is the case if and only if, for all $X \in \mathscr{G}$, the vector field X_M is locally Hamiltonian.

A *moment* of the dynamical group G is a map $J: M \to \mathscr{G}^*$ (the dual space of $\mathscr{G}$) such that, for all $X \in \mathscr{G}$

$$X_M = {}^{\#}\mathrm{d}(\langle J, X\rangle), \quad \text{or} \quad i(X_M)\omega = \mathrm{d}(\langle J, X\rangle), \tag{1.9}$$

where $\langle J, X\rangle \in \mathscr{D}(M)$ is the function $m \in M \mapsto \langle J(m), X\rangle \in \mathbb{R}$. It is clear that a moment of G exists if and only if, for all $X \in \mathscr{G}$, X_M is globally Hamiltonian; this is the case, for instance, if M is simply connected, or if $\mathscr{G} = [\mathscr{G}, \mathscr{G}]$. When it exists, a moment J is not unique: all other moments are of the form $J + \mu_0$, where $\mu_0 \in \mathscr{G}^*$ is an arbitrary constant.

From now on, we assume that $J: M \to \mathscr{G}^*$ is a moment of the dynamical group G. The importance of moments in mechanics lies in the following easy result. Let $H \in \mathscr{D}(M)$ be a Hamiltonian, and assume that H is G-invariant, that is, for all $g \in G$

$$H(g\cdot m) = H(m), \qquad m \in M. \tag{1.10}$$

Then J is constant along integral curves of the Hamiltonian vector field ${}^{\#}\mathrm{d}H$: we shall say that J is a ($\mathscr{G}^*$-valued) integral of ${}^{\#}\mathrm{d}H$.

1.3. *Action of the dynamical group G on $\mathscr{G}^*$* (Refs. [17, 18, 19])

Now let, for $g \in G$, $m \in M$,

$$\theta(g) = J(g\cdot m) - Ad^*_g \cdot J(m), \tag{1.11}$$

where Ad^* is the coadjoint representation of G in $\mathscr{G}^*$:

$$\langle Ad^*_g\cdot\mu, X\rangle = \langle \mu, Ad_{g^{-1}}\cdot X\rangle, \quad g \in G, \quad X \in \mathscr{G}, \quad \mu \in \mathscr{G}^*. \tag{1.12}$$

One can check that $\theta(g)$ does not depend on $m \in M$, and satisfies

$$\theta(gh) = \theta(g) + Ad_g^* \cdot \theta(h), \qquad g \text{ and } h \in G. \tag{1.13}$$

This means that $\theta : G \to \mathcal{G}^*$ is a 1-cocycle of G for the coadjoint representation. Let, for $g \in G$, $\mu \in \mathcal{G}^*$,

$$a_\theta(g, \mu) = Ad_g^* \mu + \theta(g). \tag{1.14}$$

We observe that $a_\theta : G \times \mathcal{G}^* \to \mathcal{G}^*$ is a left affine action of G on $\mathcal{G}^*$, such that $J : M \to \mathcal{G}^*$ is equivariant:

$$a_\theta(g, J(m)) = J(g \cdot m), \qquad g \in G, \quad m \in M. \tag{1.15}$$

Thus, if $X \in \mathcal{G}$, X_M and $X_{\mathcal{G}^*}$ being the associated vector fields on M and $\mathcal{G}^*$ respectively, defined by (1.7) and by

$$X_{\mathcal{G}^*}(\mu) = \frac{\mathrm{d}}{\mathrm{d}t}(a_\theta(\exp(-tX), \mu))|_{t=0}, \qquad \mu \in \mathcal{G}^*, \tag{1.16}$$

we see that X_M and $X_{\mathcal{G}^*}$ are J-related:

$$T_m J(X_M(m)) = X_{\mathcal{G}^*}(J(m)), \qquad m \in M.$$

Using this remark we obtain, for the Poisson bracket of the two functions $\langle J, X\rangle$ and $\langle J, Y\rangle$ (X and $Y \in \mathcal{G}$),

$$\{\langle J, X\rangle, \langle J, Y\rangle\} = \langle J, [X, Y]\rangle - \Theta(X, Y), \tag{1.17}$$

where $\Theta : \mathcal{G} \times \mathcal{G} \to \mathbb{R}$ is the bilinear form

$$\Theta(X, Y) = \langle D\theta(e) \cdot X, Y\rangle. \tag{1.18}$$

We shall also write Θ for the linear map $D\theta(e) : \mathcal{G} \to \mathcal{G}^*$. One can check that Θ satisfies, for X, Y, and $Z \in \mathcal{G}$,

$$\Theta([X, Y], Z) = \Theta(X, [Y, Z]) - \Theta(Y, [X, Z]) \tag{1.19}$$

and is skew-symmetric.

$$\Theta(Y, X) = -\Theta(X, Y). \tag{1.20}$$

Such a Θ is called a symplectic 1-cocycle of the Lie algebra $\mathcal{G}$, in its dual $\mathcal{G}^*$ (for the linear representation of $\mathcal{G}$ in $\mathcal{L}(\mathcal{G}^*, \mathcal{G}^*) : \langle X(\mu), Y\rangle = \langle \mu, [X, Y]\rangle$).

By the above formula, we see that the linear map $\mathcal{G} \to \mathcal{D}(M)$, $X \mapsto \langle J, X\rangle$, is not in general a Lie algebra homomorphism, although when composed with ${}^\#\mathrm{d}$, it gives the Lie algebra homomorphism

$X \mapsto X_M$ of $\mathscr{G}$ in $\mathscr{T}(M)$. Suppose we replace J by $J' = J + \mu_0$, with $\mu_0 \in \mathscr{G}^*$ constant. Then θ and Θ are replaced, respectively, by

$$\theta'(g) = \theta(g) + \mu_0 - Ad^*_g \mu_0,$$
$$\Theta'(X, Y) = \Theta(X, Y) + \langle \mu_0, [X, Y] \rangle.$$

This means that θ' and Θ' belong, respectively, to the cohomology classes of θ and Θ. Thus it is clear that, by an appropriate choice of the arbitrary constant $\mu_0 \in \mathscr{G}^*$, we can obtain a Lie algebra homomorphism $X \mapsto \langle J, X \rangle$ of $\mathscr{G}$ in $\mathscr{D}(M)$, if and only if the cohomology class of θ (and hence of Θ) is zero.

A case frequently encountered in mechanics, is when the 2-form ω is exact:

$$\omega = \mathrm{d}\alpha \tag{1.21}$$

and when, for all $g \in G$,

$$g^*\alpha = \alpha \qquad \text{(which implies } g^*\omega = \omega). \tag{1.22}$$

Then G has a moment J, defined by

$$\langle J(m), X \rangle = -i(X_M)\alpha, \qquad m \in M, \tag{1.23}$$

and one can check that the associated cocycle θ is identically zero.

1.4. *Orbits and integral manifolds*

Consider now the orbit ϑ_ξ of a given point $\xi \in \mathscr{G}^*$, under the action a_θ of G on $\mathscr{G}^*$. It is the maximal integral manifold through ξ of the completely integrable (involutive) distribution on $\mathscr{G}^*$ generated by the vector fields $X_{\mathscr{G}^*}$, for all $X \in \mathscr{G}$. The following important theorem is due to Kirillov [5], Souriau [19], and Kostant [6].

THEOREM 1. ϑ_ξ is a symplectic manifold, hence even dimensional, whose 2-form ω_ϑ is given by

$$\omega_\vartheta(\mu) \cdot (X_{\mathscr{G}^*}(\mu), Y_{\mathscr{G}^*}(\mu)) = -\langle \mu, [X, Y] \rangle + \Theta(X, Y). \tag{1.24}$$

Moreover, G acts transitively on ϑ_ξ, by the restriction of a_θ, as a dynamical group, and the canonical immersion of ϑ_ξ in $\mathscr{G}^*$ is a moment of this group action.

In (1.24), $\mu \in \vartheta_\xi$, X and $Y \in \mathscr{G}$; the tangent space to ϑ_ξ at μ being the subspace of $\tau_\mu \mathscr{G}^*$ generated by the $X_{\mathscr{G}^*}(\mu)$ for all $X \in \mathscr{G}$, this formula defines indeed ω_ϑ. Note also that if X and $Y \in \mathcal{G}$, $m \in M$, $\mu = J(m) \in \vartheta_\xi$, we have

$$\omega(m) \cdot (X_M(m), Y_M(m)) = \omega_\vartheta(\mu) \cdot (X_{\mathcal{G}^*}(\mu), Y_{\mathcal{G}^*}(\mu)). \quad (1.25)$$

Consider now a regular value $\xi \in \mathcal{G}^*$ of J (i.e. T_xJ surjective for all $x \in J^{-1}(\xi)$). Then $M_\xi = J^{-1}(\xi)$ is a submanifold of M, sometimes called (improperly) an integral manifold. Let

$$G_\xi = \{g\,;\, g \in G,\, a_\vartheta(g, \xi) = \xi\} \quad (1.26)$$

be the isotropy subgroup of ξ; it is closed in G, hence a Lie group; let $\mathcal{G}_\xi$ (which may be considered as a vector subspace of $\mathcal{G}$) be the Lie algebra of G_ξ. By equivariance of J, it is clear that M_ξ is invariant by restriction to G_ξ of the action of G on M. Therefore G_ξ acts on M_ξ on the left. We may consider the set $P_\xi = M_\xi / G_\xi$ of orbits of this action. Under suitable global conditions, which are assumed fulfilled, P_ξ is a manifold, the canonical projection $\pi : M_\xi \to P_\xi$ is a submersion and, for all $x \in M$, the tangent space $T_{\pi(x)}P_\xi$ to P_ξ at $\pi(x)$, is the quotient space of T_xM by the subspace ker $(T_x\pi)$ generated by the $X_M(x)$ for all $X \in \mathcal{G}_\xi$.

The next theorem, which appears under local form in the works of Cartan [3, chapter XII] (and may be even in the early works of Lagrange), has been rigorously formulated in modern language by Meyer [14]; it appears also in the works of Smale [17, 18], Marsden and Weinstein [13] and, under a different form, Souriau [19].

THEOREM 2. Under the above hypotheses, there exists on P_ξ a unique closed 2-form ω_P, such that (P_ξ, ω_P) is a symplectic manifold, and that $\pi^*\omega_P$ is the form induced on M_ξ by ω. Moreover, let $H \in \mathcal{D}(M)$ be a Hamiltonian on M, which is G-invariant (1.10). It induces a Hamiltonian $H_P \in \mathcal{D}(P_\xi)$, such that for all $x \in M_\xi$, $H(x) = H_P \circ \pi(x)$. Consider the associated Hamiltonian vector fields ${}^\#\mathrm{d}H$ on M (with respect to ω) and ${}^\#\mathrm{d}H_P$ on P_ξ (with respect to ω_P). Then, for each $x \in M_\xi$, ${}^\#\mathrm{d}H(x)$ is tangent to M_ξ, and ${}^\#\mathrm{d}H$ restricted to M_ξ and ${}^\#\mathrm{d}H_P$, are π-related.

A well-known particular case of this theorem is the reduction of the number of degrees of freedom of a Hamiltonian system by use of integrals in involution [14].

2. *A synthesis of Kirillov-Souriau-Kostant's and Meyer's theorems*

By putting together Theorems 1 and 2 above, one can obtain a new theorem which may be of interest in mechanics: Meyer's theorem is useful for the study of Hamiltonian systems whose Hamiltonian is G-invariant, while the new theorem may be applied to a larger class of Hamiltonian systems.

The hypotheses are the same as in Theorem 2. (M, ω) is a symplectic manifold, with dynamical group G and moment $J: M \to \mathscr{G}^*$ · $\xi \in \mathscr{G}^*$ is a regular value of J, ϑ_ξ the orbit of ξ under the group action a_θ, G_ξ the isotropy subgroup of ξ · $M_\xi = J^{-1}(\xi)$ is a closed submanifold of M, on which acts G_ξ, and it is assumed that the set of orbits of this action, $P_\xi = M_\xi / G_\xi$, is a manifold, the projection $\pi: M_\xi \to P_\xi$ being a smooth submersion. Now let

$$N = J^{-1}(\vartheta_\xi). \tag{2.1}$$

Clearly N is a manifold, its canonical injection in M being a smooth immersion (maybe not an embedding). We define an equivalence relation R on N by: x and $y \in N$ are R-equivalent if $J(x) = J(y)$ and if there exists $g \in G$ such that $g \cdot x = y$.

Consider now the set of R-equivalence classes N/R and the projection

$$\tilde{\pi}: N \to N/R = Q. \tag{2.2}$$

Clearly J restricted to N, induces a map $\tilde{J}$ of Q onto ϑ_ξ. Note also that G acts transitively on Q on the left, this action being noted $(g, q) \in G \times Q \mapsto g \cdot q \in Q$.

THEOREM 3. Q has a smooth manifold structure such that $\tilde{\pi}: N \to Q$ and $\tilde{J}: Q \to \vartheta_\xi$ are smooth submersions and that G acts smoothly on Q. Moreover:

(1) There exists a symplectic 2-form ω_Q on Q, such that $\tilde{\pi}^* \omega_Q$ is the 2-form induced on N by ω. G is a dynamical group of (Q, ω_Q), and $\tilde{J}: Q \to \mathscr{G}^*$ a moment of this group action.

(2) Q has a local product structure, i.e. for all $\mu \in \vartheta_\xi$, there exists an open neighborhood V of μ in ϑ_ξ, and a diffeomorphism $\hat{\varphi}$ of $\tilde{J}^{-1}(V) \subset Q$ onto $P_\xi \times V$; moreover,

$$\hat{\varphi}^*(\omega_P \times \omega_{\vartheta|V}) = \omega_{Q|\tilde{J}^{-1}(V)}. \tag{2.3}$$

Proof. Let

$$p : G \to \vartheta_\xi, \qquad p(g) = a_\theta(g, \xi),$$

ϑ_ξ being isomorphic with the homogeneous space G/G_ξ, (G, p, ϑ_ξ) is a principal bundle with structural group G_ξ. Hence, there exists an open neighborhood W of ξ in ϑ_ξ, and a map $k : W \to G$, such that

$$k(\xi) = e, \qquad p \circ k = id_W$$

and that the map of $G_\xi \times W$ in G:

$$(\gamma, z) \mapsto k(z)\gamma^{-1}, \qquad \gamma \in G_\xi, \quad z \in W$$

is a trivialization over W of the principal bundle (G, p, ϑ_ξ). Note that, if $y \in J^{-1}(W)$

$$J((k(J(y)))^{-1} \cdot y) = \xi, \text{ or equivalently } \quad (k(J(y)))^{-1} \cdot y \in M_\xi.$$

Now let $\mu \in \vartheta_\xi$ and $g \in G$ be such that $a_\theta(g, \xi) = \mu$. $a_\theta(g, W)$ is a neighborhood V of μ in ϑ_ξ. Consider the map of $J^{-1}(V)$ onto $M_\xi \times V$:

$$x \in J^{-1}(V) \mapsto (\varphi(x), J(x)) \in M_\xi \times V$$

with

$$\varphi(x) = (k(J(g^{-1} \cdot x)))^{-1} \cdot (g^{-1} \cdot x).$$

Clearly it is a diffeomorphism. Moreover, x and $y \in J^{-1}(V)$ are R-equivalent if and only if $\pi(\varphi(x)) = \pi(\varphi(y))$, where $\pi : M_\xi \to P_\xi$ is the canonical projection. Thus (φ, J) induces a bijection $(\tilde{\varphi}, \tilde{J})$ of $\tilde{J}^{-1}(V)$ onto $P_\xi \times V$. Using such maps $\tilde{\varphi} = (\tilde{\varphi}, \tilde{J})$ as charts, one obtains a manifold structure on Q, such that Q has a local product structure. It is easy to check that $\tilde{\pi} : N \to Q$ and $\tilde{J} : Q \to \vartheta_\xi$ are smooth submersions and that G acts smoothly on Q.

Now let $\dot{x} \in Q$, $\dot{u}$ and $\dot{v} \in T_{\dot{x}}Q$. Choose an $x \in \tilde{\pi}^{-1}(\dot{x}) \subset N$, u and $v \in T_x N$ such that

$$T_x\tilde{\pi}(u) = \dot{u}, \qquad T_x\tilde{\pi}(v) = \dot{v}.$$

Then we define ω_Q by

$$\omega_Q(\dot{x}) \cdot (\dot{u}, \dot{v}) = \omega(x) \cdot (u, v). \tag{2.4}$$

One must check that the right-hand side of (2.4) does not depend on the choices of $x \in \tilde{\pi}^{-1}(\dot{x})$, $u \in (T_x\tilde{\pi})^{-1}(\dot{u})$, $v \in (T_x\tilde{\pi})^{-1}(\dot{v})$. Let

$$E = \{X_M(x); X \in \mathcal{G}\} \tag{2.5}$$

$$F = \ker(T_x J) \tag{2.6}$$

E and F are vector subspaces of $T_x M$, such that

$$E + F = T_x N, \qquad E \cap F = \ker(T_x \tilde{\pi}).$$

For any vector subspace A of $T_x M$, let

$$\mathrm{orth}(A) = \{w \in T_x M;\ \omega(x) \cdot (w, z) = 0 \text{ for all } z \in A\},$$

ω being nondegenerate, if A and B are vector subspaces of $T_x M$, we have

$$\mathrm{orth}(A + B) = \mathrm{orth}(A) \cap \mathrm{orth}(B); \qquad \mathrm{orth}(\mathrm{orth}(A)) = A.$$

But by (1.9), if $X \in \mathscr{G}$ and $w \in T_x M$,

$$\omega(x) \cdot (X_M(m), w) = \langle T_x J(w), X\rangle, \quad \text{implying} \quad \mathrm{orth}(E) = F.$$

Therefore,

$$\mathrm{orth}(E + F) = \mathrm{orth}(E) \cap \mathrm{orth}(F) = E \cap F = \ker(T_x \tilde{\pi}).$$

This proves that the right-hand side of (2.4) does not depend on the choices of $u \in (T_x \tilde{\pi})^{-1}(\dot{u})$, $v \in (T_x \tilde{\pi})^{-1}(\dot{v})$, and simultaneously, that $\omega_Q(\dot{x})$ is nondegenerate (for, by the G-invariance of ω, obviously the right-hand side of (2.4) does not depend on the choice of $x \in \tilde{\pi}^{-1}(\dot{x})$). Thus (2.4) defines a 2-form ω_Q on Q. It is easy to see that ω_Q is smooth, G-invariant, and such that $\tilde{\pi}^* \omega_Q$ is the 2-form induced on N by ω. $\tilde{\pi}$ being a submersion, this proves that ω_Q is closed, hence symplectic. All other points of Theorem 3 are then easy.

Q.E.D.

COROLLARY. With the same hypotheses as in Theorem 3, assume now that there exists a closed subgroup H of G, such that either G_ξ is a normal subgroup and G a semi-direct product:

$$G = G_\xi \times_\rho H, \tag{2.7}$$

with ρ an homomorphism of H in the group of automorphisms of G_ξ, or H is a normal subgroup and G a semi-direct product:

$$G = H \times_\rho G_\xi, \tag{2.8}$$

with ρ an homomorphism of G_ξ in the group of automorphisms of H.

Then in both cases, there exists a diffeomorphism $\hat{\varphi} = (\tilde{\varphi}, \tilde{J})$ of Q onto $P_\xi \times \vartheta_\xi$, such that

$$\hat{\varphi}^*(\omega_P \times \omega_\vartheta) = \omega_Q. \tag{2.9}$$

Proof. In both cases, with notations of the proof of Theorem 3, $\hat{p} = p|_H$ is clearly an isomorphism of H onto ϑ_ξ. Thus we may define the map k not only on a neighborhood of ξ, but globally on ϑ_ξ, by $k = \hat{p}^{-1}: \vartheta_\xi \to G$. Then for all $x \in N = J^{-1}(\vartheta_\xi)$ we may define

$$\varphi(x) = (\hat{p}^{-1} \circ J(x))^{-1} \cdot x \in M_\xi,$$

and we note that (φ, J) is a diffeomorphism of N onto $M_\xi \times \vartheta_\xi$. The remainder of the proof follows easily.

Q.E.D.

Remarks. With the hypotheses of the above Corollary, we have a diffeomorphism (φ, J) of $N = J^{-1}(\vartheta_\xi)$ onto $M_\xi \times \vartheta_\xi$. Clearly, if ω_N and ω_{M_ξ} are the 2-forms (which are degenerate in general) induced by ω on N and M_ξ, we have

$$(\varphi, J)^*(\omega_{M_\xi} \times \omega_\vartheta) = \omega_N; \tag{2.10}$$

G_ξ acts on M_ξ and H on ϑ_ξ, these actions leaving ω_{N_ξ} and ω_ϑ invariant. Thus we may define an action of the direct product $G_\xi \times H$ on N, leaving ω_N invariant. Similarly, the direct product $G_\xi \times H$ acts as a dynamical group of Q. This seems to be the property tacitly used by Robinson [16], when he notes that the group of space displacements, i.e. the semi-direct product $\mathbf{R}^3 \times_\rho SO(3)$, acts on the phase space of the n-body problem, but that the direct product $\mathbf{R}^3 \times SO(3)$ also acts on this space (by translation and rotation around the center of mass). Similar properties, for the case when H is normal and abelian, appear in the book of Souriau [19, chapter III, §13], who obtains an interesting interpretation of the total mass of a mechanical system as the cohomology class of a Galilean group 1-cocycle.

The reduced symplectic manifold Q of Theorem 3 may be used for studying Hamiltonian systems, whose flow leaves N invariant and is projectible on Q. The next theorem gives an example of such Hamiltonian systems.

THEOREM 4. Under the hypotheses of Theorem 3, let $H \in \mathcal{D}(M)$ be a Hamiltonian. We assume that

$$H = H_1 + H_2 \circ J, \tag{2.11}$$

with $H_1 \in \mathcal{D}(M)$, $H_2 \in \mathcal{D}(\mathcal{G}^*)$, and H_1 G-invariant, i.e. for all $g \in G$, $m \in M$:

$$H_1(g \cdot m) = H_1(m). \tag{2.12}$$

Then the flow of the Hamiltonian vector field ${}^{\#}\mathrm{d}H$ leaves invariant $N = J^{-1}(\vartheta_\xi)$. Moreover, there exists on Q a unique Hamiltonian $\tilde{H} \in \mathcal{D}(Q)$ such that, for all $x \in N$,

$$\tilde{H}(\tilde{\pi}(x)) = H(x), \tag{2.13}$$

and the restriction of ${}^{\#}\mathrm{d}H$ to N and the vector field ${}^{\#}\mathrm{d}\tilde{H}$ on Q are $\tilde{\pi}$-related.

Proof. Let $\dot{x} \in Q$ and $x \in \tilde{\pi}^{-1}(\dot{x}) \subset N$. The tangent space T_xN is $E + F$ (Equations (2.5) and (2.6)). ${}^{\#}\mathrm{d}H_1(x)$ is in F, for J is an integral of ${}^{\#}\mathrm{d}H_1$. To prove that ${}^{\#}\mathrm{d}(H_2 \circ J)(x)$ is in $E + F$, it is sufficient to check that $T_xJ({}^{\#}\mathrm{d}(H_2 \circ J)(x))$ is tangent to ϑ_ξ. Let $X \in \mathcal{G}$. We have

$$\begin{aligned}
\langle T_xJ({}^{\#}\mathrm{d}(H_2 \circ J)(x)), X\rangle &= \omega(x) \cdot (X_M(x), {}^{\#}\mathrm{d}(H_2 \circ J)(x)) \\
&= -i(X_M(x))\,\mathrm{d}(H_2 \circ J) \\
&= -i(T_xJ(X_M(x)))\,\mathrm{d}H_2 \\
&= -i(X_{\mathcal{G}^*}(J(x)))\,\mathrm{d}H_2 \\
&= \langle {}^{\#}\mathrm{d}(H_2|_{\vartheta_\xi})(J(x)), X\rangle
\end{aligned}$$

because $X_M(x)$ and $X_{\mathcal{G}^*}(J(x))$ are J-related. This proves that $T_xJ({}^{\#}\mathrm{d}(H_2 \circ J)(x))$ is equal to the value at $J(x)$ of the Hamiltonian vector field ${}^{\#}\mathrm{d}(H_2|_{\vartheta_\xi})$ on ϑ_ξ, hence is tangent to ϑ_ξ. The remainder of the proof is easily obtained by using the following lemma.

LEMMA. Let (M, ω) be a symplectic manifold, $H \in \mathcal{D}(M)$ a Hamiltonian, and N a submanifold such that the Hamiltonian vector field ${}^{\#}\mathrm{d}H$ is tangent to N. Let (Q, ω_Q) be another symplectic manifold, and $\tilde{\pi}: N \to Q$ a submersion such that

(1) $\tilde{\pi}^*\omega_Q$ is the form induced on N by ω;

(2) there exists $\tilde{H} \in \mathcal{D}(Q)$ such that $\tilde{H} \circ \tilde{\pi} = H|_N$.

Then the restriction of ${}^{\#}\mathrm{d}H$ to N, and ${}^{\#}\mathrm{d}\tilde{H}$, are $\tilde{\pi}$-related.

Proof. Let $x \in N$, $v \in T_xN$. As $^{\#}\mathrm{d}H(x) \in T_xN$, we have

$$\begin{aligned}\omega(x)\cdot(^{\#}\mathrm{d}H(x), v) &= \omega_Q(\tilde{\pi}(x))\cdot(T_x\tilde{\pi}(^{\#}\mathrm{d}H(x)), T_x\tilde{\pi}(v)) \\ &= i(v)\,\mathrm{d}H = i(T_x\tilde{\pi}(v))\,\mathrm{d}\tilde{H} \\ &= \omega_Q(\tilde{\pi}(x))\cdot(^{\#}\mathrm{d}\tilde{H}(\pi(x)), T\tilde{\pi}(v)).\end{aligned}$$

ω_Q being nondegenerate and $T_x\tilde{\pi}$ surjective, this implies

$$T_x\tilde{\pi}(^{\#}\mathrm{d}H(x)) = {}^{\#}\mathrm{d}\tilde{H}(\tilde{\pi}(x)). \qquad \text{Q.E.D.}$$

A well-known particular case of Hamiltonian systems considered in Theorem 4, is when

$$H_2 \circ J(x) = \langle J(x), X\rangle \tag{2.14}$$

with $X \in \mathscr{G}$ fixed. By Hamilton-Jacobi theory, it appears that H_1 is the Hamiltonian of a given mechanical system relative to a fixed frame, and $H = H_1 + H_2 \circ J$ the Hamiltonian of the same mechanical system, relative to a moving frame (the relative motion of the two frames being given by the one-parameter group of diffeomorphisms of M, $\{\exp(tX);\ t \in \mathbb{R}\}$). Robbin [15] has, by these means, studied relative equilibria of mechanical systems with symmetries. Similar results appear in the earlier work of Arnold [2].

3. Groups of symplectic similarities

We consider a connected Lie group G_1, acting effectively on the left on a symplectic manifold (M, ω), by (1.6). G_1 is said to be a group of symplectic similarities of (M, ω) if, for all $g \in G_1$, there exists a $\lambda(g) \in \mathbb{R} - \{0\}$ such that

$$g^*\omega = \lambda(g)\omega. \tag{3.1}$$

Clearly, $\lambda : G_1 \to \mathbb{R} - \{0\}$ is a group homomorphism. G_1 being connected, $\lambda(g)$ is always positive.

G_1 is a group of symplectic similarities of (M, ω) if and only if, for all $X \in \mathscr{G}_1$ (the Lie algebra of G_1), there exists a $\kappa(X) \in \mathbb{R}$ such that (X_M being the vector field defined by (1.7))

$$\mathscr{L}(X_M)\omega = \kappa(X)\omega, \tag{3.2}$$

with $\kappa : \mathscr{G}_1 \to \mathbb{R}$ linear, null on $[\mathscr{G}_1, \mathscr{G}_1]$, and related to λ by

$$\lambda(\exp(tX)) = e^{-\kappa(X)t}. \tag{3.3}$$

From now on we assume that G_1 is a group of symplectic similarities of (M, ω) with λ not trivial (or, equivalently, κ not identically zero). We choose a $Z \in \mathcal{G}_1$ such that

$$\kappa(Z) = 1, \quad \text{or equivalently} \quad \mathcal{L}(Z_M)\omega = \omega. \tag{3.4}$$

We see that ω must be exact:

$$\omega = \mathrm{d}\alpha, \quad \text{with} \quad \alpha = i(Z_M)\omega. \tag{3.5}$$

Let $\mathcal{G}$ be the kernel of κ. $\mathcal{G}$ has codimension 1 in $\mathcal{G}_1$, and $[\mathcal{G}_1, \mathcal{G}_1] \subset \mathcal{G}$. Moreover, $\mathcal{G}$ is the Lie algebra of the closed normal subgroup ker (λ) of G_1 (kernel of the group homomorphism λ). Let G be the identity component in ker (λ). G is a closed, normal, connected subgroup of G_1, with codimension 1 and Lie algebra $\mathcal{G}$. We have (1.8) for all $g \in G$, so G is a dynamical group of (M, ω). From now on, we shall assume that there exists a moment $J: M \to \mathcal{G}^*$ of the dynamical group G. As ω is exact (3.5), we define $J_0: M \to \mathcal{G}^*$ by setting, for all $X \in \mathcal{G}$ and $m \in M$,

$$\langle J_0(m), X\rangle = -i(X_M(m))\alpha. \tag{3.6}$$

We observe that $\langle J - J_0, X\rangle$ is constant on M if and only if X_M leaves α invariant (i.e. $\mathcal{L}(X_M)\alpha = 0$), but we do not assume this true for all $X \in \mathcal{G}$. An easy computation shows that:

LEMMA. Let $X \in \mathcal{G}$, and take $Z \in \mathcal{G}_1$ as defined above (3.4). X_M and Z_M being the associated vector fields on M (1.7), we have (Equation (1.2))

$$[X_M, Z_M] = [X, Z]_M = {}^{\#}\mathrm{d}\langle J - J_0, X\rangle. \tag{3.7}$$

This proves that $[X, Z] = 0$ if and only if X_M leaves α invariant. When this is the case for all $X \subset \mathcal{G}$, G_1 is the direct group product $\mathbb{R} \times G$.

Results in Sections 1–3 are naturally generalized by

THEOREM 5. There exists a left affine action of G_1 on $\mathcal{G}^*$, whose restriction to G is the already defined action (1.14), such that $J: M \to \mathcal{G}^*$ is equivariant. This action is given by the formula

$$(g, \xi) \in G_1 \times \mathcal{G}^* \mapsto a_\theta(g, \xi) = \lambda(g)Ad^*_g\xi + \theta(g) \in \mathcal{G}^*, \tag{3.8}$$

where $\theta: G_1 \to \mathcal{G}^*$ is given, for any $m \in M$, by

$$\theta(g) = J(g \cdot m) - \lambda(g)Ad^*_g J(m). \tag{3.9}$$

Note that we have extended the coadjoint action of G on $\mathscr{G}^*$ by defining, for all $g \in G_1$ and $\xi \in \mathscr{G}^*$, $X \in \mathscr{G}$:

$$\langle Ad_g^* \xi, X \rangle = \langle \xi, Ad_{g^{-1}} X \rangle. \tag{3.10}$$

This is legitimate because the vector subspace $\mathscr{G}$ of $\mathscr{G}_1$ is $Ad_{g^{-1}}$-invariant for all $g \in G$.

Proof. The map $(g, \xi) \in G_1 \times \mathscr{G}^* \mapsto \lambda(g) Ad_g^* \xi \in \mathscr{G}^*$ is clearly a linear representation of G_1 in $\mathscr{G}^*$. Thus we have only to check that $\theta(g)$, defined by (3.9), does not depend on the choice of $m \in M$. For $g \in G_1$ fixed, let

$$R(m) = J(g \cdot m) - \lambda(g) Ad_g^* J(m).$$

M being connected, it is sufficient to prove that the derivative of R is zero. For $v \in T_m M$ and $X \in \mathscr{G}$, we have

$$\begin{aligned}
\langle DR(m) \cdot (v), X \rangle &= \langle DJ(g \cdot m) \circ T_m g \cdot (v), X \rangle \\
&\quad - \lambda(g) \langle Ad_g^* DJ(m) \cdot (v), X \rangle \\
&= \omega(g \cdot m) \cdot (X_M(g \cdot m), T_m g(v)) \\
&\quad - \lambda(g) \omega(m) \cdot ((Ad_{g^{-1}} X)_M(m), v) \\
&= g^* \omega(m) \cdot ((Ad_{g^{-1}} X)_M(m), v) \\
&\quad - \lambda(g) \omega(m) \cdot ((Ad_{g^{-1}} X)_M(m), v) \\
&= 0.
\end{aligned}$$

Q.E.D.

Clearly, $\theta : G_1 \to \mathscr{G}^*$ is a 1-cocycle of G_1 for the linear representation $(g, \xi) \in G_1 \times \mathscr{G}^* \mapsto \lambda(g) Ad_g^* \xi \in \mathscr{G}^*$.

The associated 1-cocycle of $\mathscr{G}_1$, $\Theta : \mathscr{G}_1 \times \mathscr{G} \to \mathbb{R}$, is

$$\Theta(Y, X) = \langle \Theta(Y), X \rangle = \langle D\theta(e) \cdot (Y), X \rangle, \qquad Y \in \mathscr{G}_1, \quad X \in \mathscr{G}. \tag{3.11}$$

An easy computation leads to the formula, valid for all $m \in M$,

$$\begin{aligned}
\Theta(Y, X) = \omega(m) \cdot (Y_M(m), X_M(m)) + \langle J(m), [Y, X] \rangle \\
+ \kappa(Y) \langle J(m), X \rangle,
\end{aligned} \tag{3.12}$$

where $\kappa : \mathscr{G}_1 \to \mathbb{R}$ is the Lie algebra homomorphism (3.2). If $Y \in \mathscr{G}$, the above formula reduces to (1.17). If $Y = Z$, we have

$$\Theta(Z, X) = \langle J(m) - J_0(m), X \rangle + \langle J(m), [Z, X] \rangle, \tag{3.13}$$

which, by (3.7), does not depend on $m \in M$.

Let $m \in M$, $\xi = J(m) \in \mathcal{G}^*$ and consider the orbit ϑ_ξ of ξ under the action a_θ of G_1 on $\mathcal{G}^*$. Two cases may be distinguished.

First case. $Z_{\mathcal{G}^*}(\xi)$ (defined as in (1.16)) belongs to the vector subspace of $T_\xi\mathcal{G}^* \equiv \mathcal{G}^*$ generated by the $X_{\mathcal{G}^*}(\xi)$, for all $X \in \mathcal{G}$. Then each connected component of ϑ_ξ is in fact an orbit of the restriction of a_θ to G. Therefore, by Theorem 1, ϑ_ξ is even dimensional and symplectic, with 2-form ω_ϑ (1.24). Clearly, G_1 is a group of symplectic similarities of $(\vartheta_\xi, \omega_\vartheta)$. Let Y be such that $Z_{\mathcal{G}^*}(\xi) = Y_{\mathcal{G}^*}(\xi)$ and define

$$\alpha' = \alpha - d\langle J, Y\rangle, \qquad Z' = Z - Y; \tag{3.14}$$

Z' and α' have the same properties (3.4) and (3.5) as Z and α, and $Z'_{\mathcal{G}^*}(\xi) = 0$. Therefore, the one-parameter group generated by Z' belongs to the isotropy subgroup $G_{1\xi}$ of ξ. When ξ is a regular value of J, $M_\xi = J^{-1}(\xi)$ is a submanifold of M, and by Theorem 2, $P_\xi = M_\xi / G_{1\xi} \cap G$, assumed a manifold, is symplectic. Then, the one-parameter group generated by Z' acts on (P_ξ, ω_P) as a group of symplectic similarities.

Second case. $Z_{\mathcal{G}^*}(\xi)$ does not belong to the vector subspace generated by the $X_{\mathcal{G}^*}(\xi)$ for all $X \in \mathcal{G}$. By Theorem 1, this subspace is even dimensional. Therefore, in this second case, ϑ_ξ is odd dimensional; it is foliated into orbits of the subgroup G, which by Theorem 1, are symplectic submanifolds of ϑ_ξ of codimension 1. But in general, there is no closed 2-form on ϑ_ξ which induces on each of these submanifolds, its symplectic 2-form. The use of contravariant 2-tensors, with Nijenhuis bracket, recently considered by Lichnerowicz for the study of contact manifolds [10], may be a better way for investigating the structure of ϑ_ξ.

4. Dynamical groups and contact transformations

1. *Preliminaries* (Ref. [1])

Let (M, ω) be a symplectic $2n$-dimensional manifold. For studying mechanical systems with a time-dependent Hamiltonian, we consider the product manifold $M_1 = M \times \mathbb{R}$. A point of M_1 is noted (x, t), with $x \in M$, $t \in \mathbb{R}$. The two projections of M_1 are noted as

$$p_1 : (x, t) \mapsto x; \qquad \bar{t} : (x, t) \mapsto t \tag{4.1}$$

There is on M_1 a closed 2-form of maximal rank $2n$:

$$\omega_1 = p_1^*\omega, \tag{4.2}$$

and E (or, when acting on a function, $\partial/\partial t$) indicates the vector field on M_1 such that

$$Tp_1(E) = 0; \qquad T\bar{t}(E) = 1. \tag{4.3}$$

Let $H \in \mathscr{D}(M_1)$ be a time-dependent Hamiltonian; for each $t \in \mathbb{R}$, the submanifold $M_t = M \times \{t\}$, with the 2-form ω_t induced by ω_1, is symplectic and naturally isomorphic with (M, ω). $H_t = H|_{M_t}$ is a Hamiltonian on M_t, and we may define the associated vector field ${}^{\#}\mathrm{d}H_t$ on M_t. ${}^{\#}\mathrm{d}H$ is defined as the vector field on M_1 which, for each $t \in \mathbb{R}$, is on M_t equal to ${}^{\#}\mathrm{d}H_t$. Clearly, it satisfies

$$T\bar{t}({}^{\#}\mathrm{d}H) = 0; \qquad i({}^{\#}\mathrm{d}H)\omega_1 = \mathrm{d}H - \frac{\partial H}{\partial t}\mathrm{d}\bar{t}. \tag{4.4}$$

Trajectories of the mechanical system defined by H are integral curves of the vector field

$${}^{\#}\mathrm{d}H + E.$$

As shown by Cartan [3], there exists on M_1 a unique 2-form, which induces ω_t on each M_t and whose inner product with $({}^{\#}\mathrm{d}H + E)$ is zero. Forms having this last property are called 'relations intégrales d'invariance' by Lichnerowicz [8], who studied their use in mechanics. Here this 2-form is

$$\omega_1 + \mathrm{d}H \wedge \mathrm{d}\bar{t}.$$

As it is closed, this 2-form is an absolute integral invariant, in the terminology of Cartan [3].

We consider now the product manifold $M_2 = M_1 \times \mathbb{R} = M \times \mathbb{R} \times \mathbb{R}$. A point of M_2 is noted (x, t, h), with $(x, t) \in M_1$ and $h \in \mathbb{R}$. The projections are

$$p_1 : (x, t, h) \mapsto x; \qquad \bar{t} : (x, t, h) \mapsto t; \qquad \bar{h} : (x, t, h) \mapsto h. \tag{4.5}$$

Given $H \in \mathscr{D}(M_1)$, consider $H_2 \in \mathscr{D}(M_2)$ defined as

$$H_2(x, t, h) = H_1(x, t) - h \tag{4.6}$$

and consider the symplectic 2-form on M_2,

$$\omega_2 = p_1^* \omega + \mathrm{d}\bar{h} \wedge \mathrm{d}\bar{t}. \tag{4.7}$$

We may then define the Hamiltonian vector field ${}^{\#}\mathrm{d}H_2$ on M_2 such that

$$i({}^{\#}\mathrm{d}H_2)\omega_2 = \mathrm{d}H_2 \tag{4.8}$$

and, as all $k \in \mathbb{R}$ are regular values of H_2, it is easy to check that:

PROPOSITION. For all $k \in \mathbb{R}$, ${}^{\#}\mathrm{d}H_2$ is tangent to the submanifold $H_2^{-1}(k)$. $(p_1, t)|_{H_2^{-1}(k)}$ is a diffeomorphism of $H_2^{-1}(k)$ onto M_1, and the vector fields $({}^{\#}\mathrm{d}H_2)|_{H_2^{-1}(k)}$ and $({}^{\#}\mathrm{d}H + E)$ are $(p_1, t)|_{H_2^{-1}(k)}$-related.

4.2. *Symplectic transformations of* (M_2, ω_2)

The use of (M_2, ω_2) instead of (M, ω) may be interesting even for the study of mechanical systems with a time-independent Hamiltonian H, because (M_2, ω_2) may have a dynamical group leaving the Hamiltonian H_2 unchanged, larger than the dynamical group of (M, ω) leaving H unchanged. As a trivial example, time translations $(x, t, h) \mapsto (x, t + t_0, h)$ form a one-parameter dynamical group of (M_2, ω_2) leaving H_2 invariant (if H is time-independent); the corresponding moment is the well-known energy integral, H itself. More interesting are the so-called canonical transformations. Such a transformation [1] is a diffeomorphism $F_1 : M_1 \to M_1$ verifying

$$\bar{t} \circ F_1 = \bar{t}, \quad \text{or equivalently} \quad F_1(x, t) = (F(x, t), t), \tag{4.9}$$

where F is a map $M_1 \to M$, and

$$F_1^* \omega_1 = \omega_1 + \mathrm{d}K \wedge \mathrm{d}\bar{t}, \tag{4.10}$$

where $K : M_1 \to \mathbb{R}$ is called the generating function of F_1.

Let $H \in \mathcal{D}(M_1)$ be a (maybe time-dependent) Hamiltonian, and $(x, t) \mapsto \varphi_t(x)$ the flow of the associated vector field $({}^{\#}\mathrm{d}H + E)$ (i.e. for $x \in M$ fixed, $t \mapsto \varphi_t(x)$ is the integral curve of $({}^{\#}\mathrm{d}H + E)$ such that $\varphi_0(x) = x$). The transformed flow by the contact transformation F_1 is defined as

$$(x, t) \mapsto \chi_t(x) = (F_t)^{-1} \circ \varphi_t \circ F_0(x), \tag{4.11}$$

where for each $t \in \mathbb{R}$, F_t is the diffeomorphism of M on itself: $x \mapsto F(x, t)$ (Equation (4.9)). By Hamilton-Jacobi theory, it is well known, and easily checked, that $(x, t) \mapsto \chi_t(x)$ is the flow of a vector field $({}^{\#}\mathrm{d}(H_F)+E)$, associated to a transformed Hamiltonian

$$H_F = K + H \circ F_1. \tag{4.12}$$

Given F_1 (and therefore F (4.9)), and chosen K (defined by (4.10) up to addition of an arbitrary function of t), we may define a diffeomorphism F_2 of $M_2 = M_1 \times \mathbb{R}$ on itself by

$$F_2(x, t, h) = (F_1(x, t), h - K(x, t)). \tag{4.13}$$

The following easy result shows that canonical transformations of M_1 appear as particular symplectic diffeomorphisms of M_2.

PROPOSITION. F_2 is a symplectic diffeomorphism of M_2 (i.e. $F_2^*\omega_2 = \omega_2$), which leaves the Hamiltonian $H_2 = H - \bar{h}$ unchanged if and only if $H_F = H$.

A connected Lie group G is called a Lie group of canonical transformations of M_1, if G acts on M_1 on the left in such a way that, for all $g \in G$, the corresponding diffeomorphism F_{1g} is a canonical transformation of generating function K_g, verifying

$$K_{g_2g_1} = K_{g_1} + K_{g_2} \circ F_1. \tag{4.14}$$

Then the above procedure extends the action of G on M_1 in an action on M_2, and G appears as a dynamical group of M_2. If there exists a moment of this group action, and if G (or at least, a subgroup of G) leaves invariant the Hamiltonian $H_2 = H - \bar{h}$ associated to a given mechanical system, one may obtain by this way integrals (generally time-dependent) of this mechanical system.

As an example, consider the case of a classical mechanical system: $M = TN$ is the tangent bundle to a Riemannian manifold (N, γ), with canonical projection $p: M \to N$, and the 2-form ω is

$$\omega = -\mathrm{d}\eta, \tag{4.15}$$

where η is the reciprocal of the canonical 1-form of the cotangent bundle T^*N, by the vector bundle isomorphism $\rho: TN \to T^*N$, $x \mapsto i(x)\gamma$, defined by the metric tensor γ. Note that if $x \in TN$, $v \in T_xTN$,

$$\eta(x)\cdot(v) = \gamma(p(x))\cdot(x, T_p(v)). \tag{4.16}$$

Let G_1 be a Lie group of isometries of N, G_2 a closed, normal, abelian subgroup of G_1, $\mathscr{G}_1$ and $\mathscr{G}_2$ their Lie algebras. For all $(X, g) \in \mathscr{G}_2 \times G_1$, we define a diffeomorphism $F_{1(X,g)}$ of $M_1 = TN \times \mathbb{R}$ onto itself by setting, for $x \in M = TN$, $n = p(x) \in N$,

$$F_{1(X,g)}(x, t) = (T_n(\exp(-tX)g) \cdot (x), t). \tag{4.17}$$

We define X_N, as usual, as the vector field on N:

$$X_N(n) = \frac{\mathrm{d}}{\mathrm{d}t}(\exp(-tX) \cdot n)|_{t=0}.$$

Then it is easy to check that:

PROPOSITION. For all $(X, g) \in \mathcal{G}_2 \times G_1$, $F_{1(X,g)}$ is a canonical transformation with generating function

$$K_{(X,g)}(x) = -\gamma(p(x)) \cdot (x, (Ad_{g^{-1}}X)_N(p(x))). \tag{4.18}$$

Moreover, consider the semi-direct product $G = \mathcal{G}_2 \times_{Ad} G_1$, i.e. the Cartesian product $\mathcal{G}_2 \times G_1$ with group composition law

$$((X_1, g_1), (X_2, g_2)) \mapsto (X_1 + Ad_{g_1}X_2, g_1 g_2).$$

Then G is a group of canonical transformations of M_1.

The above proposition explains how the Galilean group G appears as a dynamical group for classical mechanical systems. Consider, for instance, the n-body problem. N is an open set of $(\mathbb{R}^3)^n$, on which acts the group of space displacements $G_1 = \mathbb{R}^3 \times_\rho \mathrm{SO}(3)$, and the normal abelian subgroup G_2 is the group $\mathbb{R}^3$ of space translations. Then G acts as a dynamical group on $M_2 = M_1 \times \mathbb{R}$, by the transformations F_2 (4.13) associated to F_1 (4.17). But this action does not leave invariant the Hamiltonian $H_2 = H - \bar{h}$ (for the Hamiltonian $H \in \mathcal{D}(M_1)$ of the n-body problem). However, a slight change in the formula (4.17) gives a group of canonical transformations F'_1 such that the associated F'_2 leave H_2 invariant. This change is

$$\begin{aligned} F'_{1(X,g)}(x, t) = (&T_{p(x)}(\exp(-tX)g) \cdot (x) \\ &+ X_N(\exp(-tX)g \cdot (p(x))), t). \end{aligned}$$

Note that in general $F'_{1(X,g)}$ is not a canonical transformation; but in the particular case of the n-body problem, with for G_2 the group of space translations, it is a canonical transformation; this particular circumstance is due to the fact that $i(X_N)\gamma$ is a constant covector for all $X \in \mathcal{G}_2$. Therefore, the action of the Galilean group G gives rise to the classical integrals of the n-body problem.

Similarly, we remark that by considering (M_2, ω_2), in some cases one may obtain groups of symplectic similarities which do not appear when one looks only at (M, ω). For instance, in the n-body problem N is an open set in $(\mathbb{R}^3)^n$ which is homothety-invariant, i.e. for all $v \in N$ and $\lambda > 0$, $\lambda v \in N$. A point $x \in M = TN = N \times (\mathbb{R}^3)^n$ will be noted $x = (q, p)$, with $q \in N$, $p \in (\mathbb{R}^3)^n$. The Hamiltonian is

$$H(q, p) = \tfrac{1}{2}\gamma(p, p) + U(q),$$

where γ is a definite positive symmetric bilinear form on $(\mathbb{R}^3)^n$, and the potential energy U a homogeneous function of degree k:

$$U(\lambda q) = \lambda^k U(q)$$

(in celestial mechanics $k = -1$). For all $\lambda > 0$ we define a diffeomorphism of $M_2 = TN \times \mathbb{R} \times \mathbb{R}$ on itself:

$$((q, p), t, h) \mapsto ((\lambda q, \lambda^{k/2} p), \lambda^{1-k/2} t, \lambda^{-1} h).$$

It is easy to check that this is a one-parameter group of symplectic similarities of (M_2, ω_2). Therefore, the group of space similarities (semi-direct product of the group of space displacements and of this one-parameter group of homotheties) is a group of symplectic similarities of (M_2, ω_2). This appears in the works of Cartan [3], and has been more recently looked at by Losco [11].

Finally, we remark that Souriau [19] uses a similar, but slightly different approach, which may provide in mathematical physics a deeper physical insight. Instead of (M_2, ω_2) he considers the set of trajectories of the given Hamiltonian system, and shows that it is a symplectic manifold, which has essentially the same dynamical group as (M_2, ω_2).

Université de Besançon

REFERENCES

[1] Abraham, R., *Foundations of Mechanics*, W. A. Benjamin, Inc., New York, 1967.
[2] Arnold, V., 'Sur la géométrie différentielle des groupes de Lie de dimension infinie et ses applications à l'hydrodynamique des fluides parfaits', *Ann. Inst. Fourier*, Grenoble, **16**(1) (1966), 319–361.
[3] Cartan, E., *Leçons sur les invariants intégraux*, Hermann, Paris, 1958 (réédition).
[4] Godbillon, C., *Géométrie différentielle et mécanique analytique*, Hermann, Paris, 1969.
[5] Kirillov, A., *Eléments de la théorie des représentations*, Editions Mir, Moscow, 1974.

[6] Kostant, B., 'Quantization and unitary representations: part I, Prequantization', *Lectures in Modern Analysis and Applications* **III**, 87–208; *Lecture Notes in Mathematics*, 170, Springer Verlag, 1970.
[7] Libermann, P., 'Sur les automorphismes infinitésimaux des structures symplectiques et des structures de contact', *Colloque de Géométrie différentielle globale*, Bruxelles, 1958, 37–59; Gauthier Villars, Paris, 1959.
[8] Lichnerowicz, A., 'Les relations intégrales d'invariance et leurs applications à la dynamique', *Bull. Sci. Math.* **70** (1946), 82–85.
[9] Lichnerowicz, A., 'Théorèmes de réductivité sur des algèbres d'automorphismes', *Rendiconti di Matematica*, **22**(1–2) (1963), 197–244.
[10] Lichnerowicz, A., 'Variétés canoniques et transformations canoniques', *C.R. Acad. Sci. Paris*, **280A** (13 janvier 1975), 37–40.
[11] Losco, L., 'Sur un invariant intégral du problème des *n* corps conséquence de l'homogénéité du potentiel', in *The stability of the solar system and of small stellar systems* (Y. Kozai, ed.), 249–255, D. Reidel Publishing Co., 1974.
[12] Malliavin, P., *Geométrie différentielle intrinsèque*, Chapitre 4, Hermann, Paris, 1972.
[13] Marsden, J. and Weinstein, A., 'Reduction of symplectic manifolds with symmetry', *Reports on Math. Phys.*, 5, 1974.
[14] Meyer, K. R., 'Symmetries and integrals in Mechanics', in *Dynamical systems* (M. Peixoto, ed.), 259–273. Academic Press, 1973.
[15] Robbin, J. W., 'Relative equilibria in mechanical systems', in *Dynamical systems* (M. Peixoto, ed.), 425–441, Academic Press, 1973.
[16] Robinson, C., 'Fixing the center of mass in the *n*-body problem by means of a group action.' Preprint, *Colloque Symplectique*, Aix en Provence, 1974.
[17] Smale, S., 'Topology and Mechanics I', *Inventiones Math.* **10** (1970), 305–331.
[18] Smale, S., 'Topology and Mechanics II', *Inventiones Math.* **11** (1970), 45–64.
[19] Souriau, J. M., *Structure des systèmes dynamiques*, Dunod, Paris, 1970.
[20] Sternberg, S., *Lectures on Differential Geometry*, Prentice Hall, 1964.

ROGER PENROSE

ANY SPACE-TIME HAS A PLANE WAVE AS A LIMIT

It is a great pleasure for me to honour André Lichnerowicz on his 60th birthday. And since he has done so much to make relativity respectable as a branch of mathematics, I feel it is appropriate here to point out a simple, yet perhaps surprising, general mathematical property of hyperbolic normal pseudo-Riemannian manifolds which is motivated from knowledge of certain well-known exact solutions in general relativity.

The concept of tangent space at a point p in a manifold M is, of course, basic to differential geometry. Intuitively, one may envisage smaller and smaller neighbourhoods of p in M which are correspondingly scaled up by larger and larger factors. In the limit we obtain the tangent space T_p to M at p. However, it does not appear to have been noted in the literature before that a corresponding procedure can also be applied to any entire (properly embedded) *null geodesic* γ in any space-time or, more generally, in any hyperbolic normal pseudo-Riemannian manifold. But whereas T_p is, in an essential way, a flat space, the corresponding procedure applied to γ yields a curved space W_γ known as a *plane wave.*

I shall give the argument explicitly for a four-dimensional space-time, the generalization to n-dimensions being obvious. It is well-known (cf. [1], for example) that a coordinate system can be set up in some neighbourhood Q of a conjugate point-free portion γ', of γ, the metric components taking the form

$$g_{ij} = \begin{pmatrix} 0 & 1 & 0 & 0 \\ 1 & a & b_2 & b_3 \\ 0 & b_2 & -c_{22} & -c_{23} \\ 0 & b_3 & -c_{23} & -c_{33} \end{pmatrix}, \tag{1}$$

where $a, b_2, \ldots, c_{33}$ are smooth functions of the coordinates x^0, x^1, x^2, x^3, the c's forming a positive-definite matrix. (To set up such a coordinate system, we choose a null hypersurface through γ' belonging to a smooth family of null hypersurfaces labelled by $x^1 = \text{const.}$ The null geodesic generators of these hypersurfaces are to be given by

Cahen & Flato (eds.), Differential Geometry and Relativity, 271–275.

x^1, x^2, $x^3 = \text{const.}$, x^0 being an affine parameter on each one. In particular, γ' is to be given by $x^1 = x^2 = x^3 = 0$.)

Consider next, a new coordinate system (X^0, X^1, X^2, X^3) for Q

$$X^0 = x^0, \qquad \Omega^2 X^1 = x^1, \qquad \Omega X^2 = x^2, \qquad \Omega X^3 = x^3, \tag{2}$$

where Ω is a positive number, and define

$$g_{ij}(\Omega) = \begin{pmatrix} 0 & 1 & 0 & 0 \\ 1 & \Omega^2 A & \Omega B_2 & \Omega B_3 \\ 0 & \Omega B_2 & -C_{22} & -C_{23} \\ 0 & \Omega B_3 & -C_{23} & -C_{33} \end{pmatrix}, \tag{3}$$

where $A(X^0, X^1, X^2, X^3) = a(x^0, x^1, x^2, x^3)$, etc. Then

$$g_{ij}(\Omega)\,\mathrm{d}X^i\,\mathrm{d}X^j = \Omega^{-2} g_{ij}\,\mathrm{d}x^i\,\mathrm{d}x^j. \tag{4}$$

Thus the metric components (3), in the X^i-coordinate system, define a metric $g(\Omega)$ which is a multiple of (i.e. conformal to, with constant scale factor) the original space-time metric g in Q. Now let $\Omega \to 0$. We obtain, for the limiting version of (4)

$$2\,\mathrm{d}X^0\,\mathrm{d}X^1 - C_{22}(\mathrm{d}X^2)^2 - 2C_{23}\,\mathrm{d}X^2\,\mathrm{d}X^3 - C_{33}(\mathrm{d}X^3)^2, \tag{5}$$

where, because of relations (2), the coefficients C_{22}, C_{23}, C_{33} can depend on the *one* variable X^0 only. The metric form (5) is the familiar Rosen expression [2] for a plane wave – a space-time with a five-parameter group of isometries [3].

What has been established is a limiting procedure whereby the immediate neighbourhood of γ' is expanded outwards, the metric being scaled uniformly up at the same time, and an exact plane wave W_γ is the result. If desired, we may view this limiting procedure in terms of a five-dimensional manifold-with-boundary with coordinates Ω, X^0, X^1, X^2, X^3, the boundary being given by $\Omega = 0$ (cf. Geroch [4]). This boundary $\Omega = 0$ has the intrinsic metric of a plane wave, and for each $\Omega = \text{const.} > 0$ we get a space whose metric is just Ω^{-2} times that of M in Q.

The reason that the metric must be scaled up is that there is a 'boost' involved in the coordinate change which, as Ω gets smaller, corresponds to a velocity approaching the limiting value unity. A general curvature tensor will appear to 'blow up' as the coordinates (or tetrad frame) are boosted to the speed of light* (cf. [5]). So, for a finite answer, we must correspondingly scale the metric up, whence the curvature is scaled down.

There is a 'physical' interpretation of the above mathematical procedure, which is the following. We envisage a succession of observers travelling in the space-time M whose world lines approach the null geodesic γ more and more closely; so we picture these observers as travelling with greater and greater speeds, approaching that of light. As their speeds increase they must correspondingly recalibrate their clocks to run faster and faster (assuming that all space-time measurements are referred to clock measurements in the standard way [7]), so that in the limit the clocks measure the affine parameter x^0 along γ. (Without clock recalibration a degenerate space-time metric would result.) In the limit, the observers measure the space-time to have the plane-wave structure W_γ.

In view of this universal role that the plane waves have in relation to general space-times, it is worth recalling a certain peculiar feature that they possess. Throughout the history of the subject, this feature has, in its various guises, often resulted in a belief that plane waves are in some special way 'non-physical' even as idealizations of the kind that one would expect plane waves to be. For example, Rosen [2] pointed out that any non-flat vacuum metric of the form (5) necessarily encounters singularities when one attempts to extend the range of the coordinates in order to obtain a complete space-time. This led to the erroneous belief, for many years, that gravitational plane waves did not exist in general relativity. However, it was later shown by Robinson [8] (cf. Bondi *et al.* [3]) that this was merely a coordinate effect† and that a complete plane wave could be covered by using a system of overlapping coordinate neighbourhoods, in each of which the metric takes the form (5). Another way in which this peculiarity of plane waves exhibits itself

* The Weyl tensor of a generic space-time approaches a *null* type (all principal null directions coincident) when a high velocity 'boost' is applied in a general direction [5, 6]. This is consistent with the fact that all plane waves have null (or else zero) Weyl tensors. Pirani [6] established this result in the case of the Schwarzschild solution only. But as we see from the above, the phenomenon is quite general; moreover, the entire metric of a smooth plane wave arises, not just its curvature tensor at one point.

† Robinson also showed that a coordinate system for the whole of W_γ can be given for which the metric takes the form $\mathrm{d}u\,\mathrm{d}v + h_{ij}(u)x^i x^j\,\mathrm{d}u^2 - \mathrm{d}x^i\,\mathrm{d}x^i$. This form of metric is due originally to Brinkman [9] who had obtained it as a special case of a more general class of vacuum metrics known as plane-fronted waves.

is in the non-existence of a global Cauchy hypersurface and the consequent property of non-embeddability in hyperbolic normal flat spaces of arbitrarily large dimension (Penrose [10]).

In the present context this phenomenon arises because of the possible existence of pairs of conjugate points on γ [1, 11]. Coordinates for which the metric tensor components take the form (1) cannot cover a portion of γ containing a pair of conjugate points (cf. [1]). Thus, for the whole of γ, a system of overlapping coordinate neighbourhoods would be required whenever γ contains pairs of conjugate points. Furthermore [12], provided M satisfies the weak energy condition, null geodesic completeness and a (physically reasonable) generality condition, it follows that *every* null geodesic actually does contain a pair of conjugate points. Thus, the peculiar features of plane waves mentioned above are in some limiting sense already present in every such physically reasonable space-time. (This does not, however, prevent such space-times from having global coordinate systems or Cauchy hypersurfaces, or from being embeddable in hyperbolic normal flat spaces.)

It may be finally remarked that for any space-time satisfying the weak energy condition [11], every null geodesic γ gives rise to a plane wave W_γ admitting an electromagnetic field tensor which, together with the metric of W_γ constitutes a (null) solution of the Einstein-Maxwell equations (cf. Lichnerowicz [13]). This is because the metric of W_γ is governed by three free real functions of the affine parameter x^0 on γ (whose information is contained in C_{22}, C_{23} and C_{33}). Two of these functions determine the Weyl tensor of W_γ and arise from the complex gravitational null datum Ψ_0 on γ (cf. [14], take l^a tangent to γ); the third function determines the Ricci tensor of W_γ and arises from the Ricci component Φ_{00} on γ (cf. [14]). The weak energy condition gives $\Phi_{00} \geqslant 0$, so we can write $\Phi_{00} = |\varphi_0|^2$ where φ_0 is a complex function of x^0. The quantities Ψ_0 and φ_0 define the strengths of the null gravitational and null electromagnetic field, respectively, for the plane wave W_γ. They can be chosen as arbitrary functions of x^0 to give a solution of the Einstein-Maxwell equations. Note that W_γ is a solution of the vacuum equations for every γ if and only if $\Phi_{00} = 0$ along every γ, i.e. if and only if M is an Einstein space (possibly with non-zero cosmological constant).

University of Oxford

REFERENCES

[1] Penrose, R., *Techniques of Differential Topology in Relativity*, SIAM, Philadelphia, 1972.
[2] Rosen, N., *Phys. Z. Sowjet.* **12** (1937), 366.
[3] Bondi, H., Pirani, F. A. E., and Robinson, I., *Proc. Roy. Soc.* (*Lond.*) **A251** (1959), 519.
[4] Geroch, R., *Commun. Math. Phys.* **13** (1969), 180.
[5] Penrose, R., *Ann. Phys.* **10** (1960), 171.
[6] Pirani, F. A. E., *Proc. Roy. Soc.* (*Lond.*) **A252** (1959), 96.
[7] Synge, J. L., *Relativity: The Special Theory*, North-Holland Publ. Co., Amsterdam, 1956.
[8] Robinson, I., Report to the Eddington Group, Cambridge, 1956.
[9] Brinkman, H. W., *Proc. Natl. Acad. Sci.* (*U.S.*) **9** (1923), 1.
[10] Penrose, R., *Revs. Mod. Phys.* **37** (1965), 215.
[11] Hawking, S. W. and Ellis, G. F. R., *The Large Scale Structure of Space-Time*, Cambridge Univ. Press, 1973.
[12] Hawking, S. W. and Penrose, R., *Proc. Roy. Soc.* (*Lond.*) **A314** (1970), 529.
[13] Lichnerowicz, A., *Compt. Rend.* **246** (1958), 893.
[14] Newman, E. T. and Penrose, R., *J. Math. Phys.* **3** (1962), 566.

A. H. TAUB

CURVATURE INVARIANTS, CHARACTERISTIC CLASSES, AND THE PETROV CLASSIFICATION OF SPACE-TIMES*

ABSTRACT. An identity is derived which relates the integrands occurring in the theory of characteristic classes of a Lorentzian four-dimensional space with the divergence of a vector field formed from the connection 1-forms and curvature 2-forms of the manifold. The two-component spinor formalism is used to express this identity in compact form.

1. INTRODUCTION

In this paper we shall be concerned with a four-dimensional manifold and a Lorentzian metric defined over it. Such a metric determines a Riemann curvature tensor and an affine connection without torsion. Various invariants depending quadratically on the curvature tensor play a role as integrands in the theory of characteristic classes. This theory relates global properties of the manifold to integrals over the manifold of these invariants. These invariants also play a role in the discussion of the algebraically special space-times; that is, those space-times which in the Petrov classification are not of type I. It is the purpose of this paper to derive formulas relating the quadratic expressions in the curvature tensor to the divergence of vector fields formed from the connection 1-forms and curvature 2-forms defined over the manifold.

In the derivation of these formulas we shall use the two-component spinor formalism. At every point x of the manifold there is a two-dimensional complex space S_x^2. It is related to the tangent space at that point with its metric $g_{\mu\nu}$ by means of 2×2 matrices

$$q_\mu = \|\gamma^{\dot{A}}_{\mu\dot{B}}\|, \tag{1.1}$$

which satisfy the equations

$$\bar{q}_\mu q_\nu + \bar{q}_\nu q_\mu = 2g_{\mu\nu} 1_2, \tag{1.2}$$

where 1_2 is the 2×2 identity matrix, and further the γ's are such that

$$\|\gamma_{\mu\dot{A}B}\| = \bar{\varepsilon} q_\mu \tag{1.3}$$

* This paper which touches upon a number of his interests is dedicated to André Lichnerowicz. Work upon it was done in part while the author was on sabbatical leave from the University of California, Berkeley and in residence at the Collège de France, Paris and at the Département de Mécanique, Université Pierre et Marie Curie (Paris VI).

Cahen & Flato (eds.), Differential Geometry and Relativity, 277–289.

is a Hermitian matrix, the bar denotes the complex conjugate, and in every coordinate system in S_x^2

$$\varepsilon = \|\varepsilon_{AB}\| = \begin{Vmatrix} 0 & 1 \\ -1 & 0 \end{Vmatrix}. \tag{1.4}$$

The matrices $\bar{\varepsilon} q_\mu$ form a basis for all Hermitian (2×2) forms in S_x^2. Thus, for any such form

$$H = V^\mu \bar{\varepsilon} q_\mu$$

and the V^μ are the components of a vector in the tangent space at x.

The matrices

$$p_{\mu\nu} = \|p^{\ A}_{\mu\nu B}\| = \tfrac{1}{2}(\bar{q}_\mu q_\nu - \bar{q}_\nu q_\mu) = -p_{\nu\mu} \tag{1.5}$$

may be shown to satisfy

$$\operatorname{trace}(p_{\mu\nu}) = 0, \tag{1.6}$$

$$\check{p}^{\mu\nu} = \tfrac{1}{2}E^{\mu\nu\sigma\tau} p_{\sigma\tau} = g^{\mu\sigma} g^{\nu\tau} p_{\sigma\tau}, \tag{1.7}$$

where

$$E^{\sigma\tau\mu\nu} = \frac{1}{\sqrt{g}}\varepsilon^{\sigma\tau\mu\nu}; \qquad E_{\sigma\tau\mu\nu} = \sqrt{g}\,\varepsilon_{\sigma\tau\mu\nu}. \tag{1.8}$$

$\varepsilon^{\sigma\tau\mu\nu}$ is the Levi-Civita tensor density, g is the determinant of the metric tensor, and hence $E^{\sigma\tau\mu\nu}$ is a pure imaginary tensor. The $p_{\mu\nu}$ form a (redundant) basis for all traceless 2×2 matrices.

If $\Psi^{\mu\dot{A}}_B$ is a spinor field with the transformation law

$$\Psi^{\mu\dot{A}}_B = \Psi^{\nu\dot{C}}_D \frac{\partial \chi^{*\mu}}{\partial \chi^\nu} \bar{T}^{\dot{A}}_{\dot{C}} t^D_B t^\omega \bar{t}^a,$$

where

$$t = \det\|t^A_B\|, \qquad T^A_B t^B_C = \delta^A_C,$$

$$T^A_B = T^A_B(x), \qquad x^{*\mu} = x^{*\mu}(x).$$

The numbers ω and a are said to be the weight and anti-weight of the spinor Ψ. The covariant derivation of such a spinor involves the spin connection $\Gamma^A_{B\nu}$ and the Christoffel symbols of space-time, $\Gamma^\mu_{\nu\sigma}$. Thus

$$\Psi^{\mu\dot{A}}_{B;\nu} = \Psi^{\mu\dot{A}}_{B,\nu} + \Psi^{\rho\dot{A}}_B \Gamma^\mu_{\rho\nu} + \Psi^{\mu\dot{C}}_B \bar{\Gamma}^{\dot{A}}_{\dot{C}\nu}$$
$$- \Psi^{\mu\dot{A}}_C \Gamma^C_{B\nu} - \omega \Gamma^C_{C\nu} \Psi^{\mu\dot{A}}_B - a\bar{\Gamma}^C_{C\nu} \Psi^{\mu\dot{A}}_B.$$

The spinors $\gamma_B^{\mu\dot{A}}$ have the weight $\omega = -\frac{1}{2}$ and anti-weight $a = \frac{1}{2}$. The requirement that their covariant derivatives vanish as well as those of ε_{AB} and ε^{AB} enables one to determine the quantities

$$K^{A}_{B\nu} = \Gamma^{A}_{B\nu} - \tfrac{1}{2}\delta^{A}_{B}\Gamma^{C}_{C\nu}.$$

We have

$$K_\nu = \|K^{A}_{B\nu}\| = \tfrac{1}{4}\bar{q}_\mu(q^{\mu}_{,\nu} + q^{\rho}\Gamma^{\mu}_{\rho\nu}). \tag{1.9}$$

Hence

$$\operatorname{trace} K_\nu = K^{A}_{A\nu} = 0. \tag{1.10}$$

The curvature spinor is defined as

$$\mathcal{R}_{\mu\nu} = \|R^{A}_{B\mu\nu}\| = K_{\nu,\mu} - K_{\mu,\nu} + K_\mu K_\nu - K_\nu K_\mu \tag{1.11}$$

and the Ricci identity reads

$$\Psi_{A;\mu\nu} - \Psi_{A;\nu\mu} = \Psi_C R^{C}_{A\mu\nu} \tag{1.12}$$

for a spinor such that

$$\Psi_{A;\mu} = \Psi_{A,\mu} - \Psi_C K^{C}_{A\mu}. \tag{1.13}$$

The Ricci identity applied to the spinor $\gamma^{\rho\dot{A}B}$ reads

$$\begin{aligned} 0 &= \gamma^{\rho\dot{A}B}_{;\mu\nu} - \gamma^{\rho\dot{A}B}_{;\nu\mu} \\ &= -\gamma^{\sigma\dot{A}B}R^{\rho}_{\sigma\mu\nu} - \gamma^{\rho\dot{A}E}R^{B}_{E\mu\nu} - \gamma^{\rho\dot{E}B}\bar{R}^{AE}_{\mu\nu}, \end{aligned} \tag{1.14}$$

where $R^{\rho}_{\sigma\mu\nu}$ is the Riemann curvature tensor formed from the $g_{\mu\nu}$. These equations give

$$R^{\rho}_{\tau\mu\nu} = -\tfrac{1}{2}\operatorname{trace}(p^{\rho}_{\tau}\mathcal{R}_{\mu\nu} + \bar{p}^{\rho}_{\tau}\bar{\mathcal{R}}_{\mu\nu}) \tag{1.15}$$

and also

$$\mathcal{R}_{\mu\nu} = \tfrac{1}{4}p^{\sigma\rho}R_{\sigma\rho\mu\nu}. \tag{1.16}$$

Equations (1.15) and (1.16) are inverses of each other.

The quantities ε_{AB} and ε^{AB} are used to manipulate indices in the spin space. Thus

$$\Psi_A = \varepsilon_{AB}\Psi^B, \qquad \Psi_{\dot{A}} = \varepsilon_{\dot{A}\dot{B}}\Psi^{\dot{B}},$$
$$\Psi^A = \varepsilon^{BA}\Psi_B, \qquad \Psi^{\dot{A}} = \varepsilon^{\dot{B}\dot{A}}\Psi_{\dot{B}},$$

where

$$\varepsilon_{\dot{A}\dot{B}} = \bar{\varepsilon}_{AB}, \qquad \varepsilon^{\dot{A}\dot{B}} = \bar{\varepsilon}^{AB}.$$

Similar rules apply to spinors with an arbitrary number of indices. Thus we may write Equation (1.16) as

$$R_{AB\mu\nu} = \varepsilon_{AC} R^{C}_{\ B\mu\nu} = \tfrac{1}{4} p^{\sigma\rho}_{\ \ AB} R_{\sigma\rho\mu\nu}. \tag{1.17}$$

We may define

$$R_{ABCD} = R_{AB\mu\nu} p^{\mu\nu}_{\ \ CD} = \tfrac{1}{4} p^{\sigma\rho}_{\ \ AB} R_{\sigma\rho\mu\nu} p^{\mu\nu}_{\ \ CD} \tag{1.18}$$

and

$$R_{AB\dot{C}\dot{D}} = R_{AB\mu\nu} \bar{p}^{\mu\nu}_{\ \ CD} = \tfrac{1}{4} p^{\sigma\rho}_{\ \ AB} R_{\sigma\rho\mu\nu} \bar{p}^{\mu\nu}_{\ \ CD}. \tag{1.19}$$

Then

$$R_{AB\mu\nu} = \tfrac{1}{8}(R_{ABCD} p^{CD}_{\ \ \mu\nu} + R_{AB\dot{C}\dot{D}} \bar{p}^{CD}_{\ \ \mu\nu}). \tag{1.20}$$

It is evident that

$$R_{ABCD} = R_{CDAB} = R_{BACD}$$

and that

$$R_{AB\dot{C}\dot{D}} = \bar{R}_{CD\dot{A}\dot{B}} = R_{BA\dot{C}\dot{D}}.$$

The spinor R_{ABCD} may be written as

$$R_{ABCD} = C_{ABCD} - \frac{R}{6}(\varepsilon_{AD}\varepsilon_{BC} + \varepsilon_{AC}\varepsilon_{BD}), \tag{1.21}$$

where C_{ABCD} is symmetric in all its indices and R is the scalar curvature. It may be shown that

$$C_{ABCD} = \tfrac{1}{4} p^{\sigma\rho}_{\ \ AB} C_{\sigma\rho\mu\nu} p^{\mu\nu}_{\ \ CD}, \tag{1.22}$$

where $C_{\sigma\rho\mu\nu}$ is the conformal curvature tensor of space-time. Further, that

$$R_{AB\dot{C}\dot{D}} = -\gamma_{\alpha\dot{C}A}\gamma_{\beta\dot{D}B}(R^{\alpha\beta} - \tfrac{1}{4} g^{\alpha\beta} R), \tag{1.23}$$

where $R^{\alpha\beta}$ is the Ricci tensor formed from the Riemann tensor by

$$R_{\mu\nu} = R^{\rho}_{\ \mu\nu\rho}. \tag{1.24}$$

Equation (1.20) may be written as

$$R_{AB\mu\nu} = \tfrac{1}{8}\left(C_{ABCD} p^{CD}_{\ \ \mu\nu} - \frac{R}{3} p_{AB\mu\nu} + R_{AB\dot{C}\dot{D}} \bar{p}^{CD}_{\ \ \mu\nu}\right). \tag{1.25}$$

2. BASES

If φ^A and Ψ^A are a pair of spinor fields such that

$$\varphi^A\Psi_A = 1, \tag{2.1}$$

they form a basis for spinors. That is, any spinor χ_A may be written as

$$\chi_A = -(\chi_B\Psi^B)\varphi_A + (\chi_B\varphi^B)\Psi_A.$$

We have also

$$\varphi^A\Psi_B - \Psi^A\varphi_B = \delta^A_B.$$

The spinors φ^A and Ψ^A also generate bases for all symmetric two index spinors and for all Hermitian forms.

It is a consequence of the definition of the spinors $\gamma_\mu^{\dot{A}B}$ and $\gamma^\mu_{\dot{A}B}$ that

$$\gamma_\mu^{\dot{A}B}\gamma^\mu_{\dot{C}D} = -2\delta^{\dot{A}}_{\dot{C}}\delta^D_B \tag{2.2}$$

and that

$$\gamma_\mu^{\dot{A}B}\gamma^\nu_{\dot{A}B} = -2\delta^\nu_\mu. \tag{2.3}$$

It then follows that the vectors

$$\begin{aligned} 1_\mu &= \frac{1}{\sqrt{2}}\gamma_\mu^{\dot{A}B}\bar{\varphi}_A\varphi_B, \\ n_\mu &= \frac{1}{\sqrt{2}}\gamma_\mu^{\dot{A}B}\bar{\Psi}_A\Psi_B, \\ m_\mu &= \frac{1}{\sqrt{2}}\gamma_\mu^{\dot{A}B}\bar{\Psi}_A\varphi_B \end{aligned} \tag{2.4}$$

are each null, that 1_μ and n_μ are real, that m_μ is complex and that

$$g_{\mu\nu} = -1_\mu n_\nu - n_\mu 1_\nu + m_\mu\bar{m}_\nu + \bar{m}_\mu m_\nu. \tag{2.5}$$

It may be further verified that

$$\begin{aligned} p_0^{\mu\nu} &= p^{\mu\nu}_{AB}\varphi^A\varphi^B = 2(m^\mu 1^\nu - m^\nu 1^\mu) = p_0^{\upsilon\mu\nu}, \\ p_1^{\mu\nu} &= p^{\mu\nu}_{AB}\Psi^A\varphi^B = (m^\mu\bar{m}^\nu - \bar{m}^\mu m^\nu + n^\mu 1^\nu - n^\nu 1^\mu) = p_1^{\upsilon\mu\nu}, \\ p_2^{\mu\nu} &= p^{\mu\nu}_{AB}\Psi^A\Psi^B = 2(n^\mu\bar{m}^\nu - n^\nu\bar{m}^\mu) = p_2^{\upsilon\mu\nu}. \end{aligned} \tag{2.6}$$

We define

$$\begin{aligned} X^0_{AB} &= \Psi_A\Psi_B = X_{2AB}, \\ X^1_{AB} &= -(\varphi_A\Psi_B + \Psi_A\varphi_B) = -2X_{1AB}, \\ X^2_{AB} &= \varphi_A\varphi_B = X_{0AB}, \end{aligned} \tag{2.7}$$

and find that if

$$a^{ij} = X^i_{AB} X^{jAB} \tag{2.8}$$

with

$$X^{jAB} = X^j_{CD} \varepsilon^{CA} \varepsilon^{DB},$$

then

$$\|a^{ij}\| = \begin{Vmatrix} 0 & 0 & 1 \\ 0 & -2 & 0 \\ 1 & 0 & 0 \end{Vmatrix}, \qquad \|a_{ij}\| = \begin{Vmatrix} 0 & 0 & 1 \\ 0 & -\frac{1}{2} & 0 \\ 1 & 0 & 0 \end{Vmatrix}, \tag{2.9}$$

$$a = \det \|a_{ij}\| = \tfrac{1}{2}, \quad X^i_{AB} = a^{ij} X_{jAB}, \quad X_{jAB} = a_{ji} X^i_{AB}.$$

Any symmetric spinor X_{AB} may then be written as

$$X_{AB} = X_i X^i_{AB},$$

where

$$X_i = X_{AB} X_i^{AB}.$$

Thus

$$X_0 = X_{AB} \varphi^A \varphi^B, \qquad X_1 = X_{AB} \varphi^A \Psi^B, \qquad X_2 = X_{AB} \Psi^A \Psi^B.$$

Further, we have

$$X_i^{AB} X^i_{CD} = X_0^{AB} X_{2CD} - 2 X_1^{AB} X_{1CD} + X_2^{AB} X_{0CD},$$

that is

$$X_i^{AB} X^i_{CD} = \tfrac{1}{2} (\delta^A_C \delta^B_D + \delta^A_D \delta^B_C). \tag{2.10}$$

Hence

$$X_i X^i = X_{AB} X_i^{AB} X^{CD} X^i_{CD} = X_{AB} X^{AB}$$

or

$$X_i X^i = -X^A_B X^B_A = -\operatorname{trace}(X^2), \tag{2.11}$$

where

$$X = X^A_B.$$

If we define

$$C_{ij} = C_{ABCD} X_i^{AB} X_j^{CD} \tag{2.12}$$

then it follows from Equation (2.10) that

$$a^{ij}C_{ij} = C^{AB}_{AB} = 0 \tag{2.13}$$

since C_{ABCD} is symmetric in all its indices. Further, we have

$$C_{ABCD} = C_{ij}X^i{}_{AB}X^j{}_{CD}. \tag{2.14}$$

Similarly, we may define

$$s_{i\bar{j}} = R_{AB\dot{C}\dot{D}}X_i^{AB}\bar{X}_j^{CD} \tag{2.15}$$

and write

$$R_{AB\dot{C}\dot{D}} = s_{i\bar{j}}X^i{}_{AB}\bar{X}^{jCD}. \tag{2.16}$$

Equation (1.25) is then equivalent to

$$R_{i\mu\nu} = X_i^{AB}R_{AB\mu\nu} = \tfrac{1}{8}\left(\left(C_{i1} - \frac{R}{3}a_{i1}\right)p^{\,1}_{\mu\nu} + s_{i\bar{1}}\bar{p}^{\,1}_{\mu\nu}\right), \tag{2.17}$$

where

$$p^{\,1}_{\mu\nu} = a^{i1}p_{i\mu\nu}$$

and the $p_{i\mu\nu}$ are defined by Equations (2.6). It follows from the latter equations and Equation (2.5) that

$$\tfrac{1}{8}p_i^{\mu\nu}p_{j\mu\nu} = a_{ij} \quad \text{and} \quad p_i^{\mu\nu}\bar{p}_{j\mu\nu} = 0.$$

Hence

$$C_{i1} - \frac{R}{3}a_{i1} = R_{i\mu\nu}p_1^{\mu\nu}, \qquad s_{i\bar{1}} = R_{i\mu\nu}\bar{p}_1^{\mu\nu} \tag{2.18}$$

The spinor C_{ABCD} may also be expressed as

$$\begin{aligned} C_{ABCD} = {} & C_0\varphi_A\varphi_B\varphi_C\varphi_D + 4C_1\varphi_{(A}\varphi_B\varphi_C\Psi_{D)} + 6C_2\varphi_{(A}\varphi_B\Psi_C\Psi_{D)} \\ & + 4C_3\Psi_{(A}\Psi_B\Psi_C\varphi_{D)} + C_4\Psi_A\Psi_B\Psi_C\Psi_D. \end{aligned} \tag{2.19}$$

It then follows from Equation (2.12) that

$$C = \|C_{ij}\| = \left\|\begin{matrix} C_4 & -C_3 & C_2 \\ -C_3 & C_2 & -C_1 \\ C_2 & -C_1 & C_0 \end{matrix}\right\|. \tag{2.20}$$

The Petrov classification of conformal tensors is made in terms of the proper vectors of the matrix

$$\|C^i_j\| = \|a^{i1}C_{1j}\| = \begin{Vmatrix} C_2 & -C_1 & C_0 \\ 2C_3 & -2C_2 & 2C_1 \\ C_4 & -C_3 & C_2 \end{Vmatrix}. \tag{2.21}$$

The Penrose classification is made in terms of the nature of the roots of the equation

$$C_{ABCD}\chi^A\chi^B\chi^C\chi^D = 0. \tag{2.22}$$

In the Petrov classification, the quantities

$$I = C^i_jC^j_i \tag{2.23}$$

and

$$J = C^i_jC^j_kC^k_i, \tag{2.24}$$

which are, respectively, the trace of the square of the matrix C and the cube of this matrix play an important role since I and J are related to the coefficients of the characteristic equation of the matrix C. For algebraically special conformal tensors we have

$$\frac{I}{6} = \left(-\frac{J}{6}\right)^{2/3}. \tag{2.25}$$

It may be shown that

$$I = 2C^{\sigma\tau\mu\nu}(C_{\sigma\tau\mu\nu} + {}^vC_{\sigma\tau\mu\nu}), \tag{2.26}$$

$$J = 2(C^{\sigma\tau\mu\nu} + {}^vC^{\sigma\tau\mu\nu})C_{\mu\nu\lambda\rho}C^{\lambda\rho}_{\ \ \sigma\tau}, \tag{2.27}$$

where

$${}^vC_{\sigma\tau\mu\nu} = \tfrac{1}{2}E_{\sigma\tau\alpha\beta}C^{\alpha\beta}_{\ \ \mu\nu}.$$

3. The Bianchi Identities

We shall define

$$\begin{aligned} A_{0\nu} &= \varphi^A\varphi_{A;\nu}, \\ A_{1\nu} &= \varphi^A\Psi_{A;\nu} = \Psi^A\varphi_{A;\nu}, \\ A_{2\nu} &= \Psi^A\Psi_{A;\nu}. \end{aligned} \tag{3.1}$$

These $A_{i\nu}$ are twelve complex functions that Newman and Penrose have called the spin coefficients. This may be readily seen from the fact that

$$\begin{aligned} 1_{\mu;\nu} &= m_\mu \bar{A}_{0\nu} + m_\mu A_{0\nu} - 1_\nu (A_{1\nu} + \bar{A}_{1\nu}), \\ n_{\mu;\nu} &= -\bar{m}_\mu \bar{A}_{2\nu} - m_\mu A_{2\nu} + n_\mu (A_{1\nu} + \bar{A}_{1\nu}), \\ m_{\mu;\nu} &= n_\mu A_{0\nu} - 1_\mu \bar{A}_{2\nu} - m_\mu (A_{1\nu} - \bar{A}_{1\nu}), \end{aligned} \tag{3.2}$$

and may be verified by differentiating Equations (2.4).

It follows from the Ricci identity

$$\varphi_{A;\mu\nu} - \varphi_{A;\nu\mu} = \varphi_C R^C_{\ A\mu\nu}$$

by multiplying by φ^A and integrating by parts that

$$A_{0\mu;\nu} - A_{0\nu;\mu} - \varphi^A_{;\nu}\varphi_{A;\mu} + \varphi^A_{;\mu}\varphi_{A;\nu} = -\varphi^A\varphi^C R_{AC\mu\nu}.$$

If use is made of the identity

$$\delta^A_C = \varphi^A \Psi_C - \Psi^A \varphi_C,$$

we have

$$A_{0\mu;\nu} - A_{0\nu;\mu} + 2(A_{0\mu}A_{1\nu} - A_{0\nu}A_{1\mu}) = -\varphi^A\varphi^C R_{AC\mu\nu}.$$

Similar arguments show that

$$R_{i\mu\nu} = A_{i\nu;\mu} - A_{i\mu;\nu} - \sqrt{2} E_{ijk} A^j_{\ \mu} A^k_{\ \nu}, \tag{3.3}$$

where

$$A^i_{\ \mu} = a^{ij} A_{j\mu}, \qquad E_{ijk} = \sqrt{a}\,\varepsilon_{ijk},$$

and has before

$$R_{i\mu\nu} = X_i^{AB} R_{AB\mu\nu}.$$

It is an immediate consequence of Equation (3.3) that

$$R_{i\ ;\lambda}^{\ \upsilon\tau\lambda} = \tfrac{1}{2} E^{\tau\lambda\mu\nu} R_{i\mu\nu;\lambda} = -\sqrt{2} E_{ijk} R^{\upsilon j\tau\lambda} A^k_{\ \lambda}. \tag{3.4}$$

These are the Bianchi identities.

4. The Basic Identity

Consider the expression

$$\begin{aligned} (A^i_{\ \tau} R_i^{\ \upsilon\tau\lambda})_{;\lambda} &= A^i_{\ \tau;\lambda} R_i^{\ \upsilon\tau\lambda} + A^i_{\ \tau} R_{i\ ;\lambda}^{\ \upsilon\tau\lambda} \\ &= \tfrac{1}{2}(A^i_{\ \tau;\lambda} - A^i_{\ \lambda;\tau}) R_i^{\ \upsilon\tau\lambda} + A^i_{\ \tau} R_{i\ ;\lambda}^{\ \upsilon\tau\lambda}. \end{aligned}$$

It follows from the above that

$$(A^{\ i}_{\tau}R^{\ \upsilon\tau\lambda}_{i})_{;\lambda} = \tfrac{1}{2}R^{\ i}_{\lambda\tau}R^{\ \upsilon\tau\lambda}_{i} + \tfrac{1}{2}\sqrt{2}E^{ilm}A_{1\lambda}A_{m\tau}R^{\ \upsilon\tau\lambda}_{i}.$$

A straightforward calculation shows that

$$\left(\frac{\sqrt{2}}{3}E^{ilm}A_{i\mu}A_{1\nu}A_{m\tau}E^{\mu\nu\tau\lambda}\right)_{;\lambda} = -\sqrt{2}E^{ilm}A_{1\lambda}A_{m\tau}R^{\ \upsilon\lambda\tau}_{i}.$$

Hence

$$\left(A^{\ i}_{\tau}R^{\ \upsilon\tau\lambda}_{i} + \frac{\sqrt{2}}{6}E^{ilm}A_{i\mu}A_{1\nu}A_{m\tau}E^{\mu\nu\tau\lambda}\right)_{;\lambda} = -\tfrac{1}{2}R^{\ i}_{\lambda\tau}R^{\ \upsilon\lambda\tau}_{i}. \tag{4.1}$$

This is the equation referred to in the introduction for it expresses a quadratic expression in the curvature tensor in terms of the divergence of a vector field.

It may be verified from Equation (2.17), the fact that the $p^{\ 1}_{\mu\nu}$ are self-dual, the fact that $\bar{p}^{\ 1}_{\mu\nu}$ are anti-self-dual and the properties of the $p^{\ i}_{\mu\nu}$, that

$$R^{\ i}_{\mu\nu}R^{\ \upsilon\mu\nu}_{i} = \tfrac{1}{32}\left(\left(C^{i}_{j} - \frac{R}{3}\delta^{i}_{j}\right)\left(C^{j}_{i} - \frac{R}{3}\delta^{j}_{i}\right) - s_{i\bar{j}}s^{i\bar{j}}\right).$$

It then follows from Equations (2.23) and (2.26) and the fact that

$$s_{i\bar{j}}s^{i\bar{j}} = 4R^{\alpha}_{\beta}R^{\beta}_{\alpha} - R^{2},$$

where R^{α}_{β} is the Ricci tensor and R is the scalar curvature, that

$$R^{\ i}_{\mu\nu}R^{\ \upsilon\mu\nu}_{i} = \tfrac{1}{16}(C^{\ \lambda\mu}_{\sigma\tau}{}^{\upsilon}C^{\sigma\tau}_{\ \lambda\mu} + C^{\ \lambda\mu}_{\sigma\tau}C^{\sigma\tau}_{\ \lambda\mu} - 2R^{\ \mu}_{\nu}R^{\ \nu}_{\mu} + \tfrac{2}{3}R^{2}).$$

Note that the first term on the right-hand side is pure imaginary and the remaining terms are real. The first term is proportional to the integrand that determines the Pontrjagin class of a manifold and the remaining terms determine the Euler class.

Equation (4.1) may then be written as

$$\begin{aligned}\Delta &= -\tfrac{1}{32}(C^{\ \lambda\mu}_{\sigma\tau}{}^{\upsilon}C^{\sigma\tau}_{\ \lambda\mu} + C^{\ \lambda\mu}_{\sigma\tau}C^{\sigma\tau}_{\ \lambda\mu} - 2R^{\ \mu}_{\nu}R^{\ \nu}_{\mu} + \tfrac{2}{3}R^{2})\\ &= V^{\lambda}_{1;\lambda} + V^{\lambda}_{2;\lambda},\end{aligned} \tag{4.2}$$

where

$$V^{\lambda}_{1} = A^{\ i}_{\tau}R^{\ \upsilon\tau\lambda}_{i} \tag{4.3}$$

and

$$V^{\lambda}_{2} = \frac{\sqrt{2}}{6}E^{ilm}A_{i\mu}A_{1\nu}A_{m\tau}E^{\mu\nu\tau\lambda}. \tag{4.4}$$

We note that since C_{ij} is defined by Equation (2.12) we have

$$C_{ij}C^{ij} = C_{ABCD}C^{ABCD} \tag{4.5}$$

and

$$C_{ij}C^{ij} = 6C_2^2 - 8C_1C_3 + 2C_0C_4, \tag{4.6}$$

as follows from Equation (2.21). For algebraically special space-times $C_3 = C_4 = 0$, and

$$C_{ij}C^{ij} = 6C_2^2. \tag{4.7}$$

5. Evaluation of V_1 and V_2

These vectors may be studied by determining their expansion in terms of the null tetrad. Thus we may write

$$V_2^\mu = -(V_{2\rho}n^\rho)1^\mu - (V_{2\rho}1^\rho)n^\mu + (V_{2\rho}m^\rho)\bar{m}^\mu + (V_{2\rho}\bar{m}^\rho)m^\mu,$$

where of course V_2^μ is defined by Equation (4.4). Since $p_0^{\mu\nu}$ is self-dual it follows from Equation (2.6) that

$$1^\nu = -E^{\nu\mu\sigma\tau}\bar{m}_\mu m_\sigma 1_\tau.$$

Hence

$$6V_2^\mu 1_\mu = +\varepsilon^{ilm}A_{i\mu}A_{1\nu}A_{m\tau}\bar{m}^\alpha m^\beta 1^\gamma \delta^{\mu\nu\tau}_{\alpha\beta\gamma}; \tag{5.1}$$

we may also write

$$A_{i\mu} = -\alpha_i 1_\mu - \beta_i n_\mu + \gamma_i m_\mu + \delta_i \bar{m}_\mu. \tag{5.2}$$

The twelve numbers α_i, β_i, γ_i, and δ_i form a (3×4) matrix where the Greek letters α, β ... denotes the column and the index i denotes the row. The determinant of the 3×3 minor obtained by omitting the column α will be denoted by $M(\alpha)$ and in similar fashion we may define $M(\beta)$, $M(\gamma)$, and $M(\delta)$.

It follows from Equations (5.1) and (5.2) that

$$V_2^\mu 1_\mu = \sqrt{2}\,E^{ilm}\delta_i\gamma_i\beta_m = M(\alpha).$$

Similar arguments show that

$$V_2^\mu = -M(\alpha)n^\mu - M(\beta)1^\mu + M(\delta)m^\mu + M(\gamma)\bar{m}^\mu. \tag{5.3}$$

Equations (2.6) and (2.17) enable us to determine the expression for V_1^μ

in terms of the null-tetrad. It follows from the latter equations that

$$R_{i}^{\,\upsilon\tau\lambda} = \frac{1}{8}\left(\left(C_{i1} - \frac{R}{3}a_{i1}\right)p^{1\tau\lambda} - s_{i\bar{1}}\bar{p}^{1\tau\lambda}\right).$$

Hence

$$V_{1}^{\lambda} = \frac{1}{8}\left(\left(C_{i1} - \frac{R}{3}a_{i1}\right)A^{\,i}_{\,\tau}p^{1\tau\lambda} - s_{i\bar{1}}A^{\,i}_{\,\tau}\bar{p}^{1\tau\lambda}\right). \tag{5.4}$$

When Equations (5.2) and (2.6) are used one may readily write the explicit values of the coefficients in the expressions

$$\begin{aligned} A^{\,i}_{\,\tau}p^{1\tau\lambda} &= -\alpha^{i1}1^{\lambda} - \beta^{i1}n^{\lambda} + \gamma^{i1}m^{\lambda} + \delta^{i1}\bar{m}^{\lambda}, \\ A^{\,i}_{\,\tau}\bar{p}^{1\tau\lambda} &= -\alpha'^{i1}1^{\lambda} - \beta'^{i1}n^{\lambda} + \gamma'^{i1}m^{\lambda} + \delta'^{i1}\bar{m}^{\lambda}. \end{aligned} \tag{5.5}$$

These coefficients are linear functions of the quantities α^{i}, β^{i}, γ^{i}, and δ^{i}.

6. Concluding remarks

We have seen that Equation (4.1), the one we have called the basic identity, contains on the right-hand side terms simply related to the integrands occurring in the determination of the Pontrjagin and Euler classes of a manifold. The terms involving the conformal tensor are simply related to an invariant which occurs in the Petrov classification of space-times. In the algebraically special case, it is the only invariant since the one formed from a cubic combination of the components of the conformal tensor is simply related to this one.

When the right-hand side of the basic equation is integrated over all of a compact manifold as was done by Avez [1] one obtains global theorems concerning such space-times. However, we may apply Equation (4.1) to just a region of space-time (for example, one containing a singularity) and thereby obtain information about the integral of the right-hand side in terms of the boundary values of the vectors V_1 and V_2. It may be that by an appropriate choice of the region of integration significant properties of various types of singularities may be obtained.

The invariant $J^{2/3}$ when J is defined by Equation (2.24) and which is simply related to the invariant I in the algebraically special cases, is worthy of study as an integrand of an integral over a manifold. It would seem that such an integral should also relate to some global property of the manifold. As yet such integrals have not received proper attention.

University of California

REFERENCE

[1] Avez, A., *C.R. Acad. Sci. Paris* **259** (1964), 1934–37; *Proc. Natl. Acad. Sci. U.S.A.* **66** (1970), 265–268.

COLLECTED WORKS OF A. LICHNEROWICZ

1937

[1] 'Extension du théorème de Gauss-Whittaker', *C.R. Acad. Sci.* **205**, 25.

1938

[2] 'Sur les singularités du ds^2 extérieur', *C.R. Acad. Sci.* **206**, 157.
[3] 'Espaces-temps extérieurs réguliers partout', *C.R. Acad. Sci.* **206**, 313.

1939

[4] 'Sur certains problèmes globaux relatifs au système des équations d'Einstein', Thèse, Paris, Hermann; *Actual. Sci. Ind.*, fasc. 833.
[5] 'Nouvelles recherches sur les singularités du ds^2 extérieur', *C.R. Acad. Sci.* **209**, 531 (en collaboration avec M. Raymond Marrot).

1940

[6] 'Sur un théorème d'hydrodynamique relativiste', *C.R. Acad. Sci.* **211**, 117.
[7] 'Remarques sur l'équation intégro-différentielle de Boltzmann', *C.R. Acad. Sci.* **210**, 391 (en collaboration avec M. Raymond Marrot).
[8] 'Propriétés statistiques des ensembles de particules en relativité restreinte', *C.R. Acad. Sci.* **210**, 759 (en collaboration avec M. Raymond Marrot).
[9] 'Sur l'équation intégro-différentielle de Boltzmann, *C.R. Acad. Sci.* **211**, 531 (en collaboration avec M. Raymond Marrot).

1941

[10] 'Sur des théorèmes d'unicité relatifs aux équations gravitationnelles du cas intérieur', *Bull. des Sci. Math.* **65**, 54–72.
[11] 'Les espaces à connexion semi-symétrique et la Mécanique', *C.R. acad. Sci.* **212**, 328.
[12] 'Sur la définition géomètrique des processus matériels en Relativité générale,' *C.R. Acad. Sci.* **212**, 421.
[13] 'Le *H*-théorème et la notion de système isolé, *C.R. Acad. Sci.* **212**, 1074 (en collaboration avec M. Raymond Marrot).
[14] 'Opérateurs hermitiques et espaces de Riemann', *C.R. Acad. Sci.* **213**, 12.
[15] 'Sur l'intégration des équations d'Einstein', *C.R. Acad. Sci.* **213**, 516.
[16] 'Sur l'intégration des équations de la Relativité', *C.R. Acad. Sci.* **213**, 549.
[17] 'Sur l'invariant intégral de l'hydrodynamique relativiste', *Ann. Ec. Norm.* **GE**, 285–304.

1942

[18] 'Sur une généralisation des espaces de Finsler', *C.R. Acad. Sci.* **214**, 599.

Cahen & Flato (eds.), Differential Geometry and Relativity, 291–300.

1943

[19] 'Sur une extension du calcul des variations', *C.R. Acad. Sci.* **216**, 25.
[20] 'Sur l'invariant intégral des équations relativistes de l'électromagnétisme', *C.R. Acad. Sci.* **216**, 863.
[21] 'Sur les équations relativistes de l'électromagnétisme', *Ann. Ec. Norm.* **60**, 247–288.
[22] 'Sur les espaces variationnels généralisés', *C.R. Acad. Sci.* **217**, 415.
[23] 'Sur un principe de la dynamique lié à l'invariant intégral de Poincaré-Cartan', *C.R. Acad. Sci.* **217**, 260.
[24] 'L'intégration des équations de la gravitation relativiste et le problème des *n* corps', *J. Math. pures et appliquées* **23**, 37–63.
[25] 'Sur une inégalité relative aux espaces riemanniens complètement harmoniques', *C.R. Acad. Sci.* **218**, 436.
[26] 'Sur les espaces riemanniens complètement harmoniques', *C.R. Acad. Sci.* **218**, 493.
[27] 'Sur les équations de l'hydrodynamique relativiste des fluides visqueux', *C.R. Acad. Sci.* **219**, 270.
[28] 'Sur les espaces riemanniens complètement harmoniques', *Bull. Soc. Math.* **72**, 146–168.
[29] 'La composition de seconde espèce et les fonctions de Schmidt', *C.R. Acad. Sci.* **219**, 663.

1945

[30] 'Les espaces variationnels généralisés', *Ann. Ec. Norm.* **62**, 339–384.
[31] 'Sur les espaces riemanniens harmoniques de type hyperbolique normal', *C.R. Acad. Sci.* **221**, 294 (en collaboration avec M. A. C. Walker).
[32] 'Sur une proposition fondamentale de la théorie relativiste de la gravitation', *C.R. Acad. Sci.* **221**, 652.

1946

[33] 'Sur la caractère euclidien des espaces-temps extèrieurs statiques partout réguliers', *C.R. Acad. Sci.* **222**, 432.
[34] 'Sur une extension de la formule d'Allendoerfer-Weil à certaines variétés finslériennes', *C.R. Acad. Sci.* **223**, 12.
[35] 'Les relations intégrales d'invariance et leurs applications à la dynamique', *Bull. des Sci. Math.* **70**, 82–95.
[36] 'Sur la transformation des équations de la dynamique', *C.R. Acad. Sci.* **223**, 649.

1947

[37] *Algèbre et analyse linéaire*, Paris, Masson (éd.).
[38] 'Problème de calcul des variations liés à la dynamique classique et à la théorie unitaire du champ', *C.R. Acad. Sci.* **224**, 529 (en collaboration avec M. Yves Thiry).
[39] 'Sur les formes harmoniques de certains espaces fibrés', *C.R. Acad. Sci.* **224**, 1413.

1948

[40] 'Sur les nombres de Betti de certaines variétés différentielles définies comme espaces fibrés', *C.R. Acad. Sci.* **226**, 212.

[41] 'Sur un théorème global de réduction des univers statiques d'Einstein', *C.R. Acad. Sci.* **226**, 775 (en collaboration avec Mme Yvonne Choquet-Bruhat).
[42] 'Courbure et nombre de Betti d'une variété riemanienne compacte', *C.R. Acad. Sci.* **226**, 1678.
[43] 'Théorème global sur les ds^2 extèrieurs généraux d'Einstein', *C.R. Acad. Sci.* **226**, 2119 (en collaboration avec Mme Yvonne Choquet-Bruhat).
[44] 'Géomètrie différentielle et topologie', Conférence du Palais de la découverte.
[45] 'Un théorème sur l'homologie dans les espaces fibrés', *C.R. Acad. Sci.* **227**, 711.

1949

[46] 'Quelques théorèmes de géomètrie différentielle globale', *Comm. Math. Helv.*, 1949.

1950

[47] *Eléments de Calcul tensoriel*, Paris, Colin (éd.).
[48] 'Dérivation covariante et nombres de Betti', *C.R. Acad. Sci.* **230**, 1248.
[49] 'Sur certaines classes d'espaces riemanniens compacts', *C.R. Acad. Sci.* **230**, 2146.
[50] 'Courbure, nombres de Betti et espaces symétriques', Comptes Rendus du Congrès de Harvard.
[51] 'Théorème de réductibilité des variétés kählériennes et applications', *C.R. Acad. Sci.* **231**, 1280.
[52] 'Sur les variétés riemaniennes admettant une forme quadratique à dérivée covariante nulle', *C.R. Acad. Sci.* **231**, 1413.

1951

[53] 'Formes à dérivée covariante nulle sur une variété riemanienne', *C.R. Acad. Sci.* **232**, 146.
[54] 'Sur les variétés riemaniennes admettant une forme à dérivée covariante nulle', *C.R. Acad. Sci.* **232**, 677.
[55] 'Généralisations de la géomètrie kählérienne globale', Conférence du Colloque de Géomètrie différentielle de Louvain.
[56] 'Sur les formes harmoniques des variétés riemaniennes localement réductibles', *C.R. Acad. Sci.* **232**, 1634.
[57] 'Sur les variétés symplectiques', *C.R. Acad. Sci.* **233**, 723–725

1952

[58] 'Extensions de la Géomètrie kählérienne globale', Exposé ronéotypé du siménaire Bourbaki, février 1952.
[59] 'Groupes d'holonomie des variétés riemaniennes', *C.R. Acad. Sci.* **234**, 1835–1837 (en collaboration avec M. Armand Borel).
[60] 'Espaces riemanniens et hermitiens symétriques', *C.R. Acad. Sci.* **234**, 2332–2334 (en collaboration avec M. Armand Borel).
[61] 'Variétés pseudo-kählériennes à courbure de Ricci non nulle; application aux domaines bornés homogènes de C^n', *C.R. Acad. Sci.* **235**, 12–14.
[62] 'Méthodes tensorielles en Mécanique', Conférence générale aux Congrès internationaux de Mécanique, Istanbul 1953.
[63] 'The General Problem of the Transformation of the Equations of Dynamics', *J. Rat. Mech. Anal.* **1**, 499–520 (en collaboration avec M. Don Aufenkamp).

[64] 'Sur les équations relativistes de la gravitation', *Bull. Soc. Math. France* **80**, 237–251.

1953

[65] 'Etude mathématique des théories relativistes de la gravitation et de l'électromagnètisme; I: Relativité générale classique', Cours ronéotypé du Collége de France, 1952–1953.
[66] 'Espaces homogènes kählèriens', Conférence au Coll. intern. de Géomètrie diff. de Strasbourg, juin 1953, pp. 171–184.
[67] 'Sur les espaces homogènes kählèriens', *C.R. Acad. Sci.* **237**, 695–697.
[68] 'Groupes d'holonomie des variétés riemaniennes et kählériennes', Conférence au Coll. intern. de Géomètrie diff. de Venise, sept. 1953.
[69] 'Comptabilité des équations de la théorie unitaire d'Einstein-Schrödinger, *C.R. Acad. Sci.* **237**, 1383–1386.

1954

[70] 'Un théorème sur les espaces homogènes complexes', *Arch. Math.* **5**, 207–215.
[71] 'Equations de Laplace et espaces harmoniques', Coll. sur les équations aux dérivées partielles Louvain, décembre 1953, pp. 9–23.
[72] 'Compatibilité des équations de la théorie unitaire du Champ d'Einstein', *J. Rat. Mech. Anal.* **3**, 487–521.
[72ᵃ] 'Etude des équations du Champ de la Théorie unitaire d'Einstein', Rendiconti del Seminario matematico e fisica di Milano – Vol. XXV – 1953–1954.
[73] 'Groupes d'holonomie', Conférence générale du Congrès int. de Math., Amsterdam, 1954.
[74] *Les théories relativistes de la gravitation et de l'électromagnétisme: relativité générale et théories unitaires*, Masson, 1954.
[75] Sur les groupes d'automorphismes de certaines variétés kählériennes', *C.R. Acad. Sci.* **239**, 1344–1346.
[76] 'Transformations infinitésimales conformes de certaines variétés riemaniennes compactes', *C.R. Acad. Sci.* **241**, 726–729.
[76ᵃ] 'Transformations infinitésimales conformes de certaines variétés riemaniennes compactes', Dagli Atti del V Congresso dell'Unione Matematica Italiana, Pavia, Torino, 1956.
[77] 'Problèmes généraux d'intégration des équations de la relativité', Rapport du Congrès du cinquantenaire de la relativité, Berne, 1955.
[78] 'Théorie globale des connexions et des groupes d'holonomie', Cremonèse, Rome (1955).
[79] 'Transformations infinitésimales conformes de certaines variétés riemaniennes compactes', Congrès de Pavia-Torino, octobre 1955.

1956

[80] 'Sur les transformations affines d'une variété presque hermitienne compacte', *C.R. Acad. Sci.* **242**, 337–339 (en collaboration avec Mme Apte).
[81] 'Espaces homogènes riemanniens et réductibilité', *C.R. Acad. Sci.* **242**, 1410–1413.
[82] 'Sur les transformations affines des variétés riemaniennes', *C.R. Acad. Sci.* **242**, 1568–1570.
[83] 'Sur la réductibilité des espaces homogènes riemanniens', *C.R. Acad. Sci.* **243**, 640–642.

1957

[84] 'Transformations affines et holonomie', *C.R. Acad. Sci.* **244**, 1868–1870.
[85] 'Homogeneous kählerian spaces', Notes miméographiées de Princeton, janvier 1957.
[86] 'Sur les transformations analytiques des variétés kählériennes compactes', *C.R. Acad. Sci.* **244**, 3011–3013.
[87] 'Transformations analytiques d'une variété kählérienne et holonomie', *C.R. Acad. Sci.* **245**, 953–956.

1958

[88] 'Sur les ondes et radiations gravitationnelles', *C.R. Acad. Sci.* **246**, 893.
[89] 'Sur un procedé de quantification du champ de gravitation', *C.R. Acad. Sci.* **247**, 433–436.
[90] 'Transformations analytiques et isométries d'une variété kählérienne compacte', *C.R. Acad. Sci.* **247**, 855.
[90^a] 'Transformations analytiques d'une variété kählérienne compacte', *Bull. Math. de la Ste des Sci. Math. et Phys. de la R.P.R.* **2**(50) (1958), 2.
[91] 'Ondes électromagnétiques et gravitationnelles en relativité générale', Cahiers de Physique, No. 96, pp. 287–296.
[92] 'Radiations gravitationnelles et électromagnétiques en relativité générale', Centro Int. Matematico Estivo Sestrière, 20–30 juillet 1958.
[92^a] 'Radiations en relativité générale', Royaumont 21–27, juin 1959.
[93] *Géométrie des groupes de transformations*, Dunod, 1958.

1959

[94] 'Sur les transformations analytiques d'une variété kählérienne compacte', Coll. de géométrie différentielle, Bruxelles, décembre 1958.
[95] 'Isométries et transformations analytiques d'une variété kählérienne compacte', *Bull. Soc. Math. France*, **87** (1957), 427–437.
[95^a] 'Ondes et radiations électromagnétiques et gravitationnelles', *C.R. Acad. Sci.* **248**, 2728.
[96] 'Propagateurs antisymétriques en relativité générale – Quantification du champ électromagnétique dans le vide', *C.R. Acad. Sci.* **249**, 1311–1329.
[97] 'Sur la quantification du champ de gravitation pour un espace-temps à courbure constante', *C.R. Acad. Sci. Paris* **249**, 2287–2289.

1960

[98] 'Ondes et radiations électromagnétiques et gravitationnelles en relativité générale, *Annali di Matematica* **50**, 2–95.
[99] Le rôle du calcul extérieur en géométrie', Report of the second Conference on Mathematical education in South Asia, Bombay, janvier 1960, p. 73679.
[100] 'Sur la quantification du champ de gravitation', *C.R. Acad. Sci.* **250**, 3122–3124.
[101] 'Transformations affines des variétés à connexion linéaire et des variétés riemaniennes', Coll. Int. de géométrie différentielle et topologie, Zurich, 1960.

1961

[102] 'Analytic Transformations of Kähler Manifolds', Notes miméographiées de l'Université d'Illinois.

[103] 'Propagators and Commutators in General Relativity', Notes miméographiques de l'Université d'Illinois.
[104] 'Propagateurs et commutateurs en relativité générale', Publications de l'Institut des Hautes Etudes Scientiques, No. 10.
[105] 'Propagateurs et commutateurs en relativité générale', Cahiers de Physique.
[105[a]] 'Propagateurs sur une variété différentiable', Celebrazioni archimade de secolo XX, Syracusa 1962. Vol. I Conferenze gen. e symp. digeom. diff., pp. 131–138.
[106] 'Anticommutateur du champ spinoriel en relativité générale', *C.R. Acad. Sci.* **252**, 3742.
[107] 'Anticommutateur pour le champ corrispondant au spin 3/2 sur un espace d'Einstein', *C.R. Acad. Sci.* **253**, 940.
[108] 'Théorie de Petiau-Duffin-Kemmer en relativité générale', *C.R. Acad. Sci.* **253**, 983.
[109] 'Sur la réductivité de certaines algèbres d'automorphismes', *C.R. Acad. Sci.* **253**, 1302–1304.
[110] *Théorèmes de réductivité pour les algèbres d'automorphismes de certaines structures géométriques*', Rendiconti Univ. d. Roma, **61**, 43–55, 1961.
[111] 'Les spineurs en relativité générale', Rendiconti Univ. d. Bari, fasc. 79, 1962.

Article technique:

'Sur le mouvement des projectiles autour de leur centre de gravité', *Mém. de l'Artillerie française*, 1940.

[112] 'Laplaciens sur une variété riemanienne et spineurs', *Accad. Naz. dei Lincei* **33** (1962), fasc. 5, 187–191.
[113] *Propagateurs, Commutateurs et anticommutateurs en relativité générale*, Firenze, sept. 1961, Bologna, octobre 1962 – pp. 116–134.
[114] 'Propagators and commutators in general relativity', *Proc. Roy. Soc.* **270** (1962), 342.

1963

[115] 'Commutators and Anticommutators in General Relativity', Notes miméographiées de Princeton, mars 1963.
[116] 'Champs spinoriels et propagateurs en relativité générale', Cours ronéotypé de Collège de France, 1961–1962, janvier 1963.
[117] 'Opérateurs diff. invariants sur un espace symétrique', *C.R. Acad Sci.* **256** (1963), 3548.
[118] 'Spineurs harmoniques', *C.R. Acad. Sci.* **257**, 7–9.

1964

[119] 'Champs spinoriels et propagateurs en relativité générale', *Bull. Soc. Math. France* **92**, 11–100.
[120] 'Champ de Dirac, champ du neutrino et transformateurs ϵ, ρ, τ sur un espace temps courbe', *I.H.P.* **1**, 233–290.
[120[a]] 'Théorie du neutrino et transformations ε, ρ, τ sur un espace temps courbe', *C.R. Acad. Sci.* **257** (1963), 3542–3545.
[121] 'Variétés complexes et tenseur de Bergmann', *C.R. Acad. Sci.* **259**, 2737–2740.
[121[a]] 'Variétés complexes, tenseur de Bergmann et espaces homogènes', *C.R. Acad. Sci.* **261**, 2143–2145.
[122] 'Existence and Uniqueness Theorems in General Relativity', *Proc. of Symp. in Applied Math.* **17**, Ann. Math. Soc., 1964.

1965

[123] 'Opérateurs diff. invariants sur un espace homogène', *Ann. Ec. Norm.* **81**, 341–385.
[124] 'Théorème d'existence et d'unicité par un fluide thermodyn. relativiste', *C.R. Acad. Sci.* **260**, 3291–3295.
[125] 'Etude mathématique des équations de la magnétohydrodynamique relativiste', *C.R. Acad. Sci.* **260**, 4449–4453.
[126] 'Sur les transformations conformes d'une variété riemanienne compacte', *C.R. Acad. Sci.* **259** (1964), 697–700.
[127] 'Etude mathématique des fluides thermodynamiques relativistes', *Comm. Math. Phys.* **1**, 328–373.
[128] 'Variétés complexes et tenseur de Bergmann', *Ann. Inst. Fourier* **15** (1964), fasc. 2, 345–408.
[128[a]] 'Variétés analytiques complexes et tenseur de Bergmann', *Rendiconti di Matematica* (1, 2) **25**, 113–128.
[129] Ondes de choc et hypotheses de compressibilité en magnétohydrodynamique relativiste', *C.R. Acad. Sci.* (1966), 153.
[130] 'Approx. classique des équations de la magnétohydrodynamique relativiste', *C.R. Acad. Sci.*
[131] 'Sur certaines variétés kählériennes compactes', *C.R. Acad. Sci.* (1966), 570–575.
[132] 'Ondes de choc en magnétohydrodynamique relativiste', *Ann. I.H.P.* **5**, 37–75.
[132[a]] *Magnetohydrodynamics and shock waves*, Collège de France.

1967

[133] *Relativistic Hydrodynamics and Magnetohydrodynamics*, Benjamin, New York, 1967.
[134] 'Rayons en magnétohydrod. relativiste', *C.R. Acad. Sci.* **264**, 715–720.
[135] 'Théorie des rayons en hydrodynamique et magnétohydrodynamique relativistes', *Ann. I.H.P.* **7**, 271–302.
[136] 'Variétés kählériénnes et premières classes de Chern', *J. Diff. Geometry* **1**, 195–223.

1968

[137] 'Ondes de choc et hypothèses de compressibilité en magnétohydrod. relativiste', *C.R. Acad. Sci.* **266**, 888–893.
[138] 'Ondes de choc et ondes en magnétohydrod. relativistes', Coll. Int. C.N.R.S., fluides et champ gravitationnel, juin 1967.
[139] 'Sur les applications harmoniques', *C.R. Acad. Sci.* **267**, 548–553.
[140] 'Variétés kählériennes et premières classes de Chern', *Acta Cientifica Compor-tolana* **IV**, 137–148.

1969

[141] 'Th. d'existence et d'unicité des chocs en MHD relativiste', *C.R. Acad. Sci.* **268**, 256–260.
[142] 'Variétés kählériennes à premières classe de Chern positive où nulle', *C.R. Acad. Sci.* **268**, 876–880.
[143] 'Ondes de choc et hypothèses de compressibilité en MHD relativistes', *Comm. Math. Phys.* **12**, 145–174.

[144] 'Ondes de choc et ondes en MHD relativiste', Fluides et champ gravitationnel en relativité générale, C.N.R.S., 1969.
[145] 'Applications harmoniques dans un tore', *C.R. Acad. Sci.* **269**, 912–916.

1970

[147] *Applications harmoniques et variétés kählériennes*, Istituto Nazionale di alta Matematica, Symp. Math. Vol. III, 1970.
[148] *La Magnétohydrodynamique classique et relativiste*, Lille, 16–20 juin 1969, Colloq. Intern. du C.N.R.S.
[149] *Applications harmoniques et variétés kählériennes*, Symp. Math., Vol. 3, Bologne, 1970.
[150] 'Shock Waves in Relativistic Magnetohydrodynamics', *Physica Scripta* **2** (1970), 221–225.
[151] 'Variétés riemaniennes à tenseur C non négatif', *C.R. Acad. Sci. Paris* **271** (1970), 650–653.
[152] 'Point-selle et systèmes différentiels', *C.R. Acad. Sci. Paris* **271** (1970), 1123–1128.

1971

[153] 'Espaces Fibres et Espaces-temps', *General Relativity and Gravitation* **1** (1971), 235–245.
[154] 'Magnétohydrodynamique Relativiste et ondes de choc', *Actes Congrès Intern. Math.* **3** (1970), 47–56.
[155] 'Sur les ondes de choc gravitationnelles', *C.R. Acad. Sci. Paris* **273** (1971), 528–532.
[156] 'Sur les transformations holomorphes des variétés kählériennes compactes', *C.R. Acad. Sci. Paris* **273** (1971), 1148–1151.
[157] 'Ondes de choc, ondes infinitésimales et rayons en hydrodynamique et magnétohydrodynamique relativistes', C.I.M.E., Roma, 1971.
[158] 'Variétés kählériennes à première classe de Chern non négative et variétés riemaniennes à courbure de Risci généralisée non négative', *Journ. of diff. Géom.* **6**(1) (1971), 47–94.

1972

[159] 'Zéros des vecteurs holomorphes sur une variété kählérienne à première classe de Chern non négative', *Diff. Geom. Kinokuniya*, Tokio, 1972, pp. 253–266.
[160] 'Sur l'algèbre de Lie des automorphismes symplectiques', *C.R. Acad. Sci.* **274** (1972), 1177–1180.
[161] 'Sur les idéaux de l'algèbre de Lie des automorphismes d'une variété symplectique', *C.R. Acad. Sci.* **274** (1972), 1494–1498.
[162] 'Dérivations et premier groupe de cohomologie pour des algèbres de Lie attachées à une variété symplectique', *C.R. Acad. Sci.* **275** (1972), 113–118.
[163] 'Dérivations de l'algèbre de Lie des automorphismes infinitésimaux d'une structure de contact', *C.R. Acad. Sci.* **275** (1972), 709–714.
[164] 'Sur l'algèbre de Lie des automorphismes infinitesimaux d'une structure de contact', *C.R. Acad. Sci.* **275** (1972), 825–829.

1973

[165] 'Dérivations de l'algèbre de Lie des automorphismes infinitésimaux d'une variété unimodulaire', *C.R. Acad. Sci.* **276** (1973), 55–60.

[166] 'Sur l'algèbre de Lie des automorphismes infinitésimaux d'une variété unimodulaire', *C.R. Acad. Sci.* **276** (1973), 199–203.
[167] 'Algébres de Lie et groupes de difféomorphismes pour les structures classiques', *C.R. Acad. Sci.* **276** (1973), 1112–1118.
[168] 'Variétés kählériennes à première classe de Chern non négative et situation analogue dans le cas riemannien', *Symposia Matematica* **10** (1972).
[169] 'Ondes de choc gravitationnelles et électromagnétiques', *C.R. Acad. Sci.* **276** (1973), 1385–1389.
[170] 'Cohomologie 1-différentiable d'algèbres de Lie associées à une variété symplectique', *C.R. Acad. Sci. Paris* **277** (1973), 215–219.
[171] 'Tenseurs holomorphes sur une variété kählérienne compacte', *C.R. Acad. Sci. Paris* **277** (1973), 801–805 (en collaboration avec R. J. Ibrahim).
[172] 'Ondes de choc gravitationnelles et électromagnétiques', Symposia Math. Ist. Naz. di Alta Matematica, Rome, février 1972, **12** (1973), 93–110.

1974

[173] 'Algèbre de Lie dynamique d'une variété unimodulaire', *C.R. Acad. Sci. Paris* **278** (1974), 431–434.
[174] 'Sur l'algèbre des automorphismes infinitésimaux d'une variété symplectique', *J. of Diff. Geometry* **9** (1974), 1–40 (en collaboration avec A. Avez et A. Diaz-Miranda).
[175] 'Algèbre de Lie des automorphismes infinitésimaux d'une structure de contact', *J. de Math. pures et appl.* **52** (1973), 473–508.
[176] 'Automorphismes infinitésimaux d'une structure subordonnée à une structure unimodulaire', *C.R. Acad. Sci. Paris* **278** (1974), 1425.
[177] 'Algèbre de Lie des automorphismes infinitésimaux d'une structure unimodulaire', *Ann. Inst. Fourier* **24** (1974), 219–266.
[178] 'Cohomologie 1-différentiable des algèbres de Lie attachées à une variété symplectique ou de contact', *J. de Math. pures et appl.* **53** (1974), 459–484.
[179] 'Dérivations et cohomologies des algèbres de Lie attachées à une variété symplectique et à une variété de contact', Colloque Int. C.N.R.S., Aix, juin 1974.
[180] 'Déformations 1-différentiables d'algèbres de Lie attachées à une variété symplectique ou de contact', *C.R. Acad. Sci. Paris* **279**, 877–881 (en collaboration avec M. Flato et D. Sternheimer).
[181] 'Déformations 1-différentiables d'algèbres de Lie attachées à une variété symplectique ou de contact', *Comp. Mat.* **31**, 47–82, 1975 (en collaboration avec M. Flato et D. Sternheimer).
[182] 'Variétés symplectiques, variétés canoniques et systèmes dynamiques' (volume en mémoire de E. T. Davis) (Rund. Forbes Ed.).
[182[a]] *Topics in Differential Geometry*, Academic Press, New York, 1976.
[182[b]] 'Gravitational Waves' in *Mathematical Physics and Physical Mathematics* (eds. K. Maurin and R. Rączka), D. Reidel, Dordrecht, 1976.

1975

[183] 'Variétés canoniques et transformations canoniques', *C.R. Acad. Sci. Paris* **280** (1975), 37–40.
[184] 'Variétés symplectique et dynamique associées à une sous-variété', *C.R. Acad. Sci. Paris* **280** (1975), 523–527.
[185] 'Cohomologie 1-différentiable et déformations de l'algèbre de Lie dynamique d'une variété canonique', *C.R. Acad. Sci. Paris* **280** (1975), 1217–1220.
[186] 'Algébres de Lie attachées à une variété canonique', *J. de Math pures et appl.* **54** (1975), 445–480 (en collaboration avec M. Flato et D. Sternheimer).

[187] 'Les variétés de Poisson et leurs algèbres de Lie associées', *J. of diff. geom.*

[188] 'Structures de contact et formalisme hamiltonien invariant', *C.R. Acad. Sci. Paris* **281** (1975), 171–175.

[189] 'Algèbre de Lie des champs de vecteurs: cohomologie 1-différentiable et déformations', *C.R. Acad. Sci. Paris* **281**, 507–512.

[190] 'Ondes de choc, sous des hypothèses générales, en magnétohydrodynamique', *C.R. Acad. Sci. Paris* **281** (1975), 723–728.

[191] 'Sur l'algèbre de Lie des champs de vecteurs', *Comm. Math. Helv.*

[192] 'Shock waves in Relativistic Magnetohydrodynamics under general assumption', *Journ. of Math. Physics*

[193] 'Variétés symplectiques et variétés canoniques', Trends in Applications Pure Mathematics to Mechanics, Lecce, mai 1975, Pickman Publ., London.

[194] 'Ondes de choc gravitationnelles et ondes de choc', *C.R. Acad. Sci. Paris* **281** (1975), 929–934.

1976

[195] 'Fibrés canoniques, structures, structures unimodulaires exactes de Lie d'automorphismes infinitésimaux', *C.R. Acad. Sci. Paris* **282** (1976), 283–287.

[196] 'Fibrés vectoriels, structures unimodulaires exactes et automorphismes infinitésimaux', *J. de Math. Pures et appl.*

[197] 'Deformations of Poisson brackets, Dirac brackets and applications', *J. of Math. Phys.* **17**, 1754–1762 (en collaboration avec M. Flato and D. Sternheimer).

[198] 'Poisson manifolds and Dirac brackets', Journées Relativistes de Bruxelles, avril 1976 et Symposium Varsovie, juin 1976.

INDEX OF NAMES

GPSR Compliance

The European Union's (EU) General Product Safety Regulation (GPSR) is a set of rules that requires consumer products to be safe and our obligations to ensure this.

If you have any concerns about our products, you can contact us on ProductSafety@springernature.com

In case Publisher is established outside the EU, the EU authorized representative is:

Springer Nature Customer Service Center GmbH
Europaplatz 3
69115 Heidelberg, Germany

Batch number: 10607282

Printed by Printforce, the Netherlands